VOLUME ONE HUNDRED AND FIFTY ONE

PROGRESS IN MOLECULAR BIOLOGY AND TRANSLATIONAL SCIENCE

Approaches to Understanding Breast Cancer

VOLUME ONE HUNDRED AND FIFTY ONE

PROGRESS IN MOLECULAR BIOLOGY AND TRANSLATIONAL SCIENCE

Approaches to Understanding Breast Cancer

Edited by

Rajkumar Lakshmanaswamy

Center of Emphasis in Cancer Research, Paul L. Foster School of Medicine, Graduate School of Biomedical Sciences, Texas Tech University Health Sciences Center El Paso, El Paso, TX, United States

ELSEVIER

ACADEMIC PRESS

An imprint of Elsevier

Academic Press is an imprint of Elsevier
125 London Wall, London EC2Y 5AS, United Kingdom
525 B Street, Suite 1800, San Diego, CA 92101-4495, United States
50 Hampshire Street, 5th Floor, Cambridge, MA 02139, United States
The Boulevard, Langford Lane, Kidlington, Oxford OX5 1GB, United Kingdom

First edition 2017

ISBN: 978-0-12-812772-8
ISSN: 1877-1173

For information on all Academic Press publications
visit our website at https://www.elsevier.com/

Publisher: Zoe Kruze
Acquisition Editor: Ashlie M. Jackman
Editorial Project Manager: Joanna Collett
Production Project Manager: Vignesh Tamil
Designer: Mark Rogers

Typeset by Thomson Digital

CONTENTS

CONTRIBUTORS

Shrikant Anant
University of Kansas Medical Center, Kansas City, KS, United States

Roy A. Jensen
University of Kansas Medical Center, Kansas City, KS, United States

Jeronay King
Morehouse School of Medicine, Atlanta, GA, United States

Naveen Krishnegowda
The University of Texas Health Science Center at San Antonio, San Antonio, TX, United States

Rajkumar Lakshmanaswamy
Center of Emphasis in Cancer Research, Paul L. Foster School of Medicine, Graduate School of Biomedical Sciences, Texas Tech University Health Sciences Center El Paso, El Paso, TX, United States

Sridhar Mani
Albert Einstein College of Medicine INC, Bronx, NY, United States

Charmaine Martin
Texas Tech University Health Sciences Center, El Paso, Paul L Foster School of Medicine, El Paso, TX, United States

Hina Mir
Morehouse School of Medicine, Atlanta, GA, United States

Daniel Murphy
Texas Tech University Health Sciences Center, El Paso, Paul L Foster School of Medicine, El Paso, TX, United States

Sushmita Bose Nandy
Center of Emphasis in Cancer Research, Paul L. Foster School of Medicine, Texas Tech University Health Sciences Center El Paso, El Paso, TX, United States

Jiangjiang Qin
University of Houston, Houston; Texas Tech University Health Sciences Center, Amarillo, TX, United States

Kumaraguruparan Ramasamy
The University of Texas Health Science Center at San Antonio, San Antonio, TX, United States

Rajeshwar R. Tekmal
The University of Texas Health Science Center at San Antonio, San Antonio, TX, United States

Randi Ryan
University of Kansas Medical Center, Kansas City, KS, United States

Cathy Samayoa
The University of Texas Health Science Center at San Antonio, San Antonio, TX, United States

Navkiran K. Shokar
Texas Tech University Health Sciences Center, El Paso, Paul L Foster School of Medicine, El Paso, TX, United States

Shailesh Singh
Morehouse School of Medicine, Atlanta, GA, United States

Ramadevi Subramani
Center of Emphasis in Cancer Research, Paul L. Foster School of Medicine, Texas Tech University Health Sciences Center El Paso, El Paso, TX, United States

Ossama Tawfik
University of Kansas Medical Center, Kansas City, KS, United States

Wei Wang
University of Houston, Houston; Texas Tech University Health Sciences Center, Amarillo, TX, United States

Stella Winters
Texas Tech University Health Sciences Center, El Paso, Paul L Foster School of Medicine, El Paso, TX, United States

Ruiwen Zhang
University of Houston, Houston; Texas Tech University Health Sciences Center, Amarillo, TX, United States

PREFACE

Breast cancer is one of the major cancers in women and is also a leading cause of cancer-related deaths. In the past several decades there has been tremendous advancements in breast cancer prevention and treatment. The purpose of this book is to organize a comprehensive compilation of recent advances in breast cancer. The ambitious task of compiling such a comprehensive volume was possible by bringing together materials from recognized authorities in their respective fields. This book will enable the reader appreciate the various aspects of advancement in breast cancer research. Each chapter in the book refers to selected journal articles, monographs, or chapters from other resources. In essence, this volume will address different aspects of breast cancer from epidemiology to complementary and alternate medical treatments.

The book has been organized into nine chapters. The first chapter deals with breast cancer epidemiology, prevention, and screening. The second chapter reviews the current approaches available to diagnose and treat early breast cancer lesion, such as ductal carcinoma in situ. The third chapter talks about how pregnancy alters the risk of breast cancer. The fourth chapter describes the major associations between key cytokines and chemokines in the development of breast cancer. The fifth chapter defines the role of cancer stem cells in breast cancer metastasis. The sixth chapter describes how estrogen receptor beta is gaining importance as a preventive and therapeutic target for breast cancer. The seventh chapter gives an appraisal of molecular pathways involved in breast cancer with a particular emphasis on the NFAT–MDM2–p53 pathway. The eighth chapter deals with the current trend of how microbiota influences breast cancer. Finally, the ninth chapter reviews the complementary and alternative medical approaches available for breast cancer patients.

I am very appreciative of all the contributors of this volume for devoting their time in compiling the latest advances in breast cancer that will enable a better understanding of the approaches available today to deal with breast cancer. I gratefully acknowledge the opportunity provided to me to put together various facets of breast cancer with a goal of making the reader appreciate the current developments in breast cancer prevention and treatment.

Rajkumar Lakshmanaswamy
Editor

CHAPTER ONE

Breast Cancer Epidemiology, Prevention, and Screening

Stella Winters, Charmaine Martin, Daniel Murphy, Navkiran K. Shokar[1]
Texas Tech University Health Sciences Center, El Paso, Paul L Foster School of Medicine, El Paso, TX, United States
[1] Corresponding author. E-mail address: Navkiran.shokar@ttuhsc.edu

Contents

Abstract

Globally, breast cancer is both the most commonly occurring cancer and the commonest cause of cancer death among women. Available data suggest that incidence and mortality in high-resource countries has been declining whereas incidence and mortality in low-resource countries has been increasing. This pattern is likely to be due to changing risk factor profiles and differences in access to breast cancer early

Progress in Molecular Biology and Translational Science, Volume 151
ISSN 1877-1173
http://dx.doi.org/10.1016/bs.pmbts.2017.07.002

detection and treatment. Risk factors for breast cancer include increasing age, race, menarche history, breast characteristics, reproductive patterns, hormone use, alcohol use, tobacco use, diet, physical activity, and body habitus. Mutations in the BRCA 1 and BRCA 2 tumor suppressor genes are significantly associated with the development of breast and ovarian cancer by the age of 70. Survival depends on both stage and molecular subtype. As there are few signs and symptoms early on, early detection is an important strategy to improve outcomes. Major professional organizations in the United States and elsewhere recommend screening with mammography with appropriate follow up for an abnormal screening test, although they differ somewhat by recommended ages and frequency of screening. Studies suggest a 15%–40% mortality reduction secondary to screening, however, there are also concerns about harms, such as overdiagnosis (5%–54%) and overtreatment leading to long term complications, and false negatives (6%–46%). Identification of women at risk for BRCA1 and BRCA 2 mutations is also recommended with referral for genetic testing. Preventive interventions, such as lifestyle, medical, and surgical options are available for women testing positive for BRCA mutations.

1. INTRODUCTION

Breast tissue is composed of lobules that are glands involved in milk production, ducts that connect the lobules to the nipple and connective tissue, fatty tissue, and lymphatic tissue. Breast cancer occurs when there is an unregulated growth of cells within any of the components of the breast, although this occurs most commonly in the lobules. Although significantly less common, breast cancer can also occur among males. Breast cancer develops over time and may go through an in situ phase. Whether in situ or invasive, it may be detected during routine self-breast exams, mammographic screening, or once signs or symptoms have developed. Initially, there are no signs or symptoms associated with breast cancer until a palpable, or visible lump develops within the breast. The most common physical sign that can be appreciated is a painless breast lump. Swollen and enlarged lymph nodes may be present within the axillary region during the early stages of metastasis. Evidence of bloody nipple discharge, heaviness, redness, swelling, breast deformity, or retractions are less common signs, but a substantial indicator of breast malignancy and maybe more evident with advanced stages of disease. As, there are few early signs, it is recommended that all females follow the current breast cancer screening guidelines to identify those with abnormal breast tissue during the early evolutionary stages.

A majority of breast cancers encompass the invasive type. These malignancies have extended beyond the ducts and glands into surrounding tissues and lymph nodes. Currently, there are 21 histologically distinct breast cancers and four different molecular variations. Each subtype differs in their overall presentation and type of treatment required[1]. Invasive malignancies are evaluated by molecular techniques to determine if hormone receptors are present, and if there is an excessive concentration of human epidermal growth factor receptor 2. These molecular techniques have proven that a majority (74% in the USA) of invasive breast malignancies are either estrogen or progesterone receptor positive, but human epidermal growth receptor 2 negative (luminal A). These malignancies are characteristically slow growing and less aggressive. This is contrast to triple negative cancers (12% of cases in the USA), which have the poorest short term prognosis largely because there are no targeted therapies. Breast malignancies that are hormone receptor negative, but human epidermal growth factor receptor 2 positive (HER2-enriched, 4% of cases in the USA) are also more aggressive and have a poorer short-term prognosis than hormone positive types, although recent targeted therapies are reversing this trend.[1] Treatment options for breast cancer include surgery, chemotherapy, radiation therapy, and hormone therapy, and depend on the stage, as well as histological and molecular subtypes. Survival depends on stage of diagnosis, but overall, survival in the USA is 89% at 5 years, 83% at 10 years, and 78% at 15 years. Survival rates are highest for early stages of diagnosis (99% 5 year survival in the USA) versus late stages (26% 5 year survival in the USA).[2]

2. BREAST CANCER EPIDEMIOLOGY

Breast cancer is the most frequently diagnosed cancer in women in all world regions, regardless of per capita income.[3,4] Worldwide about 1.7 million cases of breast cancer are diagnosed every year, or approximately one new case recognized every 18 s.[3,4] However, incidence rates vary nearly fourfold (Fig. 1), with higher incidence documented in higher income regions (92 per 100,000 in North America) and lower incidence documented in lower income regions (27 per 100,000 in Middle Africa and Eastern Asia).[4–6] In the United States, breast cancer accounts for 14% of newly identified cancers.[7,8] One in eight US women will be diagnosed with breast cancer in their lifetime, with the highest percentage occurring in the 55–64 age group.[7,8]

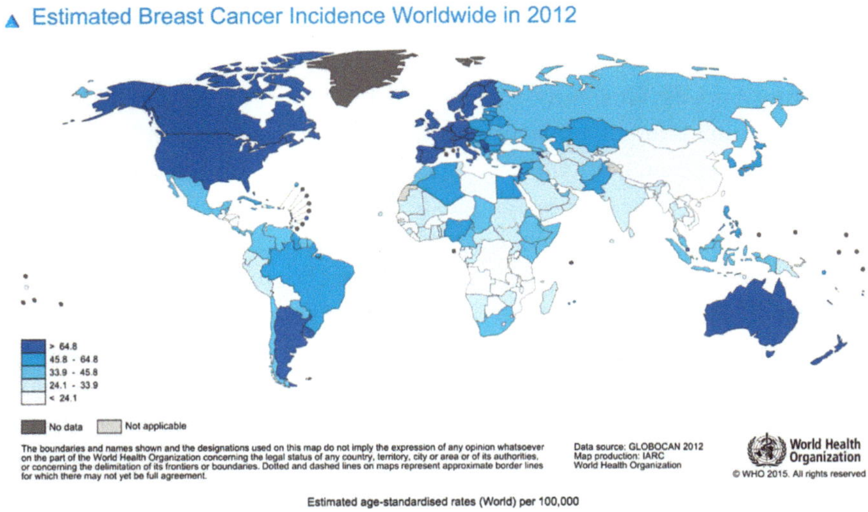

Fig. 1 Estimated breast cancer incidence worldwide. *From Ferlay J, Soerjomataram I, Ervik M, Dikshit R, Eser S, Mathers C, Rebelo M, Parkin DM, Forman D, Bray F. GLOBOCAN 2012 v1.0, Cancer Incidence and Mortality Worldwide: IARC CancerBase No. 11 [Internet]. Lyon, France: International Agency for Research on Cancer; 2013. Available from: http://globocan.iarc.fr/Pages/fact_sheets_cancer.aspx.*[4]

In addition to being most common, breast cancer is also the leading cause of cancer death in women worldwide.[3–6] There were approximately 0.5 million deaths worldwide from breast cancer in 2013, which approximates to almost one death per minute and accounts for 15% of all cancer deaths.[4] Mortality rates vary even more widely than incidence, with breast cancer survival being significantly lower in low- and middle-income countries compared to high-income countries.[6] According to a recent study that examined 25 population-based cancer registries from 12 different countries in diverse world regions, survival for women with localized disease and regional disease was reported to be around 90% and 75%, respectively for countries with highly developed health services, whereas survival in women in countries with less-developed health services with localized disease and regional disease was 76% for regional disease and 47%, for regional disease[9] (Fig. 2) In the United States, it is estimated that 40,450 women will die from this disease in 2016.[8] Breast cancer is the second leading cause of cancer death in women of all age groups, and the leading cause of cancer death in women aged 20–59 years.[10,11] Breast cancer mortality is highest in the southern belt region and along the Mississippi river.[12]

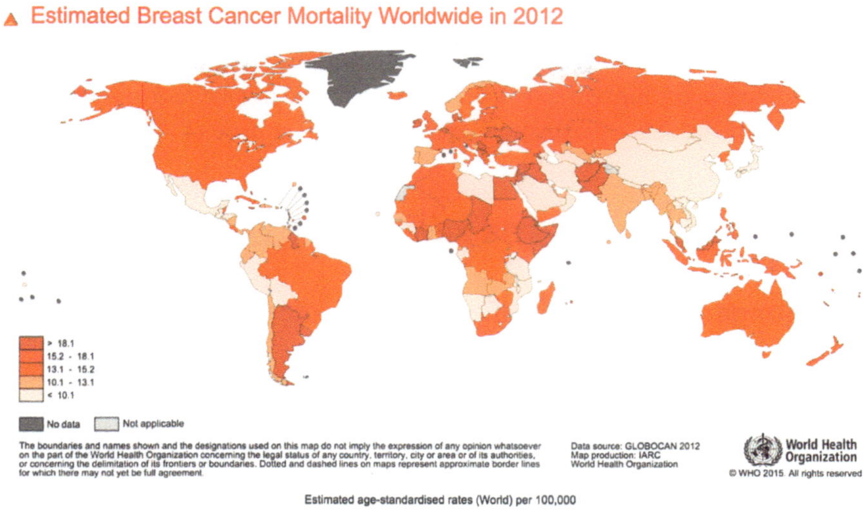

Fig. 2 Estimated breast cancer mortality worldwide. *From Ferlay J, Soerjomataram I, Ervik M, Dikshit R, Eser S, Mathers C, Rebelo M, Parkin DM, Forman D, Bray F. GLOBOCAN 2012 v1.0, Cancer Incidence and Mortality Worldwide: IARC CancerBase No. 11 [Internet]. Lyon, France: International Agency for Research on Cancer; 2013. Available from: http://globocan.iarc.fr/Pages/fact_sheets_cancer.aspx.*[4]

2.1 Trends Over Time

Globally, breast cancer incidence has been on the rise and that trend is expected to continue. There has been a 3.1% annual rate of rise in global breast cancer incidence beginning with 641,000 cases in 1980 to over 1.6 million in 2010.[13] By 2030, the number of breast cancer diagnoses worldwide is expected to increase to almost 3.2 million per year.[14]

In the United States, incidence has been stable since 2005 for women in their 40s and 50s, but incidence has increased for women and aged 60 or older.[10] For reasons not fully understood, breast cancer cases in the United States are expected to increase by 64% between 2011 and 2030 with a rising prevalence of ER positive in situ tumors and a shift in age of diagnosis to women 70–84 years of age.[15,16] Despite this daunting increase in numbers, significant progress has been achieved over the past 30 years in improving survival rates following a diagnosis of invasive breast cancer.[8,12,17] United States breast cancer mortality rates have decreased 36% from 1989 to 2012[8,12] and estimates suggest that 249,000 breast cancer deaths have been averted among US women within that time period.[7] Death rates from breast cancer have decreased in all age groups, most notably in women ages 20–39 years.[7]

Relative survival rates for women diagnosed with breast cancer are 89% at 5 years after diagnosis, 83% after 10 years, and 78% after 15 years. Five year survival rates were highest in women with localized disease and in those with small tumor size (i.e., less than or equal to 2.0 cm) at time of diagnosis.[7,15,18]

3. RISK FACTORS FOR BREAST CANCER

3.1 Nongenetic, Nonmodifiable Risk Factors

3.1.1 Age

Breast cancer incidence and mortality increase proportionally with age. Throughout the world, this disease peaks around age 60 with a sharp incline beginning at age 40.[4,8] There are, of course, some exceptions. For example, in Latin America, 20%–30% of breast cancers are diagnosed between the ages of 20 and 44, roughly doubling the corresponding proportion observed in the combined USA and Canadian registries.[19] In a similar vein, the peak age at diagnosis was 40–50 years in Asian and African Countries versus 60–70 years in Western countries.[20,21] In the United States, the median age at diagnosis is 61, with the median age being younger for black women (58 years) compared to non-Hispanic white women (62 years)[22]. The median age at death due to breast cancer is 68 years overall, with the median age at death of 62 years for black women and 69 years for non-Hispanic white women.[8]

3.1.2 Race

There are significant disparities in long-term breast cancer survival between black and white women in the United States.[8,10] Although the overall incidence of breast cancer is slightly lower in black women than in white women, black women have the highest 5 year breast cancer mortality rate for each known stage at diagnosis.[7,22–27] As of 2012, the breast cancer death rate was 42% higher in blacks than in whites .[7] Black women were less likely than white women to survive their breast cancer despite uniform treatment, even after controlling for stage of disease, tumor characteristics, follow-up, and other breast cancer risk factors.[22,24,28] In addition to obvious socioeconomic disparities that occur in some black communities, black women in general are more likely to have a genetically more aggressive type of cancer.[7] Black women have the largest proportion of HR−/HER2 breast cancers compared with women of other race/ethnicities.[7,22,23] This subtype of breast cancer is associated with a poorer prognosis.[29,30] Additionally, black women have the smallest proportion of localized breast cancers (53%) and the largest proportion of regional (35%) and distant (8%) stage disease compared to other races/ethnicities in the United States.[7]

In contrast, Asian/Pacific Islander women have the lowest incidence and mortality in the United States, with Alaskan natives, American Indians, and Hispanic women closely behind.[7,8]The reason for lower breast cancer incidence in these minority groups is thought to reflect variation of reproductive patterns, including greater number of children, first childbirth at a younger age, and longer duration of breastfeeding.[24,31] Asian/Pacific Islander women, specifically, are also less likely to consume alcohol and have lower rates of obesity both of which are associated with lower breast cancer risk.[32,33]

3.1.3 Early Menarche/Late Menopause

Younger age at menarche and older age at menopause increases breast cancer risk.[34–37] Breast cancer risk is approximately 20% higher among girls that begin menstruating before age 11 compared to those that begin at age 13.[38,39] In addition, women who experience menopause at age 55 or older have about a 12% higher risk compared to those who do so between ages 50–54.[38] There is also a disproportionate prevalence of estrogen receptor positive and lobular carcinomas in this population,[38] suggesting that both ovarian synthesis of endogenous hormones, as well as a lengthening in the total number of reproductive years play a contributory role in breast cancer development.[38,40]

3.1.4 Breast Characteristics

Women with a history of breast cancer are at higher risk of developing the disease again, especially if the initial age at diagnosis was less than 40 years of age.[41] The risk of reoccurrence in women with an initial age at diagnosis greater than 40 years was 1.5-fold compared to a 4.5-fold increase in women diagnosed with breast cancer at age less than 40.[41] Similarly, women with a diagnosis of a cancer precursor, such as ductal carcinoma in situ or lobar carcinoma in situ are 8–10 and 7–12 times, respectively, more likely to develop a new invasive breast cancer compared to women without a previous diagnosis.[42,43] Other significant associations with increased risk are proliferative lesions with atypia (i.e., atypical ductal hyperplasia and atypical lobular hyperplasia), which are associated with 4–5 times higher than average breast cancer risk and proliferative lesions without atypia (i.e., ductal hyperplasia and fibroadenoma), which have a 1.5–2 times higher than average risk.[44] Lastly, dense breast tissue increases the risk of breast cancer.[45,46] Compared to women with 11%–25% breast density, those with 26%–50% or greater than 50% breast density have about a 1.6 or 2.3 times, respectively, higher risk of breast cancer.[47,48] Complicating this issue is the fact that denser breasts impair mammographic detection of breast cancer, which couples higher risk with a more elusive diagnosis.[49]

3.2 Nongenetic, Modifiable Risk Factors

3.2.1 Income, Education, and Insurance Status

Throughout the world, there exists a disparity in breast cancer survival based on the median income of each country, as well as their *Human Development Index*, which is a measure of national well-being calculated from average life expectancy, mean years of schooling, and gross national income per capita. Studies have shown that the mortality-to-incidence ratio of breast cancer was found to be significantly lower in countries with a very high human development index compared to those with high-, medium-, and low-human development indices.[50,51]

Similar disparities exist in the United States. Although incidence of breast cancer increases among women living in areas with better socioeconomic conditions,[52–55] the 5 year survival rates for each stage of diagnosis are lower among breast cancer patients who reside in lower-income areas.[52,54–62] Moreover, the level of educational attainment was found to be inversely related to poor breast cancer outcomes.[56] Specifically, women who resided in zip codes with the lowest educational attainment were more likely to present with metastatic breast cancer, large tumor size, triple negative tumors, or high-grade breast cancer compared to women residing in areas of higher educational attainment.[56]. Finally, payer status may affect access to health care and influence breast cancer stage at diagnosis. Reduced access to health care has also been linked to poor outcomes. The highest mortality bracket includes those women with no insurance or those with Medicaid only.[63–67]

3.2.2 Changing Reproductive Patterns

Delayed childbearing, smaller family size, decreased duration of breastfeeding, and hormonal contraceptives have been linked to increased breast cancer risk. On the contrary, women having a first child at a younger age and having a greater number of children is associated with decreased breast cancer risk.[48,68,69] There seems to be particular importance in timing of pregnancy relative to menarche, with a smaller time gap associated with decreased risk (i.e., 50% reduced lifetime risk in women having first child before age 20).[18,70–72] Another study showed that the risk of premenopausal breast cancer is increased by 5% for each additional year that first birth is delayed.[73]

Increased parity is, overall, a long-term protective feature. Each full term pregnancy leads to a 12% reduction of postmenopausal breast cancer risk.[73] Although there is a transient (approximately 10 years duration) increased risk after childbirth, the cumulative incidence of breast cancer is lower in multiparous women.[74–76] Inherent to this protection is a time-sensitive caveat. Compared to nulliparous females, those with first full-term birth after age 35 were shown to have a 1.26 greater risk of developing the disease within 5 years after delivery, although after 15 years, they were found to have lower risk than their nulliparous counterparts.[68,69,74,77] Furthermore, the parity-specific effects on breast cancer risk seem to be limited to ER positive/PR positive histologic types.[78]

Breastfeeding for 1 or more years reduces a woman's overall risk of breast cancer (reduced by 4% for every 12 months of breastfeeding).[79,80] Regarding hormonal contraceptives, there is a general increased risk associated with earlier age at onset of use and higher estrogen formulations.[81–88]

3.2.3 Menopausal Hormone Use

There are two general formulations of hormonal therapy for controlling menopausal symptoms: combined estrogen and progestin versus estrogen-only therapy. Combined estrogen and progestin therapy increases the risk of developing breast cancer.[89,90] Higher risk correlated most with longer use and starting therapy soon after the onset of menopause.[89–92] However, the increased risk appears to be temporary and goes back to baseline risk after 5 years of discontinuation.[91,93,94] The studies analyzing risk of breast cancer associated with estrogen-only hormonal therapy have yielded mixed results. Although one large RCT study found that women who used estrogen-only therapy for an average of 6 years had a 23% lower risk of developing breast cancer,[95] other studies looking at estrogen-only therapy have found a positive correlation with breast cancer risk, specifically when used by lean women and when starting therapy soon after the onset of menopause.[91,94,96,97]

3.2.4 Tobacco

Based on current available information there is evidence to suggest that smoking prior to menopause increases breast cancer risk. Greater risk was seen in women who started smoking before their first pregnancy (up to 21% higher risk) and in those who were heavy, long-term smokers (i.e., more than 40 pack years).[98–103] On the contrary, some studies show that women who took up smoking after menopause had a significant decrease in breast cancer risk, likely due to tobacco's antiestrogenic effect.[99,102,104] There have been no conclusive studies regarding the association of second hand smoke and breast cancer risk, however some studies have found a connection with premenopausal breast cancer.[99,105,106]

3.2.5 Alcohol

Alcohol consumption has been linked to increased breast cancer risk. Some studies have found a dose-dependent relationship,[107–112] while others found an association with cumulative average alcohol consumption over long periods of time.[113,114] Specifically, consumption of 5–9.9 g/day (i.e., 3–6 glasses of wine per week) was associated with up to a 10% increased risk.[113] Others found an association with binge-drinking (i.e., 1.5 RR for 10–15 drinks per weekend and 2.5 RR for 16–21 drinks per weekend).[115–117] There was no difference whether alcohol consumption was prevalent before age 40 or after age 40.[115] There was also no difference based on type of alcohol (i.e., beer, wine, or liquor).[108,109,114,118] The mechanism behind these findings is likely related to the ability of alcohol to increase levels of estrogen in the blood. As such, the majority of breast cancers linked to alcohol consumption are of the ER+ subtype.[89,119–122]

3.2.6 Fitness and Nutrition

Breast cancer risk is decreased in women performing regular physical activity and in those who consume high amounts of fruits and vegetables. Women who walk at least 7 h per week have a 10%–25% lower risk of breast cancer compared to women who are inactive, with an even greater protective effect in postmenopausal women.[123–126] Fruit and vegetable consumption, as measured by plasma carotenoid levels, was inversely related to breast cancer risk, especially among nonobese women. Furthermore, these women tended to have less aggressive cancers.[115,127] One study found that plasma alpha carotene levels above the lowest quartile were associated with over 40% lower risk of invasive breast cancer.[127]

Elevated BMI and breast cancer risk is contingent on menopausal status.[95,128–130] Obesity seems to have a protective effect in development of premenopausal breast cancer, however, is linked to increased breast cancer development after menopause.[131,132] Overweight and obese premenopausal women (40 and 49 year old) had a 14% and 26%, respectively lower risk of developing breast cancer compared to premenopausal women who were normal weight.[48] This is likely due to the lack of ovulation commonly experienced by overweight and obese women.[133–135] On the contrary, there is an increased risk of breast cancer in postmenopausal overweight and obese women.[95,128,129,133,134,136] Risk is approximately 1.5 times higher in overweight postmenopausal women and 2.1 times higher in obese postmenopausal women compared to postmenopausal lean women.[95,134] This is likely explained by higher levels of circulating estrogen released from adipose tissue in elderly women with elevated BMI.[134] Elevated BMI is also highly correlated with type II diabetes, which is an independent risk factor for breast cancer development.[137,138]

The relationship between weight gain and weight loss with breast cancer risk is less clear, likely because these changes, especially the latter, are not sustained and therefore difficult to study prospectively. However, a recent metaanalysis found that each 5 kg (about 11 pounds) gained during adulthood increases the risk of postmenopausal breast cancer by 11%.[139] There is currently inconclusive evidence regarding the role of weight loss and breast cancer risk.[140–143]

3.3 Genetic Risk Factors

3.3.1 BRCA1 and BRCA 2 Mutations

Hereditary breast and ovarian cancer syndrome is usually caused by genetic mutations in one of two BReast CAncer (BRCA) susceptibility genes, BRCA1 and BRCA2. Mutations in these genes can cause cancer in the breasts and in other organs, such as the ovaries, pancreas, and prostate. In the United States, 1 in 500 may carry one of these mutations.[144,145] Approximately 3% of breast cancers and 10% of ovarian cancers are caused by these gene mutations.[146] Women who carry these mutations are more likely to develop cancer by age 70 compared with the general population, 65% with BRCA1 and 45% with BRCA 2.[147] Women with close relatives that have triple negative breast cancer, premenopausal breast cancer, and ovarian cancer, bilateral breast cancer, or breast cancer in a male relative or multiple relatives with breast cancer is likely to have a gene mutation in BRCA1 and BRCA2. Ashkenazi Jews or Eastern European descent have a much higher risk (1 in 40).[146]

Tumor suppressor genes are essential to the normal biophysiology of human function and development. Tumor suppressor genes are key to regulating and inhibiting cell growth and proliferation. Specific proteins are created from these genes that have numerous functions and specialize in DNA repair, initiating apoptosis, block tumor cell adhesion, and preventing metastasis. Mutations can occur within these genes leading to inappropriate and dysfunctional proteins that are unable to perform effectively. Once mutations occur, there is an increased risk of tumor formation and malignancy. Specific tumor suppressor genes, including the breast cancer genes, BRCA1 and BRCA2, are fundamental in cellular processes through genomic integrity and transcriptional regulation.[148] Mutations within these two genes are linked to the development of breast and/or ovarian cancer. These mutations can occur through an inherited, or sporadic process leading to malignancy.

Those with inherited BRCA mutations also have an inactivated BRCA allele. During puberty, the estrogen dependent–breast epithelium undergoes rapid cell replication and proliferation. These changes exert a stress on the DNA repair system that is incapable of repairing damaged DNA, therefore resulting in cell death. During this process, a small population of cells with damaged DNA will escape cellular death resulting in protein production based on mutated and damaged DNA. Multiple DNA regions that are damaged will often affect genes responsible for cell cycle checkpoint activation. Once a cell is able to bypass this checkpoint, it will permanently escape cell death and will undergo uncontrolled cell proliferation. With the loss of BRCA1 mammary cells, this typically produces mutated p53 checkpoint genes leading to mammary tumors.[149]

Sporadic BRCA mutations are similar to the pathophysiology seen with an inherited process. With the rapid proliferation within breast epithelium stimulated by estrogen, the rate of mutations that occur within the BRCA gene is increased. Inactivation of either BRCA alleles, or a hypofunctional active allele are required in sporadic malignancy. Clinical symptoms and presentation typically present later in life as mutations and cell inactivation increase.[148]

Transcriptional silencing has also been considered a factor contributing to BRCA expression. The BRCA alleles are dependent on regulatory proteins. Inactivation of positively regulated proteins results in decreased expression of these tumor suppressor cells, whereas an increased expression of negative regulatory proteins, decreases gene expression resulting in an increased risk of malignancy.[148]

More than 1800 mutations have been identified within the BRCA1 gene, which increases the risk of breast cancer in both men and women. These mutations that occur are present within every cell in the body, and therefore can be passed on to future generations. Statistically, women are 35%–60% more likely to develop ovarian cancer in comparison to the 1.6% seen in the general population.[150]

3.3.2 BRCA Mutation Screening

It has been clearly defined and proven that BRCA1 and BRCA2 function as tumor suppressor genes. Of the mechanisms described earlier, the commonality seen amongst each possible mechanism is an increased prevalence of breast and/or ovarian malignancy. Risk factors for potentially harmful BRCA mutations include breast cancer diagnosed before the age of 50, bilateral breast malignancy, presence of both breast and ovarian cancer, presence of breast cancer in at least one male–related family member, multiple cases of breast cancer, and Ashkenazi Jewish ethnicity. Women who have at least one family member diagnosed and treated for BRCA-related breast or ovarian cancer should be considered for genetic testing. Several tools are available to assess the risk of having a BRCA mutation and this guides whether referral for consideration of testing should occur. The United States Preventive Service Task Force has evaluated and recommended the use of several assessment tools to identify patients at risk. These tools include the Ontario Family History Assessment Tool, Manchester Scoring System, Referral Screening Tool, Pedigree Assessment Tool, or the Family History Screening 7.[151]

BRCA genetic testing should be recommended in individuals who have had breast and/or ovarian cancer, or a family history of BRCA related malignancies. It is important for patients to be evaluated by a professional who is trained in interpreting the results. Individuals who have a family history of specific, known mutations or from specific ethnicities, may be tested for those specific mutations, or mutations known to be present with certain ethnic groups. If there is no specific history or ethnic background, a more comprehensive BRCA genetic testing is recommended. Results of genetic testing are reported as positive, variants of uncertain clinical significance, uninformative-negative, or true negative. It is important that these results are discussed with a genetic counselor, or a professional trained in interpreting the results.[151]

Screening for BRCA mutations may begin once the individual has reached an age of consent, with periodic follow-up evaluations to determine if any changes have occurred with the patient or within the family.

4. BREAST CANCER PREVENTION

Interventions are available exist for those with positive genetic testing for BRCA mutation(s). Options should be discussed with the patient, including adverse effects of each treatment, and potential long-term consequences.

4.1 Lifestyle/Behavior Change Interventions

More than 2/3 of the U.S. adult population is either overweight or obese. As females transition through menopause, the increase adipose tissue increases the amount of circulating estrogen levels putting individuals at an increased risk of breast cancer by 12%. In comparison, those who lost and maintained more than 10 kg of weight demonstrated more than 50% reduction in breast cancer risk.[152]

Moderate to vigorous physical activity has demonstrated an overall reduction in breast cancer risk. Women who engage in recreational physical activity, who achieve greater than 42 MET-h/week, have demonstrated an overall breast cancer risk reduction of at least 25%. Women who walked at least 7 h/week had a 14% reduced risk of postmenopausal breast cancer in comparison to those who only walked less than three hours per week.[153]

In addition to poor physical activity, as discussed earlier, consumption of alcohol increases the risk of breast cancer. During 1980–2008, a large prospective, cohort study was conducted on 105,986 women enrolled in the Nurses' Health Study. This study identified a modest, but significant increase in breast cancer risk with low level alcohol consumption. Low-level consumption was defined as 5–9.9 g/day, or 3–6 glasses of wine per week (multivariate RR, 1.15; 95% CI, 1.06–1.24; 333 cases/100,000 person-years). Women who consumed at least two alcoholic drinks per day, or more than 30 g daily, had a greater risk of developing breast cancer (RR 1.51; 95% CI, 1.35–1.70; 413 cases/100,000 person-years).[154] In this sample, a combination of weight reduction, physical activity, and limited alcohol consumption reduced the overall risk of breast malignancy.

4.2 Chemoprevention

Chemoprevention has demonstrated benefits in reducing the risk of breast cancer. Selective estrogen receptors modulators, such as Tamoxifen and Raloxifene, have demonstrated evidence for the reduction of breast cancer risk. In 2013, the American Society of Clinical Oncology, recommended that physicians discuss the use of Tamoxifen as an option to reduce the risk of invasive breast cancer, especially those with estrogen receptor positive breast cancer, premenopausal, or postmenopausal women older than 35 at an increased risk, or those with lobular carcinoma in situ. Raloxifene has similar recommendations, but should not be used in premenopausal women. The STAR trial, which compared Tamoxifen with Raloxifene, is the only randomized controlled trial comparing breast cancer risk reduction between the two medications. The STAR trial demonstrated an overall risk reduction between 31% and 67%. After a follow-up period of 4.6 years, the efficacy of Tamoxifen and Raloxifene were equal. At 6 years, women taking a Raloxifene were 24% more likely to develop invasive breast cancer in comparison to women taking Tamoxifen (RR 1.24; 95% CI, 1.05–1.47).[155] Physicians should engage in a shared conversation and decision when considering the use of Tamoxifen or Raloxifene in women with an increased risk of breast cancer.

The use of chemoprevention is dependent on several factors, such as age, race, breast cancer risk, the presence of a uterus, and the type of medication being considered. The International Breast Cancer Intervention Study-1 demonstrated an increased risk of endometrial cancer and venous thromboembolism. This randomized controlled trial was conducted on premenopausal and postmenopausal women aged 35–70 years of age. There was an increase in endometrial cancer, but this was not statistically significant. Unfortunately, there was a significant increase in deep venous thrombosis during the same study. Fifty of the 3,579 (1.4%) women receiving Tamoxifen developed a deep venous thrombosis in comparison to 29 (0.8%) women who were not on Tamoxifen therapy [OR 1.73(95% CI 1.07–2.85)].[156]

In comparison, Raloxifene has a lower risk of cataracts, endometrial cancer, and thromboembolic disease. Women younger than age 50, and young black women may benefit from the use of Tamoxifen, whereas women greater than age 50 would benefit from using Raloxifene if the patient still has a uterus. The STAR trial demonstrated significantly lower risk of thromboembolic disease, uterine cancer, uterine hyperplasia, and cataracts. Raloxifene also demonstrated a significantly lower incidence of ovarian cysts, endometrial polyps, hot flashes, vaginal discharge, and vaginal bleeding during a retrospective analysis of data from the NSABP-P1 and STAR trials).[155]

In addition to the use of selective estrogen receptor modulators, chemoprevention can also be implemented with the use of aromatase inhibitors. Exemestane, which functions to inhibit the conversion of androgens to estrogen, is recommended by the American Society of Clinical Oncology. Postmenopausal women who are at an increased risk of breast cancer, have a history of atypical hyperplasia, or lobular carcinoma in situ are encouraged to discuss with their physician the benefits of starting Exemestane for chemoprevention.[152]

Bisphosphonates, such as Zoledronate, have a threefold benefit in treating individuals with breast cancer. Bisphosphonates aid in reducing skeletal related events, such as pathological fractures, bone pain, and hypercalcemia related to breast cancer metastasis. The use of aromatase inhibitors decreases the amount of circulating estrogen leading to decreased bone strength. The use of bisphosphonates can help with stabilizing the bone during treatment. In addition, bisphosphonates have demonstrated a decrease in bone recurrence, and breast cancer–related death in postmenopausal patients. The Early Breast Cancer Trialists' Collaborative Group included 5 oral clodronate trials ($N = 5{,}053$), and 17 aminobisphosphonates trials ($N = 12{,}738$).[157] This metaanalysis concluded that treatment with bisphosphonates reduced the risk of bone recurrence by 34%, and the risk of breast cancer death by 17% in postmenopausal women.

4.3 Surgery

Mutations in BRCA1 and BRCA2 tumor suppressor genes increase the risk of breast cancer. The option of prophylactic mastectomy should be discussed and offered to those with this mutation(s). Body image is a concern and must be discussed when individuals are considering a mastectomy. Genetic risk and psychological assessment should be completed, along with an evaluation by a breast surgeon. The option for reconstructive surgery is necessary to maintain a patient's self-identify and confidence.[158]

Overall, several different preventive options are available, and each should be discussed thoroughly with the patient. Comprehension of the disease pathophysiology and potential side effects of preventive treatment, will allow the patient to make an informed decision based on what is most appropriate.

5. BREAST CANCER EARLY DETECTION AND SCREENING

Screening tests detect life-threatening diseases at early stages. Earlier diagnosis of disease can lead to improved outcomes in those with disease and lessen the incidence of a disease presenting at later stages. It is important to have discussions with patients to compare potential benefits and harms.

5.1 History of Screening

In 1913, Albert Salomon, a German surgeon, performed a roentgeno-histological study on 3000 mastectomies. Through X-rays he was able to find differences in normal and cancerous breast tissue. In 1930, Stafford Warren described and categorized the appearance of normal breasts. He also discussed the importance of comparing the right and left breasts in side-by-side views.[159] In the 1950's, after more than a decade of research, Gershon-Cohen and Strickler published a report on the variations of a normal mammogram. Gershon-Cohen and colleagues went on to define the criteria for normal and characteristic findings of benign and malignant disease. In 1965, through the work of Robert Egan, the American College of Radiology sponsored a conference that led to recommended dosages and reproducible techniques for diagnostic examinations of the breasts.[159]

In the 1950s mammography was gaining usefulness as a diagnostic modality and 20 years later, Betty Ford, the First Lady of the United States, and Rose Kushner, an American journalist, was diagnosed with breast cancer and the two became prominent advocates for breast cancer awareness in the United States. Betty Ford openly discussed her disease in public and used her illness to help " . . . save other lives . . . " Rose Kushner adamantly refused to have the "one-step" procedure. This procedure involved an immediate mastectomy of the breast with biopsy proven cancer. In 1979, the National Institutes of Health conference on breast cancer treatment concluded that "the 1-step procedure was no longer appropriate and other less disfiguring surgeries combined with radiotherapy, chemotherapy, or both afforded the same survival."[160] Years later Nancy Reagan, former First Lady of the United States, and Nancy Brinker, sister of Susan G. Komen, began to increase public awareness of breast cancer even further. This led to "legislative, regulatory, and funding changes" that have brought breast cancer into mainstream and greatly improved research into the diagnosis and treatment of this disease.[161]

5.2 Screening Benefits and Harms

Breast cancer will affect the majority of the population through patient, personal, and family or friend experiences. Screening mammography in women aged 40 years and older during the 1980s–90s in the United States led to a large increase in cases of early-stage cancer diagnoses (112–234 cancers per 100,000), a decrease in late stage cancer diagnoses (102 to 94 cases per 100,000 women), and a decrease in the death rate from breast cancer (71 to 51 deaths per 100,000 women).[162] Case-control studies, RCTs, and observational studies have also shown a decrease in breast cancer mortality secondary to screening mammograms.[163,164] These studies have shown 15%–40% reduction in mortality from breast cancer secondary to screening mammograms. Contributing to this reduction is a combination of advanced screening, better treatments, and underreporting of breast cancer as a cause of death.[163,165] Breast cancer screening does save lives, however, as with any screening test, there is the potential risk of harm from screening.

Harms associated with screening mammograms include false negative results, overdiagnosis, radiation-induced cancer, and false positives. False-negative mammograms result in undetected breast cancer and estimates range from 6% to 46% of women screened, and they may be more common in younger women with denser breasts.[2] It can result in women erroneously confident that they do not have breast cancer and this can cause late stage diagnosis of breast cancer or death. Radiation-induced cancers from digital mammograms are rare and are estimated to be 0.4–1.2 per 10, 000 women screened over a lifetime. Screening mammograms may also lead to false-positive results.[1,147,164] Women with false-positive results may require further imaging and often a breast biopsy.[122] Fifty percent of women screened annually for 10 years will experience a false positive and 7%–17% will have biopsies.[2,18] On average, of 10% of women recalled for further testing, 5 out of 100 women will have cancer.[2] Screening low-risk women at younger ages may cause unnecessary further imaging and biopsies. This particular harm needs to be discussed with women prior to initiating screening. Women may experience anxiety and psychological distress for up to 2 years after a false positive result.[122,147] Another important potential harm is overdiagnosis, which is the diagnosis and treatment of noninvasive and invasive breast cancer that would ultimately not cause symptoms, threaten life or cause death. Women may be exposed to disfiguring surgery, radiation, hormone therapy or chemotherapy, and the adverse consequences of these treatments (scarring, cardiac toxicity, and lymphedema).[164] Observing trends using the surveillance, epidemiology, and end results data from 1976 through 2008, the estimated proportion of overdiagnosed breast cancers varies widely between 5% and 54%.[1,2,162,164] It is estimated that 1 in 8 women diagnosed with screening mammograms from the ages 50 to 75 years are overdiagnosed,[122] see Table 1.

Table 1 Lifetime Benefits and Harms of Biennial Screening Mammography, 1000

Variable	Ages 40–74 years	Ages 50–74 years
False-positive test results (*n*)	1529 (1100–1976)	953 (830–1325)
Unnecessary breast biopsies (*n*)	213 (153–276)	146 (121–205)
Overdiagnosed breast tumors, (*n*)	21 (12–38)	19 (11–34)

Women screened: model results compared with no screening.
Adapted with permission of the Final Recommendation Statement—Breast Cancer: Screening. U.S. Preventive Services Task Force. November 2016. https://www.uspreventiveservicestaskforce.org/Page/Document/RecommendationStatementFinal/breast-cancer-screening1[166].

5.3 Screening Recommendations

In the USA, the U.S. Preventive Services Task Force (USPSTF) and other professional organizations make recommendations for prevention and screening based on periodic assessments of the evidence base, and also evaluate the level of the recommendation based on the strength of the evidence, see Table 2 for the level of evidence recommendation categories for the USPSTF.[122,166] The breast cancer screening guidelines also address women at an increased risk for breast cancer, women who request screening outside of the standard age ranges and the use of breast cancer risk screening tools to obtain a thorough family and personal history to assist the clinician and patient to identify individual risks and make informed decisions for breast cancer screening.[122]

5.4 Breast Cancer Screening in Average Risk Women

The USPSTF recommends screening asymptomatic women aged 50–74 years old with biennial screening mammograms (B recommendation).[122,166] Screening women aged 40–49 years with mammograms is a C recommendation.[122,166] This C recommendation indicates that healthcare providers discuss the pros and cons of screening to ensure patients understand the potential harms associated with screening mammograms, such as false-positives, false-negatives, and overdiagnosis so that they can make an informed decision about whether to get screened. In RCTs, screening mammograms are shown to be effective in women aged 40–49 in averting breast cancer

Table 2 United States Preventive Services Task Force Grades and Suggestions for Practice

Grades	Definitions	Suggestions for Practices
A	Net benefit, recommend service	Offer or provide this service
B	Moderate net benefit, recommend service	Offer or provide this service
C	Small net benefit, selectively provide service	Offer service for select patients
D	No net benefit, do not recommended	Discourage use of service
I statement	Current evidence is insufficient to assess net benefit	If service is offered, patients should understand the uncertainty about harms and benefits

Adapted with permission of the Final Recommendation Statement—Breast Cancer: Screening. U.S. Preventive Services Task Force. November 2016. https://www.uspreventiveservicestaskforce.org/Page/Document/RecommendationStatementFinal/breast-cancer-screening1[166].

death [RR 0.92 (0.75–1.02)], but to a lesser degree compared to women aged 60–69 years [RR 0.67 (0.54–0.83)].[147] Women aged 40–49 years, do receive a greater number and proportion of negative additional imaging and breast biopsies compared to older women[122] (Table 1). Women aged 60–69 years benefit the most from mammographic screening and are most likely to avoid death from breast cancer. Screening younger women is based on individual risk factors and requires a thorough family history of certain cancer syndromes. The USPSTF has no recommendations for screening women 75 years and older (I statement). The USPSTF also supports that women have an awareness of changes in their bodies (i.e., breasts) but does not recommend teaching breast self examinations.[122,166]

Other national guidelines are similar to the USPSTF. The American College of Radiology (ACR), Appropriateness Criteria,[167] and the American College of Obstetricians and Gynecologists (ACOG) recommend that women start annual screening at age 40.[168,169] The ACR defines women of average risk as having a <15% lifetime risk of breast cancer and breast that are not classified as dense. The health provider should discuss the benefits and potential harms of screening starting at aged 40 years.[122] ACOG also recommends clinical breast exams in women aged 20–39 years old every 1–3 years and annually in women aged 40 years. ACOG, USPSTF, and ACS promote "breast self-awareness"[18,166,169]; ACOG includes instruction for and encourages self-breast examinations for high-risk women.[169] The American Cancer Society (ACS) has a strong recommendation that women with average risk for breast cancer should undergo screening for breast cancer: women aged 45–54 years should begin annual screening for breast cancer (qualified recommendation).[18] Women 55 years and older should transition from annual to biennial screening or can continue with annual screening after discussion with health provider (qualified recommendation). Women aged 40–44 years should have the opportunity to start annual screening (qualified recommendation). Finally, women should continue mammograms as long as they are healthy and there is at least a 10 year life expectancy (qualified recommendation). The ACS does not recommend clinical breast exam for any patient (qualified recommendation).[18]

5.5 Genetic Mutations, Breast Cancer Screening Tools, and Screening Methods in High-Risk Women

As discussed previously mutations in BRCA1 and BRCA2 can cause cancer in the breasts and in other organs, such as the ovaries, pancreas, and prostate. In the United States, approximately 3% of breast cancers and

10% of ovarian cancers are caused by these gene mutations.[145] The USPSTF recommends screening women to identify relatives with these hereditary syndromes and refer them for genetic counseling.[166] Women referred to genetic counseling discuss possibilities for genetic testing and receive counseling on early mammographic screening with adjunctive imaging (MRI), medications, chemotherapy, and risk reducing surgery (mastectomy and oophorectomy).

There are various online clinical screening tools that can assist the clinician in stratifying patients into different risk categories. The National Cancer Institute has developed the Breast Cancer Risk Assessment Tool (BCRAT), a validated calculator to estimate invasive breast cancer risk for women in the United States (https://www.cancer.gov/bcrisktool/), which is based upon the Gail Model that has been validated for white women.[18] Family history is a key element of this tool as with most other breast cancer screening tools. The BCRAT is not meant for women with a diagnosis of breast cancer, DCIS, LCIS, or for women who may have received radiation to the chest for Hodgkin lymphoma. Women with LCIS can use the IBIS Breast Cancer Risk Evaluation Tool (www.ems-trials.org). For women with known BRCA1 or BRCA2 gene mutations, the BOADICEA (Breast and Ovarian Analysis of Disease Incidence and Carrier Estimation Algorithm) model (www.ccge.medschl.cam.ac.uk) can be used to estimate their breast cancer risk. This model is a computer program that calculates the risk of breast and ovarian cancer in women based on their family history. There are also tools available for screening and providing information on BRCA gene mutations. A BRCA decision support tool can be found at www.KnowBRCA.org and through a digital media campaign that provides information on (http://www.cdc.gov/cancer/breast/young_women/bring/your/brave)[145] hereditary breast and ovarian cancer syndrome.[145] Women who screen positive would benefit from a referral to a breast specialty center for genetic counseling, further testing, and management.

According to ACR guidelines, women deemed high risk should be screened earlier; ages 25–30 or 10 years earlier (whichever is later) for women with a first-degree relative with premenopausal breast cancer; ages 25–30 years for women who are carriers of the BRCA gene 1, or having a lifetime risk of 20% of developing breast cancer on the basis of family history[122,158]; 8 years after radiation therapy but not before age of 25 years for those who have received mantle radiation between 10 and 30 years of age; and any age for women with biopsy proven lobular neoplasia, atypical ductal hyperplasia, DCIS, or invasive breast cancer.[158,168]

The ACS also includes personal and family history of BRCA1 and BRCA2 gene mutations, Li-Fraumeni syndrome (autosomal dominant disorder also known as the Sarcoma, Breast, Leukemia and Adrenal Gland cancer syndrome) or Cowden syndrome (a germline mutation in the phosphatase and tensin homolog (PTEN) gene with harmartomatous tumor development) as risk factors for the development of breast cancer. Women with any of these conditions are encouraged to begin mammographic screening at age 30 years.[1]

5.6 Screening Modalities

As stated earlier, mammography is a procedure that allows visualization of the internal structures of the breasts. Across all ages screening mammogram has a sensitivity of 77%–95% and a specificity of 94%–97%.[166] Three types of mammography are film, digital, and breast tomosynthesis. Film uses X-ray equipment to record images of the breasts and digital mammography (DM) uses specialized computer equipment and delivers lower doses of radiation. DM has mainly replaced film mammography.[18]

Dense breasts are a risk factor for breast cancer.[122] A classification system for dense breasts from the American College or Radiology's Breast Imaging Reporting and Data System (BI-RADS) is a four-category scale (Table 3).

Dense breast tissue reduces mammography sensitivity from 87% to 73% and specificity from 96% to 90%.[166] Many states in the United States require patient notification of breast density status when a mammogram is performed. Some states require that adjunctive screening should at least be considered.[166] According to data from BSCS, 25 million women aged 40–74 years have a BI-RADS Class D. Relative risk for developing invasive breast cancer in women with extremely dense breasts is, 1.23 aged 40–49 years, 1.29 for women aged 50–64 years, and 1.30 for women aged 65–74 years.[122] DM accounts for the majority of mammography screening in the United States and has a higher sensitivity that increases with age and decreases with denser breast tissue.[147] The USPSTF does not recommend

Table 3 Breast Tissue[167]

Breast Composition Categories
a. The breasts are almost entirely fatty
b. There are scattered areas of fibroglandular density
c. The breasts are heterogeneously dense, which may obscure small masses
d. The breasts are extremely dense, which lowers the sensitivity of mammography

Permission granted from the ACR Appropriateness Criteria.

use of adjunctive imaging, such as breast MRI, breast ultrasound, or digital breast tomosynthesis (DBT) in women with a normal screening mammogram.[166]

In 2011, the FDA approved use of DBT or 3-dimensional (3-D) mammography. A 3-D image of the breast is constructed with multiple high-resolution X-rays and can be used in combination with a 2-D or DM with single breast compression. DBT allows for the display of breast tissue in 1 mm sections, which reduces superimposition of breast tissue.[168] DBT with digital mammogram results in less recall rates for women to come in for additional imaging compared to conventional screening (7.8% vs. 12.3% recall). Cancer detection is 5.9 per 1000 with 3-D compared to 4.4 per 1000 with 2D alone. The combined radiation does is higher but still within acceptable limits and is still within the FDA approved limit.[170]

Breast MRIs use magnetic fields instead of X-rays to produce very detailed, cross sectional images of the body. There is use of contrast, either gadolinium or DTPA that is injected into a vein. Breast MRI is complementary imaging to mammography for women who are high risk.[1] In high-risk women, breast MRI has a higher sensitivity than mammography and combined with mammography has the highest sensitivity (92.7%).[168] If a woman cannot undergo an MRI then an ultrasound is recommended.[168] It is important to note that nephrogenic systemic fibrosis is related to underlying severe renal disease with the administration of gadolinium-based contrast agents. It occurs primary in patients with estimated glomerular filtrations rates of <30 mL/min/1.73 m^2

5.7 Breast Ultrasound

Ultrasound of the breasts can detect more cancer than mammogram alone especially in women with dense breast tissue. Ultrasound should not be used alone to screen for breast cancer.[18] It can be used on women with inconclusive mammogram findings, to guide for different breast biopsies, and to distinguish a solid from a nonsolid mass.[169] It can also be used in high-risk women who need adjunctive imaging and cannot tolerate an MRI.[168]

Breast cancer screening saves lives. The recommendations for average-risk women follow age-based protocols. Women undergoing screening should routinely be assessed for personal and family history and other risk factors that place them at an increased risk of breast cancer. There are calculators that estimate a women's risk for having gene certain mutations (BRCA1 and BRCA2) and a women's risk of developing breast cancer based on family history. When screening women, especially younger women, it is important to discuss the limitations and possible harms from mammography screening, such as overdiagnosis and false positives that can lead to additional imaging, biopsies, and even treatment. DM is the foremost screening tool, however, breast DBT in addition to DM results in less recalls for additional imaging, better detection of cancer, and less false positives in some cases. MRI of the breasts is a complementary test and is usually reserved for high-risk women based on their family history. Healthcare providers should strongly consider referral for women considered high risk for developing breast cancer.

REFERENCES

1. *Breast Cancer Facts & Figures*. American Cancer Society; 2017
2. *National Cancer Institute*. Breast Cancer Screening (PDQ)—Health Professional Version. National Cancer Institute; 2017
3. Ferlay J, Soerjomataram I, Dikshit R, et al. Cancer incidence and mortality worldwide: sources, methods and major patterns in GLOBOCAN 2012. *Int J Cancer.* 2015;136(5): E359–E386.
4. Ferlay J, Soerjomataram I, Ervik M, Dikshit R, Eser S, Mathers C, Rebelo M, Parkin DM, Forman D, Bray F. *GLOBOCAN 2012 v1.0, Cancer Incidence and Mortality Worldwide: IARC CancerBase No. 11 [Internet]*. Lyon, France: International Agency for Research on Cancer; 2013. Available from: http://globocan.iarc.fr
5. Torre LA, Siegel RL, Ward EM, Jemal A. Global cancer incidence and mortality rates and trends—an update. *Cancer Epidemiol Biomarkers Prev.* 2016;25(1):16–27.
6. Ginsburg O, Bray F, Coleman MP, et al. The global burden of women's cancers: a grand challenge in global health. *Lancet.* 2016;389:847–860.
7. DeSantis CE, Fedewa SA, Goding Sauer A, Kramer JL, Smith RA, Jemal A. Breast cancer statistics, 2015: convergence of incidence rates between black and white women. *Cancer J Clin.* 2016;66(1):31–42.
8. Howlader NNA, Krapcho M, Miller D, Bishop K, Altekruse SF, Kosary CL, Yu M, Ruhl J, Tatalovich Z, Mariotto A, Lewis DR, Chen HS, Feuer EJ, Cronin KA. *Cancer Statistics Review, 1975–2013—SEER Statistics*. Based on November 2015 SEER data submission; 2016. Available from: https://seer.cancer.gov/csr/1975_2013/
9. Sankaranarayanan R, Swaminathan R, Brenner H, et al. Cancer survival in Africa, Asia, and Central America: a population-based study. *Lancet Oncol.* 2010;11(2): 165–173.
10. DeSantis C, Jemal A, Ward E. Disparities in breast cancer prognostic factors by race, insurance status, and education. *Cancer Causes Control.* 2010;21(9):1445–1450.
11. Siegel RL, Miller KD, Jemal A. Cancer statistics, 2015. *Cancer J Clin.* 2015;65 (1):5–29.
12. Mokdad AH, Dwyer-Lindgren L, Fitzmaurice C, Stubbs RW, Bertozzi-Villa A, Morozoff C, Charara R, Allen C, Naghavi M, Murray C. Trends and patterns of disparities in cancer mortality among US counties, 1980–2014. *JAMA.* 2017;317 (4):388–406.
13. Bray F, Ferlay J, Laversanne M, et al. Cancer incidence in five continents: inclusion criteria, highlights from volume X and the global status of cancer registration. *Int J Cancer.* 2015;137(9):2060–2071.
14. American Cancer Society. *Cancer Statistics Presentation*; 2005. Available from: www. cancer.org

15. Rosenberg PS, Kimberly BK, Anderson WF, Government of Canada. National Research Council Canada. *Estrogen Receptor Status and the Future Burden of Invasive and In Situ Breast Cancers in the United States*; 2011
16. Smith BD, Smith GL, Hurria A, Hortobagyi GN, Buchholz TA. Future of cancer incidence in the United States: burdens upon an aging, changing nation. *J Clin Oncol.* 2009;27(17):2758–2765.
17. Horner MJ, Ries LAG, Krapcho M, Aminou R, Howlader N, Altekruse SF, Feuer EJ, Huang L, Mariotto A, Miller BA, Lewis DR, Eisner MP, Stinchcomb DG, Edwards BK. *SEER Cancer Statistics Review 1975–2006—Previous Version—SEER Cancer Statistics*; 2009. Available from: https://seer.cancer.gov/archive/csr/1975_2006/
18. American Cancer Society. *Breast Cancer Facts & Figures*. American Cancer Society; 2017
19. Franco-Marina F, López-Carrillo L, Keating NL, Arreola-Ornelas H, Marie Knaul F. Breast cancer age at diagnosis patterns in four Latin American populations: a comparison with North American countries. *Cancer Epidemiol.* 2015;39(6):831–837.
20. Leong SP, Shen ZZ, Liu TJ, et al. Is breast cancer the same disease in Asian and Western countries? *World J Surg.* 2010;34(10):2308–2324.
21. Abdulrahman GO, Rahman GA. Epidemiology of breast cancer in europe and Africa. *J Cancer Epidemiol.* 2012;2012:915610.
22. Iqbal J, Ginsburg O, Rochon PA, Sun P, Narod SA. Differences in breast cancer stage at diagnosis and cancer-specific survival by race and ethnicity in the United States. *JAMA.* 2015;313(2):165–173.
23. Tao L, Gomez SL, Keegan TH, Kurian AW, Clarke CA. Breast cancer mortality in african-american and non-hispanic white women by molecular subtype and stage at diagnosis: a population-based study. *Cancer Epidemiol Biomarkers Prev.* 2015;24 (7):1039–1045.
24. Chlebowski RT, Chen Z, Anderson GL, et al. Ethnicity and breast cancer: factors influencing differences in incidence and outcome. *J Natl Cancer Inst.* 2005;97(6): 439–448.
25. Maskarinec G, Sen C, Koga K, Conroy SM. Ethnic differences in breast cancer survival: status and determinants. *Womens Health.* 2011;7(6):677–687.
26. Silber JH, Rosenbaum PR, Clark AS, et al. Characteristics associated with differences in survival among black and white women with breast cancer. *JAMA.* 2013;310 (4):389–397.
27. Gomez SL, Clarke CA, Shema SJ, Chang ET, Keegan TH, Glaser SL. Disparities in breast cancer survival among Asian women by ethnicity and immigrant status: a population-based study. *Am J Public Health.* 2010;100(5):861–869.
28. Albain KS, Unger JM, Crowley JJ, Coltman CA, Hershman DL. Racial disparities in cancer survival among randomized clinical trials patients of the Southwest Oncology Group. *J Natl Cancer Inst.* 2009;101(14):984–992.
29. Blows FM, Driver KE, Schmidt MK, et al. Subtyping of breast cancer by immunohistochemistry to investigate a relationship between subtype and short and long term survival: a collaborative analysis of data for 10,159 cases from 12 studies. *PLoS Med.* 2010;7(5):e1000279.
30. Perou CM, Børresen-Dale AL. Systems biology and genomics of breast cancer. *Cold Spring Harb Perspect Biol.* 2011;3(2):a003293.
31. American College of Obstetricians and Gynecologists. *Routine cancer screening*; 1997
32. American Gastroenterological Association. *New National Colorectal Cancer Practice Guidelines Recommend Life-Saving Tests*; 2000
33. Edelist D. Does fecal occult blood screening reduce colorectal cancer morbidity? *J Fam Pract.* 2002;51:412–415.
34. Heflin M, Oddone E, Pieper C, Burchett B, Cohen H. The effect of comorbid illness on receipt of cancer screening by older people. *J Am Geriatr Soc.* 2002;50:1651–1658.

35. Frew E, Wolstenholme J, Whynes D. Mass population screening for colorectal cancer: factors influencing subjects' choice of screening test. *J Health Serv Res Policy.* 2001;6:85–91.
36. Roetzheim R, Pal N, Gonzalez E, Ferrante J, Van Durme D, Krischer J. Effects of health insurance and race on colorectal cancer treatments and outcomes. *Am J Publ Health.* 2000;90:1746–1754.
37. Nichols S, Koch E, Lallemand R, et al. Randomised trial of compliance with screening for colorectal cancer. *Br Med J.* 1986;293:107–110.
38. Duggan C, Coronado G, Martinez J, et al. Cervical cancer screening and adherence to followup among Hispanic women study protocol: a randomized controlled trial to increase the uptake of cervical cancer screening in Hispanic women. *BMC Cancer.* 2012;12(1):170.
39. Masi C, Blackman D, Peek M. Interventions to enhance breast cancer screening, diagnosis, and treatment among racial and ethnic minority women. *Med Care Res Rev.* 2007;64(suppl 5):195S–242S.
40. Schueler K, Chu P, Smith-Bindman R. Factors associated with mammography utilization: a systematic quantitative review of the literature. *J Womens Health.* 2008;17(9):1477–1498.
41. Esserman L, Joe BN. *Diagnostic Evaluation and Initial Staging Work-Up of Women With Suspected Breast Cancer*; 1998. Available from: http://www.uptodate.com/home/index.html
42. Goel MS, O'Conor R. Increasing screening mammography among predominantly Spanish speakers at a federally qualified health center using a brief previsit video. *Patient Educ Couns.* 2016;99(3):408–413.
43. *Screening and Risk Factors for Texas Counties*; 2008–2010. Available from: https://statecancerprofiles.cancer.gov/map/map.withimage
44. Sanchez N, Stierman B, Saab S, Mahajan D, Yeung H, Francois F. Physical activity reduces risk for colon polyps in a multiethnic colorectal cancer screening population. *BMC Res Notes.* 2012;5(1):312.
45. Lindberg NM, Stevens VJ. Immigration and weight gain: Mexican-American women's perspectives. *J Immigr Minor Health.* 2011;13(1):155–160.
46. Baron R, Rimer B, Breslow R, et al. Client-directed interventions to increase community demand for breast, cervical, and colorectal cancer screening a systematic review. *Am J Prev Med.* 2008;35(suppl 1):S34–S55.
47. Wilson TE, Fraser-White M, Feldman J, et al. Hair salon stylists as breast cancer prevention lay health advisors for African American and Afro-Caribbean women. *J Health Care Poor Underserved.* 2008;19(1):216–226.
48. Thrush RS, Lanese RR. The use of printed material in diabetes education. *Diabetes.* 1962;11:132–137.
49. Baron R, Melillo S, Rimer B, et al. Intervention to increase recommendation and delivery of screening for breast, cervical, and colorectal cancers by healthcare providers a systematic review of provider reminders. *Am J Prev Med.* 2010;38(1):110–117.
50. Wallace D, Hunter J, Papenfuss M, De Zapien JG, Denman C, Giuliano AR. Pap smear screening among women >/=40 years residing at the United States-Mexico border. *Health Care Women Int.* 2007;28(9):799–816.
51. Mendez LE, Manci N, Cantuaria G, et al. Expression of glucose transporter-1 in cervical cancer and its precursors. *Gynecol Oncol.* 2002;86(2):138–143.
52. Invasive Cancer Incidence & Mortality Rates in Texas Counties; 2009–2013. Available from: http://www.cancer-rates.info/tx/
53. Miles L. Physical activity and health. *Nutr Bull.* 2007;32(4):314–363.
54. http://muafind.hrsa.gov/index.aspx

55. Pérez A, Reininger BM, Aguirre Flores MI, Sanderson M, Roberts RE. Physical activity and overweight among adolescents on the Texas-Mexico border. *Rev Panam SaludP°blica.* 2006;19:244–252.
56. Jacobs F. The Patients Per Doctor Map of the World. *BigThink*; 2007. Available from: http://bigthink.com/ideas/21237
57. Jylhä M. What is self-rated health and why does it predict mortality? Towards a unified conceptual model. *Soc Sci Med.* 2009;69(3):307–316.
58. Suarez L, Pulley L. Comparing acculturation scales and their relationship to cancer screening among older Mexican-American women. *J Natl Cancer Inst Monogr.* 1995;(18):41–47.
59. Adams EK, Breen N, Joski PJ. Impact of the National Breast and Cervical Cancer Early Detection Program on mammography and Pap test utilization among white, Hispanic, and African American women: 1996-2000. *Cancer.* 2007;109(suppl 2):S348–S358.
60. Ferrante JM, Piasecki AK, Ohman-Strickland PA, Crabtree BF. Family physicians' practices and attitudes regarding care of extremely obese patients. *Obesity.* 2009;17(9):1710–1716.
61. Scarinci IC, Garcia FA, Kobetz E, et al. Cervical cancer prevention: new tools and old barriers. *Cancer.* 2010;116(11):2531–2542.
62. Mier N, Ory MG, Zhan D, Wang S, Burdine JN. Levels and correlates of exercise in a border mexican american population. *AmJ HealthBehav.* 2007;31(2):159–169.
63. Veach E, Xique I, Johnson J, et al. Race matters: analyzing the relationship between colorectal cancer mortality rates and various factors within respective racial groups. *Front Public Health.* 2014;2:239.
64. El Paso University Medical Center Cancer Registry. *Cervical Cancer Cases 2006–2010*; 2012
65. Usher-Smith JA, Walter FM, Emery J, Win AK, Griffin SJ. Risk prediction models for colorectal cancer: a systematic review. *CancerPrev Res.* 2015;9:13–26.
66. Census 2000 summary file 3 (SF 3)—Sample data: P19. *Age By Language Spoken at Home by Ability to Speak English for the Population 5 Years and Over.* Available from: http://factfinder.census.gov/servlet
67. Society of teachers of family medicine. *Family Medicine Digital Resources Library*; 2012. Available from: http://www.fmdrl.org/
68. Jandorf L, Bursac Z, Pulley L, Trevino M. Breast and cervical cancer screening among Latinas attending culturally specific educational programs. *Progr Commun Health Partnersh.* 2008;2(3):195–204.
69. Hashim MJ, Prinsloo A, Leduc C, Raasch BA, Mirza DM. What community-based preceptors want in teaching medical students: findings from a mixed methods study. *Educ Health.* 2010;23(2):366.
70. Warner E, Plewes DB, Hill KA, et al. Surveillance of BRCA1 and BRCA2 mutation carriers with magnetic resonance imaging, ultrasound, mammography, and clinical breast examination. *JAMA.* 2004;292(11):1317–1325.
71. Wujcik D, Fair A. Barriers to diagnostic resolution after abnormal mammography: a review of the literature. *CancerNurs.* 2008;31(5):E16–E30.
72. Anonymous. *2015 El Paso Community Cancer Needs Assessment—A Comprehensive Report of the State of Cancer Support Care in Our Region.* El Paso, Texas: Rio Grande Cancer Foundation; 2015
73. Cuzick J, Szarewski A, Mesher D, et al. Long-term follow-up of cervical abnormalities among women screened by HPV testing and cytology—results from the Hammersmith study. *IntJ Cancer.* 2008;122(10):2294–2300.
74. Taplin SH, Barlow WE, Ludman E, MacLehos R, Meyer DM, Seger D, Herta D, Chin C, Curry S. Testing Reminder and Motivational Telephone Calls to Increase Screening Mammography: a Randomized Study. *J Natl Cancer Inst.* 2000;92(93):233–242.

75. Ghaddar S, Brown CJ, Pagán JA, Díaz V. Acculturation and healthy lifestyle habits among Hispanics in United States-Mexico border communities. *Rev Panam Salud P°blica.* 2010;28:190–197.
76. Warburton DE, Nicol CW, Bredin SS. Health benefits of physical activity: the evidence. *CMAJ.* 2006;174:801–809.
77. Wareham NJ, van Sluijs EM, Ekelund U. Physical activity and obesity prevention: a review of the current evidence. *Proc Nutr Soc.* 2005;64:229–247.
78. Anonymous. *Medically Underserved Areas*; 2009. Available from: http://www.cprit.state.tx.us/images/uploads/muas.pdf
79. Engelstad LP, Stewart SL, Nguyen BH, et al. Abnormal Pap smear follow-up in a high-risk population. *Cancer Epidemiol Biomarkers Prev.* 2001;10(10):1015–1020.
80. Dohan D, Schrag D. Using navigators to improve care of underserved patients: current practices and approaches. *Cancer.* 2005;104(4):848–855.
81. Shantakumar S, Terry MB, Paykin A, et al. Age and menopausal effects of hormonal birth control and hormone replacement therapy in relation to breast cancer risk. *Am J Epidemiol.* 2007;165(10):1187–1198.
82. Nelson HD, Zakher B, Cantor A, et al. Risk factors for breast cancer for women aged 40 to 49 years: a systematic review and meta-analysis. *Ann Intern Med.* 2012;156 (9):635–648.
83. Faupel-Badger JM, Arcaro KF, Balkam JJ, et al. Postpartum remodeling, lactation, and breast cancer risk: summary of a National Cancer Institute-sponsored workshop. *J Natl Cancer Inst.* 2013;105(3):166–174.
84. Bassuk SS, Manson JE. Oral contraceptives and menopausal hormone therapy: relative and attributable risks of cardiovascular disease, cancer, and other health outcomes. *Ann Epidemiol.* 2015;25(3):193–200.
85. Althuis MD, Brogan DR, Coates RJ, et al. Hormonal content and potency of oral contraceptives and breast cancer risk among young women. *Br J Cancer.* 2003;88 (1):50–57.
86. Beaber EF, Buist DS, Barlow WE, Malone KE, Reed SD, Li CI. Recent oral contraceptive use by formulation and breast cancer risk among women 20 to 49 years of age. *Cancer Res.* 2014;74(15):4078–4089.
87. Li CI, Beaber EF, Tang MT, Porter PL, Daling JR, Malone KE. Effect of depo-medroxyprogesterone acetate on breast cancer risk among women 20 to 44 years of age. *Cancer Res.* 2012;72(8):2028–2035.
88. Strom BL, Berlin JA, Weber AL, et al. Absence of an effect of injectable and implantable progestin-only contraceptives on subsequent risk of breast cancer. *Contraception.* 2004;69(5):353–360.
89. Coughlin SS, Uhler RJ, Richards T, Wilson KM. Breast and cervical cancer screening practices among Hispanic and non-Hispanic women residing near the United States-Mexico border, 1999–2000. *Fam Commun Health.* 2003;26(2):130–139.
90. Brandsma LL. Physician and patient attitudes toward obesity. *Eat Disord.* 2005;13: 201–211.
91. Watlington AT, Byers T, Mouchawar J, Sauaia A, Ellis J. Does having insurance affect differences in clinical presentation between Hispanic and non-Hispanic white women with breast cancer? *Cancer.* 2007;109(10):2093–2099.
92. Byrd TL, Wilson KM, Lee Smith J, et al. A multi-city, muli-component cervical cancer prevention trial among Mexcican-American women. *AMIGAS*; 2012
93. Cameron AJ, Shaw JE, Zimmet PZ. The metabolic syndrome: prevalence in worldwide populations. *Endocrinol Metab Clin North Am.* 2004;33(2):351–375.
94. Ahmad F, Cameron JI, Stewart DE. A tailored intervention to promote breast cancer screening among South Asian immigrant women. *Soc Sci Med.* 2005;60(3): 575–586.

95. Center for Substance Abuse Treatment. *Enhancing Motivation for Change in Substance Abuse Treatment*. Rockville, MD, USA: Substance Abuse and Mental Health Services Administration; 1999. [vol series].
96. Slater JS, Finnegan JR, Madigan SD. Incorporation of a successful community-based mammography intervention: dissemination beyond a community trial. *Health Psychol.* 2005;24(5):463–469.
97. Hollingsworth A, Stough R. The emerging role of breast magnetic resonance imaging. *J Okla State Med Assoc.* 2003;96(7):299–307.
98. Laukkanen JA, Laaksonen DE, Niskanen L, Pukkala E, Hakkarainen A, Salonen JT. Metabolic syndrome and the risk of prostate cancer in Finnish men: a population-based study. *Cancer Epidemiol Biomarkers Prev.* 2004;13(10):1646–1650.
99. Alberti KG, Eckel RH, Grundy SM, et al. Harmonizing the metabolic syndrome: a joint interim statement of the International Diabetes Federation Task Force on Epidemiology and Prevention; National Heart, Lung, and Blood Institute; American Heart Association; World Heart Federation; International Atherosclerosis Society; and International Association for the Study of Obesity. *Circulation.* 2009;120(16): 1640–1645.
100. Association of American Medical Colleges, Center for Workforce Studies. *State Physician Workforce Data Release.* Association of American Medical Colleges; 2011
101. Worden JK, Solomon LJ, Flynn BS, McVety DJ, Dorwaldt AL, Geller BM. Community-based promotion of breast screening using small group education. *J Public Health Manag Pract.* 1999;5(3):54–62.
102. CfHS. *Texas Behavioral Risk Factor Surveillance System Survey Data*; 2010. Available from: https://www.dshs.state.tx.us/Layouts/ContentPage.aspx?pageid=35474&year=10&risk=170&area=ElPaso
103. Kolonel LN, Henderson BE, Hankin JH, et al. A multiethnic cohort in Hawaii and Los Angeles: baseline characteristics. *Am J Epidemiol.* 2000;151(4):346–357.
104. Slater JS, Ha CN, Malone ME, et al. A randomized community trial to increase mammography utilization among low-income women living in public housing. *Prev Med.* 1998;27(6):862–870.
105. Stager JL. The comprehensive breast cancer knowledge test: validity and reliability. *J Adv Nurs.* 1993;18(7):1133–1140.
106. Gill JM, Cooper AR. Physical activity and prevention of type 2 diabetes mellitus. *Sports Med.* 2008;38:807–824.
107. Stocks T, Lukanova A, Johansson M, et al. Components of the metabolic syndrome and colorectal cancer risk; a prospective study. *Int J Obes.* 2008;32(2):304–314.
108. HealthyPeople.gov. *General Health Status—Healthy People* 2020; 2014. Available from: http://healthypeople.gov/2020/about/genhealthabout.aspx#physically
109. Ganesh A, Ganesh G. Reflective writing by final year medical students: lessons for curricular change. *Natl Med J India.* 2010;23(4):226–230.
110. Hoskins KF, Zwaagstra A, Ranz M. Validation of a tool for identifying women at high risk for hereditary breast cancer in population-based screening. *Cancer.* 2006;107 (8):1769–1776.
111. Sorkin DH, Biegler K, Billimek J. Differences in self-reported physical activity and body mass index among older Hispanic and non-Hispanic white men and women: findings from the 2009 California Health Interview Survey. *J Am Geriatr Soc.* 2015;63: 2158–2163.
112. United States Census Bureau. *2005–2009 American Community Survey*; 2011. Available from: http://2020.census.gov/2010census/date
113. Byrd TL, Wilson KM, Smith JL, et al. Using intervention mapping as a participatory strategy: development of a cervical cancer screening intervention for Hispanic women. *Health Educ Behav.* 2012;39:603–611.

114. Legler J, Meissner H, Coyne C, Breen N, Chollette V, Rimer B. The effectiveness of interventions to promote mammography in women with historically lower rates of screening. *Cancer Epidemiol Biomarkers Prev.* 2002;11:59–71.
115. Byrd TL, Chavez R, Wilson KM. Barriers and facilitators of cervical cancer screening among Hispanic women. *Ethn Dis.* 2007;17(1):129–134.
116. Chukwueke I, Firek A, Beeson L, et al. The En Balance Spanish diabetes education program improves apolipoproteins, serum glucose and body composition in Hispanic diabetics. *Ethn Dis.* 2012;22(2):215–220.
117. Kriege M, Brekelmans CT, Boetes C, et al. Efficacy of MRI and mammography for breast-cancer screening in women with a familial or genetic predisposition. *N Engl J Med.* 2004;351(5):427–437.
118. Domchek SM, Rebbeck TR. Preventive surgery is associated with reduced cancer risk and mortality in women with BRCA1 and BRCA2 mutations. *LDI Issue Brief.* 2010;16 (2):1–4.
119. National Center for Health Statistics.*Health, United States, 2011: With Special Feature on Socioeconomic Status and Health.* Hyattsville, MD: National Center for Health Statistics; 2012.
120. Nelson HD, Smith ME, Griffin JC, Fu R. Use of medications to reduce risk for primary breast cancer: a systematic review for the U.S. Preventive Services Task Force. *Ann Intern Med.* 2013;158(8):604–614.
121. Frankel SD, Sickles EA, Curpen BN, Sollitto RA, Ominsky SH, Galvin HB. Initial versus subsequent screening mammography: comparison of findings and their prognostic significance. *AJR Am J Roentgenol.* 1995;164(5):1107–1109.
122. Siu AL. Force USPST.Screening for breast cancer: U.S. preventive services task force recommendation statement. *Ann Intern Med.* 2016;164(4):279–296.
123. Markides K, Coreil J. The health of Hispanics in the southwestern United States: an epidemiologic paradox. *Public Health Rep.* 1986;101(3):253–265.
124. Buki LP, Jamison J, Anderson CJ, Cuadra AM. Differences in predictors of cervical and breast cancer screening by screening need in uninsured Latino women. *Cancer.* 2007;110(7):1578–1585.
125. Taplin SH, Barlow WE, Ulcickas-Yood M, et al. Re: breast cancer screening comes full circle. *J Natl Cancer Inst.* 2005;97(6):461.
126. Marmot M. Social determinants of health inequalities. *Lancet.* 2005;365 (9464):1099–1104.
127. *Census 2000 Summary File 3 (SF 3) Sample Data, Place of Birth for the Foreign Born Population*; 2008. Available from: http://factfinder.census.gov/servlet/DTTable?bm+y&context=dt&-ds_name=DEC2000
128. Wee CC, Huang A, Huskey KW, McCarthy EP. Obesity and the likelihood of sexual behavioral risk factors for HPV and cervical cancer. *Obesity.* 2008;16(11):2552–2555.
129. Hernandez BY, McDuffie K, Wilkens LR, Kamemoto L, Goodman MT. Diet and premalignant lesions of the cervix: evidence of a protective role for folate, riboflavin, thiamin, and vitamin B12. *Cancer Causes Control.* 2003;14(9):859–870.
130. Texas Department of State Health Statistics. *Behavioral Risk Factor Surveillance Query System*; 2008–2012. Available from: http://www.dshs.state.tx.us/chs/brfss/query/brfss_form.shtm
131. Jia H, Muennig P, Lubetkin EI, Gold MR. Predicting geographical variations in behavioural risk factors: an analysis of physical and mental healthy days. *J Epidemiol Community Health.* 2004;58(2):150–155.
132. Tan D, Zwar NA, Dennis SM, Vagholkar S. Weight management in general practice: what do patients want? *Med J Aust.* 2006;185:73–75.
133. Friedenreich CM, Neilson HK, Lynch BM, Vagholkar S. State of the epidemiological evidence on physical activity and cancer prevention. *Eur J Cancer.* 2010;46:2593–2604.

134. Ford ES, Caspersen CJ. Sedentary behaviour and cardiovascular disease: a review of prospective studies. *Int J Epidemiol.* 2012;41:1338–1353.
135. Affairs TDoR. Medically under-served areas. *Medically Under-Served Status is Designated to Areas or Populations Having a Shortage of Personal Health Services According to U.S. Department of Health and Human Services Rules*; 2009
136. Walboomers JM, Jacobs MV, Manos MM, et al. Human papillomavirus is a necessary cause of invasive cervical cancer worldwide. *J Pathol.* 1999;189(1):12–19.
137. Hariri S, Unger ER, Sternberg M, et al. Prevalence of genital human papillomavirus among females in the United States, the National Health And Nutrition Examination Survey, 2003–2006. *J Infect Dis.* 2011;204(4):566–573.
138. Baron R, Rimer B, Coates R, et al. Client-directed interventions to increase community access to breast, cervical, and colorectal cancer screening a systematic review. *Am J Prev Med.* 2008;35(suppl 1):S56–S66.
139. 2015 SGKCPR. *Lubbock Area*; 2015
140. Brown ER, Wyn R, Teleki S. *Dispairities in Health Insurance and Access to Care for Residents across U.S. Cities.* Los Angeles, CA: UCLA Center for Health Policy Research; 2000.
141. University of Wisconsin Population Health Institute. *County Health Rankings 2013*; 2013. Available from: www.countyhealthrankings.org
142. Borgstrom E, Cohn S, Barclay S. Medical professionalism: conflicting values for tomorrow's doctors. *J Gen Intern Med.* 2010;25(12):1330–1336.
143. New C, Xiao L, Ma J. Acculturation and overweight-related attitudes and behavior among obese Hispanic adults in the United States. *Obesity.* 2013;21(11): 2396–2404.
144. Moyer VA. Risk assessment, genetic counseling, and genetic testing for BRCA-related cancer in women: U.S. Preventive Services Task Force recommendation statement. *Ann Intern Med.* 2014;160(4):271–281.
145. Rodriguez JL, Thomas CC, Massetti GM, et al. CDC grand rounds: family history and genomics as tools for cancer prevention and control. *Morb Mortal Wkly Rep.* 2016;65 (46):1291–1294.
146. Escribà-Agüir V, Rodríguez-Gómez M, Ruiz-Pérez I. Effectiveness of patient-targeted interventions to promote cancer screening among ethnic minorities: a systematic review. *Cancer Epidemiol.* 2016;44:22–39.
147. Nattinger AB, Mitchell JL. Breast cancer screening and prevention. *Ann Intern Med.* 2016;164(11):ITC81–ITC96.
148. Welcsh PL, King MC. BRCA1 and BRCA2 and the genetics of breast and ovarian cancer. *Hum Mol Genet.* 2001;10(7):705–713.
149. Xu X, Wagner KU, Larson D, et al. Conditional mutation of Brca1 in mammary epithelial cells results in blunted ductal morphogenesis and tumour formation. *Nat Genet.* 1999;22(1):37–43.
150. *Conditions GHRYGtUG. BRCA1 Gene.* National Institute of Health; 2017
151. Moyer VA. Force USPST.Risk assessment, genetic counseling, and genetic testing for BRCA-related cancer in women: U.S. Preventive Services Task Force recommendation statement. *Ann Intern Med.* 2014;160(4):271–281.
152. Colditz GA, Bohlke K. Priorities for the primary prevention of breast cancer. *Cancer J Clin.* 2014;64(3):186–194.
153. Hildebrand JS, Gapstur SM, Campbell PT, Gaudet MM, Patel AV. Recreational physical activity and leisure-time sitting in relation to postmenopausal breast cancer risk. *Cancer Epidemiol Biomarkers Prev.* 2013;22(10):1906–1912.
154. Chen WY, Rosner B, Hankinson SE, Colditz GA, Willett WC. Moderate alcohol consumption during adult life, drinking patterns, and breast cancer risk. *JAMA.* 2011;306(17):1884–1890.

155. Visvanathan K, Hurley P, Bantug E, et al. Use of pharmacologic interventions for breast cancer risk reduction: American Society of Clinical Oncology clinical practice guideline. *J Clin Oncol.* 2013;31(23):2942–2962.
156. Cuzick J, Sestak I, Cawthorn S, et al. Tamoxifen for prevention of breast cancer: extended long-term follow-up of the IBIS-I breast cancer prevention trial. *Lancet Oncol.* 2015;16(1):67–75.
157. Mathew A, Brufsky A. Bisphosphonates in breast cancer. *Int J Cancer.* 2015;137(4): 753–764.
158. Eeles RA. Future possibilities in the prevention of breast cancer: intervention strategies in BRCA1 and BRCA2 mutation carriers. *Breast Cancer Res.* 2000;2(4):283–290.
159. Bassett LW, Gold R. The evolution of mammography. *J Roentgenol.* 1988;150:493–498.
160. Lerner BH. No shrinking violet: Rose Kushner and the rise of American breast cancer activism. *West J Med.* 2001;174(5):362–365.
161. Braun Susan. The history of breast cancer advocacy. *Breast J.* 2003;9(suppl 2): S101–S103.
162. Bleyer A, Welch HG. Effect of three decades of screening mammography on breast-cancer incidence. *N Engl J Med.* 2012;367(21):1998–2005.
163. Puliti D, Zappa M. Breast cancer screening: are we seeing the benefit? *BMC Med.* 2012;10(106):1741–7015.
164. Oeffinger K, Fontham. Elizabeth. Etzioni. Ruth. et al. Breast cancer screening for women at average risk 2015 guideline update from the American Cancer Society. *JAMA.* 2015;314(15):1599–1614.
165. Autier P, Boniol M, Smans M, Sullivan R, Boyle P. Observed and predicted risk of breast cancer death in randomized trials on breast cancer screening. *PLoS One.* 2016;11(4): e0154113.
166. Final Recommendation Statement—Breast Cancer: Screening. U.S. Preventive Services Task Force. November 2016. https://www.uspreventiveservicestaskforce.org/Page/Document/RecommendationStatementFinal/breast-cancer-screening1
167. ACR American College of Radiology; 2016. Available from: https://www.acr.org/Quality-Safety/Appropriateness-Criteria
168. Mainiero MB, Lourenco A, Mahoney MC, et al. ACR appropriateness criteria breast cancer screening. *J Am Coll Radiol.* 2016;13(11S):R45–R49.
169. ACOG Practice Bulletin.Clinical management guidelines for obstetrician-gynecologists. Number 42, April 2003. Breast cancer screening. *Obstet Gynecol.* 2003;101 (4):821–831.
170. Durand MA, Haas BM, Yao X, et al. Early clinical experience with digital breast tomosynthesis for screening mammography. *Radiology.* 2015;274(1):85–92.

CHAPTER TWO

Current Approaches to Diagnosis and Treatment of Ductal Carcinoma In Situ and Future Directions

Randi Ryan[1], Ossama Tawfik, Roy A. Jensen, Shrikant Anant
University of Kansas Medical Center, Kansas City, KS, United States
[1] Corresponding author. E-mail address: rryan2@kumc.edu

Contents

Progress in Molecular Biology and Translational Science, Volume 151
ISSN 1877-1173
http://dx.doi.org/10.1016/bs.pmbts.2017.08.001

Abstract

The presentation and treatment of ductal carcinoma in situ (DCIS) has changed substantially over the years. While previously an incidental pathologic finding in more advanced, palpable tumors, the institution of screening mammography has repositioned this disease entity as one largely diagnosed as a non-palpable lesion, often prior to any invasive disease. As DCIS is a precursor to invasive carcinoma, evolution in the approach to treatment has followed in the footsteps of that for invasive disease, including breast conservation therapy, adjuvant radiation, and use of antihormonal therapy. Survival outcomes for DCIS are very high and more recent literature has investigated tailoring therapeutic approaches to avoid overtreatment. Two important areas of ongoing clinical debate concerning overtreatment include use of preoperative MRI and the role of adjuvant radiation. The heterogeneity of the disease makes it difficult to differentiate lesions that would benefit from more aggressive treatment from those in which overtreatment could be avoided. Clinical characteristics, such as histologic appearance, age at diagnosis, and margin status at tumor excision have been established as moderate predictors of disease recurrence, but none has provided strong enough evidence as to guide consensus decisions on adjuvant therapy. Continuing research seeks to define the genetic and molecular characteristics that can predict disease course and serve as the potential targets for novel therapeutic agents. While several markers have shown promise in differentiating tumor aggressiveness, there is still much to be discovered about the precise mechanisms of disease progression and how this can be applied clinically to optimize treatment.

1. INTRODUCTION AND BACKGROUND

Ductal carcinoma in situ (DCIS) is a noninvasive and malignant proliferation of ductal epithelial cells confined by the basement epithelium. Its origin is believed to be at the terminal ductal lobular unit based on pathologic studies.[1] Over time, our understanding of DCIS has evolved drastically, and it is now recognized as a heterogeneous disease in which some, but not all lesions progress to invasion. DCIS has an excellent overall survival rate. Much of current the debate related to DCIS is related to striking a balance between overtreatment of a low-risk lesion and preventing recurrence of invasive carcinoma. With continued research, discovery of genetic, and/or molecular markers that indicate risk of progression to invasion may allow us to better stratify lesions for more tailored, patient-specific therapies.

2. EPIDEMIOLOGY

The introduction of screening mammography had a profound impact on the incidence, presentation, and treatment of this disease. Previously, DCIS made up only a small proportion of all diagnosed breast cancers, with most sources reporting a rate less than 5%.[2,3] Screening mammography resulted in a staggering increase in the proportion of new breast cancer diagnoses that were DCIS, ranging from 20%–30%.[4–6] This is likely the result of lesions being discovered prior to becoming invasive. An estimated more than 63,000 new cases of in situ breast carcinoma in 2017 according to data from the Surveillance, Epidemiology and End Results Program, making this the fourth most common cancer diagnosis in women after invasive breast, lung, and colorectal.[7] The risk factors for DCIS are similar to those for invasive breast cancer and include increasing age, nulliparity, first birth after the age of 30, family history of breast cancer, and BMI 25 kg/m^2 or higher.[8–10]

3. NATURAL HISTORY

The natural history of DCIS is poorly understood. This is because in the prescreening era, DCIS most commonly presented as a palpable mass and was treated by mastectomy. There is some limited data from follow-up studies of small numbers of patients treated with biopsy alone. A study from 1978 reported that long-term follow-up of 10 women treated at Memorial Sloan Kettering Cancer Center found to have undiagnosed DCIS from a cohort treated between 1940 and 1950. Seven of these 10 women developed carcinoma in the same breast.[11] Another study from 1968 reported that 15 patients out of a cohort of 267 diagnosed with breast cancer that previously had benign biopsies from the same site. Review of these biopsies revealed that 10 out of 15 actually had areas of low-grade DCIS.[12] Black *etal.* reviewed pathology slides of 93 patients with previously benign breast biopsies who later developed cancer. Seventeen of the 93 (18%) original benign biopsies were reclassified as in situ carcinoma, although whether this was ductal or lobular was not specified.[13] A more recent study reported 30-year follow-up of 28 women with low-grade DCIS, originally diagnosed as benign and treated with biopsy alone. In this small cohort, 11 women (39.3%) developed invasive breast carcinoma in the same quadrant of the initial lesion. Eight of the recurrences were within 15 years of the initial biopsy, and five of these women died of metastatic disease at the time of last follow-up.[14] Page *et al.* reported a series of 25 women found to have DCIS on biopsies originally diagnosed as benign. Seven of these women (28%) developed invasive carcinoma with an average follow up of 16 years.[15] These studies support the association of DCIS as a precursor lesion to invasive breast carcinoma. Based on these limited reports, the estimated rate of DCIS progressing to invasive disease is 25%–35%.[16] This data is limited by the fact that the number of cases is small, all are retrospective reviews, and nearly all missed lesions were low-grade DCIS, which is known to follow a much different disease course than that of high-grade lesions. The true proportion of all DCIS lesions that will eventually progress to invasive carcinoma will likely never be known given that today essentially all known cases are treated.

Overall prognosis of DCIS is excellent. Data has shown a 20-year breast cancer-specific mortality of 3.3% in women diagnosed with DCIS before the age of 70.[17] Understanding the prognostic factors that influence recurrence after treatment of DCIS is important, as roughly 50% of recurrences are invasive cancer[18] and invasive recurrence is associated with an increased mortality risk.[19] Younger age at diagnosis is associated with poorer outcomes. In a large retrospective review of women diagnosed with DCIS, the overall risk of death from breast cancer was 1.8 times that of the general population, but in the cohort of women diagnosed before the age of 35, this risk increased to 17 times and was 7.3 in women aged 35–39.[17]

4. DIAGNOSIS

The gold standard for the diagnosis of any concerning breast lesion involves a triple assessment, including clinical exam, imaging, and biopsy. Clinical exam is important in assessing patients presenting with palpable abnormalities and should include assessment of the axillary and clavicular lymph node basins. In the current era, DCIS is most often diagnosed as a nonpalpable mass on screening images. Screening mammography is the most common method of detecting DCIS, however, ultrasound and MRI are additional imaging modalities sometimes used. Lastly, tissue diagnosis can be achieved via fine needle aspiration or core needle biopsy. In current practice, fine needle aspiration is no longer recommended, as this does not demonstrate tissue architecture and therefore is unable to differentiate preinvasive from invasive disease. Core needle biopsy is the standard therapy used for tissue diagnosis today, and we describe some of the improvements in this method later.

5. IMAGING

5.1 Mammography

The institution of mammography as a screening modality in the 1980s had a substantial impact on breast cancer mortality. In a large, randomized controlled trial comparing screened versus unscreened populations in Sweden, mammography was estimated to reduce breast cancer mortality by approximately 30%.[20] This is similar to the estimates in the United States based on 30 years of data.[21] Screening mammography also drastically impacted DCIS as a disease entity. The incidence of DCIS was reported to increase more than 500% in the initial years of screening mammography.[22] Today, the overwhelming majority of DCIS diagnoses are now made preclinically, with 75% of cases presenting as microcalcifications on mammography.[23] The sensitivity of mammography for detection of DCIS is 86% based on results from a large cohort of screened patients in the UnitedStates.[24]

Microcalcifications associated with DCIS are a result of cellular necrosis. Certain mammographic patterns of calcifications can be suggestive of architectural type of DCIS as well, although with limited accuracy. The calcifications of comedo carcinoma are often fine, linear, and branching, sometimes referred to as "casting calcifications." Calcifications associated with noncomedo and low-grade carcinoma are described as dystrophic, granular, clustered, or having the appearance of "crushed stone," representing intraductal cellular debris or calcified secretions into cribriform spaces.[16,25–28]

Mammography has been shown to underestimate the size of DCIS lesions in some cases,[29,30] due to its inability to detect noncalcified disease. In a large retrospective trial of over 2000 women in the United Kingdom with DCIS, 30% of patients had underestimation of disease on mammography and required additional surgery as a result.[29]

Mammography is more accurate in assessing extent of DCIS in high-grade lesions compared to low-grade. This is in part attributed to the amount of necrosis in the lesion, which is apparent as calcifications on mammography. In cases of comedonecrosis, the calcifications on mammography more closely match the extent of the lesions, whereas noncomedo lesions may have more extensive in situ carcinoma than is depicted on mammography[23]

5.2 Ultrasonography

Ultrasonography of the breast has been around as early as the 1950s and serves as an adjunct to mammography.[31] Breast ultrasound is useful in further characterizing the suspicious lesions or calcifications detected on screening mammography and for image-guided biopsy. In women with dense breast tissue, ultrasound in addition to mammography improves detection rates.[32,33] Pure DCIS is detected on ultrasonography in up to 85% of cases and up to 15% of DCIS visible on ultrasound is not apparent on mammography.[34] Findings of DCIS on ultrasound vary based on mammographic presentation. For DCIS without calcifications, DCIS often appears as architectural distortion, isoechoic to hypoechoic heterogenous multilobulated mass, and ductal extension. In the case of DCIS presenting with microcalcifications on mammobraphy, ultrasound often shows nonmass hypoechoic abnormalities.[34]

5.3 MRI

MRI was initially investigated as a diagnostic tool in breast imaging in 1982,[35] with the recognition a few years after that gadolinium contrast improved identification of malignant lesions and differentiating them from benign findings based on morphology, enhancement patterns, and enhancement kinetics.[36–38] Carcinomas tend to have low signal intensity on T1-weighted images and variable signal intensity on T2.[39] Enhancement of lesions with contrast is a result of tumor angiogenesis.[40–43] Morphologic patterns typical of malignancy on MRI are similar to those seen on mammography, including irregular, spiculated borders, or microlobulated borders. Early rim enhancement and peripheral washout are other features suggestive of malignancy.[39,44,45] Presentation of pure DCIS is more variable on MRI compared to invasive carcinoma. The most common presentation of DCIS on MRI is a nonmass enhancement[46,47] and can have a clumped or linear pattern that follows a ductal distribution.[48,49]

MRI has been shown to improve sensitivity in identifying breast malignancies, especially with improved imaging techniques overtime. A recent metaanalysis on the diagnostic accuracy of MRI pooled data from 44 eligible studies and reported an estimated sensitivity of 90% and specificity of 72% for diagnosis of malignant breast lesions.[50] A metaanalysis of 19 studies looking at the role of MRI in preoperative staging of the affected breast found that MRI resulted in detection of additional disease in 16%, with 2/3 of these findings shown to be malignant on histology.[51] These studies include patients with invasive carcinoma and DCIS. In the case of pure DCIS, previous reports showed lower sensitivity of MRI in diagnosis of this compared to invasive carcinoma.[39,52–54] In several cases, patients with positive mammography findings and negative MRI was attributed to limited ability of MRI to detect microcalcifications. More recent studies, however, have shown sensitivities ranging from 70%–100% for identification of DCIS on MRI.[55–57] Several studies have also demonstrated that sensitivity of MRI for detecting DCIS differs based on grade, with higher sensitivity for increasing histologic grade.[55,58] Improved ability of MRI to diagnose DCIS has been attributed to improved imaging techniques with use of 3D volume whole-breast techniques and background suppression.[48,49,59]

The specificity of MRI in detecting malignant lesions has often been lower than the sensitivity, with reports ranging from 37%–75%.[52,59,60] The low specificity and increased false positive rate is a result of benign lesions that have also been shown to enhance on MRI, including fibroadenomas, radial scars, fibrocystic disease, lymph nodes, areas of inflammation, and physiological parenchymal enhancement due to hormonal stimuli.[37,38,41,59,61] Some reports have shown improved specificity on MRI when enhancement kinetics are considered to differentiate benign from malignant lesions. For example, malignant lesions tend to have early, rapid uptake of contrast; whereas fibroadenomas often enhance later.[37,62] In another study, use of a new model with expanded architectural criteria yielded a specificity of 80%.[44] However, while this has worked for more malignant phenotype, further enhancement is required for increasing specificity for DCIS.

Current routine use of MRI is only recommended as a *screening* modality in select populations. This includes patients with a personal history of or first-degree relative with a known BRCA mutation, a lifetime risk of breast cancer 20% or greater, a personal history of lobular carcinoma in situ, a history of radiation to the chest between 10 and 30 years of age, or a personal history of or a first-degree relative with Li–Fraumeni syndrome, Cowden syndrome, or Bannayan–Riley–Ruvalcaba syndrome.[63,64]

Despite the fact that no guidelines exist for its routine is in other scenarios, the use of MRI as an adjunct in preoperative evaluation for breast has increased.[65,66] Other situations in which breast MRI could be considered according to guidelines published by the American College of Radiology include: investigation of lesions for which previous imaging is inconclusive, evaluation of neoadjuvant therapy, evaluation of contralateral breast cancer, treatment planning after positive margins, evaluation of reconstructed breasts, patients with isolated axillary disease, surveillance for high risk patients (described earlier).[41,67] The role of MRI in preoperative assessment of patients with DCIS has been a subject of controversy. Because of its high sensitivity and reports that it better characterizes extent of disease the use of MRI in preoperative planning for patients with DCIS has the proposed advantages of decreasing the positive margin rate, identifying patients who would benefit from mastectomy as an initial operation due to disease extent, and identifying synchronous contralateral disease requiring bilateral treatment.[68]

Positive margins have been shown to significantly impact local failure rates of DCIS.[18,69–73] Reducing the positive margin rate could impact disease-free survival, as well as reducing the number of surgeries for reexcision. A recently published metaanalysis evaluated the effect of preoperative MRI on rates of positive margins and mastectomy rates in patients with DCIS. The final analysis included two randomized controlled trials and seven retrospective cohort studies. Results showed no significant difference in rates of positive margins or surgical reexcision between patients who underwent preoperative MRI versus those who did not. There were significantly higher rates of mastectomy as the initial procedure in patients who underwent preoperative MRI, but the overall mastectomy rate, which included initial mastectomy and mastectomy after BCS with positive margins, was not statistically different based on preoperative MRI.[74] This pattern seen with mastectomy was demonstrated in a recently published randomized controlled trial as well.[75] Other studies have failed to show any significant effect of MRI on the decreasing the positive margin rate as well.[65,76] The effect of MRI on mastectomy has been proposed as an advantage in that identification of more extensive disease prior to surgery allows selection of those patients who would have otherwise failed breast conserving therapy (BCT) and required mastectomy as a second procedure.[77] There is some speculation that mastectomy, as a result of occult disease on MRI is an overtreatment as these patients may have had similar outcomes with whole breast radiation to treat the occult disease after lumpectomy. In a retrospective review of 414 patients, Miller *et al.* found that preoperative MRI was associated with significantly increased rates of mastectomy in patients with positive *and* negative MRI findings.[78]

Identification of synchronous contralateral breast cancer is another proposed advantage of preoperative MRI in patients diagnosed with DCIS.

Detection of synchronous disease would serve to decrease the potential number of treatment interventions, as well as identifying in situ cancers at an early stage that could otherwise become invasive. In a retrospective review of over 200 cases, 19% of which were DCIS, findings on MRI changed surgical management in 43 patients (20% of the cohort). The affected surgical plan was deemed favorable/somewhat favorable in 24 patients, and involved either change in plan from lumpectomy to mastectomy or widening of excision area in breast conserving surgery. In 14 patients the affect of MRI on treatment was determined unfavorable, and included 10 who underwent additional biopsies that were negative and 3 who underwent mastectomy with final pathology showed lumpectomy would have been adequate.[79] There were 6 patients in this cohort with DCIS in which MRI changed management, and in 5 of these cases, this change was deemed strongly favorable. In a study of 136 women diagnosed with DCIS who underwent preoperative bilateral breast MRI, 21 patients (18%) were found to have more extensive disease based on MRI findings and 4 (4%) were found to have contralateral DCIS, resulting in alterations in treatment recommendations.[80]

Pilewskie *et al.* performed a retrospective review of a prospectively maintained database looking at the impact of preoperative breast MRI on local recurrence and contralateral breast cancer. There were 2321 women over a 13-year period included in the analysis. The results showed no significant decrease in local recurrence or rates of contralateral breast cancer with the use of preoperative MRI versus conventional imaging.[81] Solin *et al.* also retrospectively examined outcomes of patients who underwent BCT that had preoperative MRI (28%) versus those who didn't (72%). The results on 8-year follow-up of all patients showed no significant differences in local recurrence, overall-survival, or cause-specific survival. For the subset of patients with DCIS in this cohort, the local recurrence rate was 6% in both the MRI and non-MRI groups ($P = 0.58$).[82] There have also been a few randomized controlled trials evaluating the effects of preoperative MRI in breast cancer. The Comparative Effectiveness of MRI in Breast Cancer Trial (COMICE) was a multicenter study that included 1623 patients from the United Kingdom with biopsy proven cancer randomized to preoperative MRI or conventional imaging. Preoperative MRI was not shown to have a significant effect on the rate of reoperation. The majority of patients (90%) in this study had invasive cancer.[83] The MR Mammography of Nonpalpable Tumors (MONET) study randomized patients with BIRADS 3–5 lesions to preoperative MRI or conventional imaging/workup. In this study, close to 50% of patients in each cohort had DCIS. Results unexpectedly showed that preoperative MRI was associated with a significantly higher reexcision rate as a result of positive margins.[84] A more recent randomized controlled trial in Sweden compared the effect of preoperative MRI on surgical management in women under the age of 56 diagnosed with breast cancer. There were significantly fewer reoperations in the group who underwent preoperative MRI (5% vs. 15%; $P < 0.001$). Preoperative MRI resulted in alteration of the surgical plan in 18% of patients; although, there is no comment on whether these changes were all deemed favorable.[75] There was no report of the proportion of patients in this study with DCIS.

Fischer *et al.* reported significantly lower local recurrence rate in patients diagnosed with breast cancer who underwent preoperative MRI versus those who did not (1.2% vs. 6.8%; $P < 0.001$).[85] However, this study was retrospective and most patients had invasive ductal carcinoma, with pure DCIS making up only a minority of the cohort.

It is important to note that the majority of these studies are retrospective, with the few prospective randomized trials including patients with invasive and in situ cancer. To date, there has not been a prospective, randomized controlled trial looking specifically at the effects of preoperative MRI in patients with DCIS. Preoperative MRI in the setting of a diagnosis of DCIS is not without its disadvantages. MRI has been shown in several studies to overestimate tumor size, which could lead to unnecessary biopsies and potentially overtreatment.[51] Additional considerations are the cost of MRI and the burden of this cost to the patient should insurance coverage be denied. Despite a wealth of research on the use of MRI in preoperative assessment, there has been no strong evidence to support or refute its routine use in patients with DCIS. The NCCN recommendation regarding preoperative assessment of DCIS using MRI is that this is optional. As a result, clinicians must advise patients about the available data and present the pros and cons of preoperative MRI, allowing patients to be involved in an informed decision.

6. BIOPSY

While the gold standard for biopsy of suspicious breast lesions is surgical excision, this has largely been replaced by less invasive methods. Stereotactic core needle biopsy for use on breast lesions was first implemented in the late 1980s.[86] This less invasive means to establish a diagnosis allows patients with benign lesions to avoid undergoing surgical intervention.[87] Initially, this method involved use of an automated core needle biopsy gun to sample lesions under mammographic guidance. Studies of long-term follow-up have shown excellent sensitivity and specificity in detecting carcinoma on stereotactic biopsy. For DCIS in particular, the specificity of core needle biopsy is slightly lower due to cases of missed invasive disease. A large cohort of patients from a single institution study reported a sensitivity of 97.3% and specificity of 99.7% for detection of carcinoma. For detection of DCIS, sensitivity was 65.4% and specificity was 97.7%. The lower sensitivity in this case is a result of discovery of invasive carcinoma on subsequent excision specimen.[87] If lesions are visible on ultrasonography, this is the preferred method of image-guided biopsy given its decreased cost, improved patient comfort, and avoidance of radiation exposure.[88,89] Vacuum-assisted stereotactic biopsy involves use of suction to pull breast tissue into the biopsy device. Reports have demonstrated that this method results in a lower rate underestimations of lesions[90,91] due to its ability to obtain more tissue, produce specimens that are more contiguous with less blood in the sample, and require less precision in localization of the lesion compared to core needle biopsy.[92,93] This can be performed under stereotactic or ultrasound guidance and has replaced core needle biopsy at many major centers.

7. HISTOLOGIC CLASSIFICATION

Histologic classification of DCIS has evolved over time as our understanding of significant pathologic features has evolved. Traditional classification of DCIS was primarily according to architecture. Architectural types include comedo, solid, cribriform, micropapillary, and papillary.[94,95] The comedo pattern is characterized by ductal distention with a solid growth of cancerous cells with high grade nuclei surrounding a central zone of necrosis.[96,97] Solid DCIS is characterized by neoplastic cells that fill the duct space.[97] Cribriform DCIS is evidenced by cuboidal cells arranged around multiple small microlumina with smooth edges.[94,96] These microlumina give a "punched out" appearance. Micropapillary DCIS contains small stalks of cells that protrude into the lumen of the ducts varying from very short to more extensive processes.[94,96] Pure micropapillary tumors are more commonly associated with more extensive and multicentric disease.[98–100] Micropapillary and cribriform DCIS are the most common histological subtypes of noncomedo DCIS.[23] These architectural patterns often coexist in the same lesion and are more commonly associated with low nuclear grade.[15,95,96] Papillary DCIS is differentiated from micropapillary by the presence of fibrovascular cores within papillae of malignant cells (Fig. 1).[94–97]

Pathological characterization by architecture alone is problematic given the incidence of multiple architectural types within a single specimen,[99,101] high degree of interobserver variability,[102] and lack of prognostic significance.[16,103] The exception to this is the presence of necrosis, often termed comedonecrosis, which has been shown to be associated with a high risk of recurrence,[3,104–106] as well as high nuclear grade.[98,101] As a result, there has been an emergence in most classification systems a categorization of lesions as comedo or noncomedo, with noncomedo lesions to include the other architectural types listed before.

Histologically, necrosis is defined as the presence of ghost cells and karyorrhectic debris. It has an eosinophilic, granular appearance as a result of hematoxylin staining of nuclear debris.[107] This is distinguished from single, apoptotic cells, which are not considered necrosis.[96] The terms comedo and comedonecrosis are often used interchangeably depending on the resource consulted. According to the Consensus Conference on

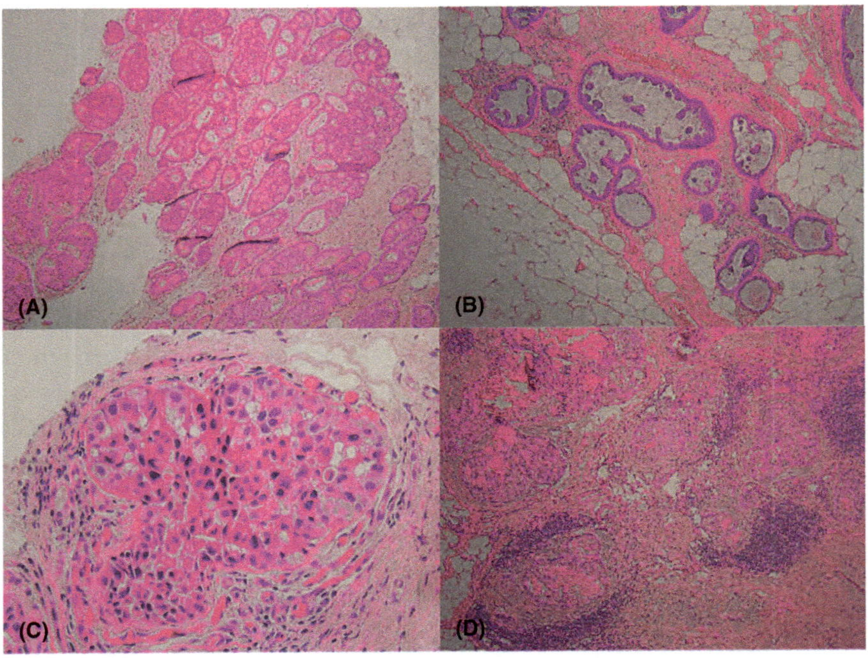

Fig. 1 Architectural patterns of ductal carcinoma in situ (DCIS), including (A) cribriform, (B) micropapillary, (C) solid, and (D) comedo.

Classification of DCIS, the term comedo refers to "solid intraepithelial growth within the basement membrane with central necrosis."[108] Other sources apply this term to tumors containing necrosis, as well as high nuclear grade.[109,110] This type of necrosis forms a linear pattern, which can appear as linear or branching calcifications on mammography.[96] Necrosis may also be present in a punctate pattern, which is distinct from the comedo pattern.[96]

Although necrosis is an important marker for high-risk lesions, this feature alone oversimplifies the categorization of DCIS lesions, signified by evidence of high-risk lesions with noncomedo architecture. Nuclear grade has emerged as a strong predictor of local recurrence (Fig. 2).[100,106,111] As a result, the general consensus among experts is that nuclear grade is the most important feature governing histologic classification.[108] The characteristics used to classify low and high nuclear grade are listed in Table 1, with intermediate grade reserved for specimens with characteristics

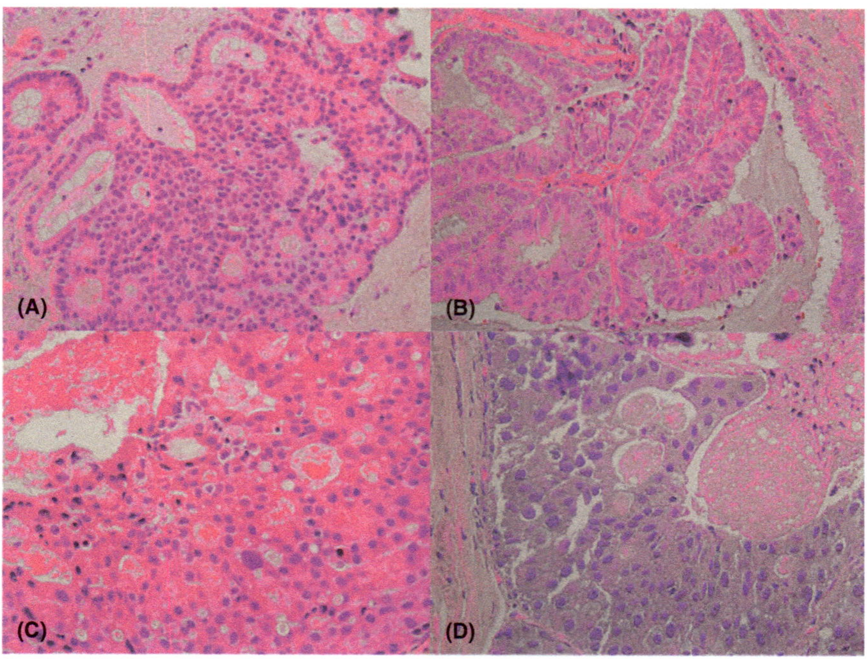

Fig. 2 Nuclear grade is an important predictor of prognosis, and is now a component of more recent histologic classification systems. (A–B) Low nuclear grade is characterized by nuclear monotony and polarization with diffuse chromatin. (C–D) High nuclear grade is evident by pleomorphic, nonpolarized nuclei with irregular chromatin, and prominent nucleoli.

Table 1 Characteristics Used to Classify Nuclear Grade in DCIS. Intermediate Grade Assigned to Specimens Not Suited to Either Low or High Grade

Nuclear Characteristics	Low Grade	High Grade
Appearance	Monotonous	Pleomorphic
Size compared to normal RBC or duct epithelial cell nucleus	1.5–2x	>2.5x
Chromatin	Diffuse	Vesicular, irregularly dispersed
Nucleoli	Occasional	Multiple, prominent
Mitoses	Occasional	May be frequent
Orientation	Polarized toward luminal spaces	Not polarized

Information adapted from the Consensus Conference on the Classification of Ductal Carcinoma In Situ.[108]

that are not consistent with low or high grade. In addition to nuclear grade, other histologic features noted on pathological examination of DCIS include presence of necrosis, cell polarization, and architecture.[108] Studies have shown less heterogeneity when classifying lesions according to nuclear grade compared to architectural pattern[109] and improved inter-observer reliability.

There is currently no single standard classification system agreed upon for DCIS and therefore a number of different systems have been published by different groups. The most commonly recognized classification systems share important similarities. All recognize the significance of nuclear grade in prognosis and risk of recurrence, making this the primary feature of each histologic classification. Additionally, the majority has a three-tiered system for histologic grading, which allows for an intermediate category. This is important given the notorious heterogeneity of DCIS as a disease entity. Many lesions with overlapping features that are not obviously high or low grade are more accurately categorized as "intermediate" lesions. The value of a classification system is in its ability to stratify lesions according to the important prognostic characteristics to predict risk of recurrence, as well as its reproducibility among pathologists. An early study published by Lagios *et al.* was the first to prioritize cytological features in classification of DCIS. This grading system incorporated nuclear grade, architecture, and presence/absence of necrosis into a four-tiered grading system. Review of specimens from 79 patients who underwent excision alone for DCIS showed the rate of recurrence was higher in patients with high nuclear grade and the presence of necrosis.[106] The modified Lagios classification combined the two groups with high nuclear grade into one high risk group, resulting in a three-tiered system that had improved concordance between observers.[112] The European Pathologists Working Group Classification, published by Holland, is primarily based on nuclear grade with incorporation of cell polarity in a three-tiered system.[113] Alternatively, the European Commission Breast Screening Pathology Working Groups use the system proposed by Holland, but only considers nuclear grade, and does not incorporate cell polarity into histologic classification[94] (National Coordinating Group for Breast Screening Pathology: Pathology Reporting in Breast Cancer Screening, second edition. Sheffield, United Kingdom, Breast Screening Publications, 1995). The Nottingham system categorizes lesions based solely on the presence or absence of necrosis.[114] The Van Nuys histologic classification was published in 1995 and has been shown in multiple subsequent publications to predict prognosis of lesions. This classification categorizes lesions based on nuclear grade and the presence or absence of comedonecrosis. Lesions with high nuclear grade are categorized as high-grade lesions. Lesions with low or intermediate nuclear grade were categorized as intermediate if comedonecrosis was present and low grade if comedonecrosis was absent. This classification was applied to 238 patients with DCIS treated by excision with or without radiation. At a median follow up of 6.5 years, there was a significant difference in local recurrence rates according to histologic grade, with increasing rates of recurrence as histologic grade increased.[111] Other studies have demonstrated a significant correlation between Van Nuys histologic grade and recurrence as well.[103] Several studies have shown a good rate of interobserver reliability using the Van Nuys classification,[115–118] however, there does exist some disagreement in determining lesions that fall within the intermediate group. A more recent classification system published by Tawfik *et al.* in 2007 uses a more objective approach to increase reproducibility.[119] This classification is based on nuclear grade, but incorporates automated evaluation of proliferation in place of manual mitotic count. Nuclear grade was determined according to that used for the Van Nuys. Proliferation was determined objectively by evaluation of Ki-67, which is a cellular protein whose expression correlates with tumor cell proliferation. Ki-67 staining was performed using MIB-1, an anti-Ki-67 antibody. Grade I included lesions with nuclear grade 1 or 2 and MIB-1 $<25\%$. Grade II included lesions with nuclear grade 3 or MIB-1 $>25\%$. Grade III lesions contained nuclear grade 3 and MIB-1 $>25\%$. This classification, termed the Nuclear Grade and Proliferation Index (N+P), significantly correlated with overall survival when used to classify patients with invasive breast carcinoma.[119] This classification was compared to the Van Nuys Index for evaluation of 162 DCIS lesions. In this analysis, 94% of lesions classified as grade II according to the Van Nuys system were down graded to grade I using the N+P system. Additionally, 40% of lesions classified as Van Nuys grade III were down graded to grade II according to the N+P system. Application of the N+P grading system to in situ lesions correlated with other known prognostic markers for DCIS, including p53, Her-2, Bcl-2, and epidermal growth factor receptor (EGFR).[120]

8. TREATMENT

8.1 Mastectomy

As mentioned previously, mastectomy was historically the treatment of choice for DCIS when its most common presentation was as a palpable mass. Outcomes after mastectomy for DCIS are excellent, with studies showing 98%–99% disease-free survival.[17,121,122] The excellent survival rates for DCIS has led to more conservative therapy, with rates of mastectomy decreasing in favor of breast conservation therapy.[17,123,124] Currently, mastectomy is still recommended in the case of multicentric disease, lesions too extensive to achieve clear margins with good cosmetic result, and patients with high grade lesions who cannot receive radiation, including those who are pregnant or have a history of previous radiation to the chest.[125]

8.2 BCT

Today, almost 75% of newly diagnosed patients with DCIS are treated with BCT.[123] BCT involves lumpectomy of the lesion with or without adjuvant radiation therapy to the breast. Because the majority of DCIS lesions are detected mammographically and therefore not palpable, lesions must be localized under imaging guidance prior to excision. This can be done by wire or radioactive seed localization. Wire-guided localization involves placement of two wires crossing at the site of the lesion. This can be performed under stereotactic or ultrasound guidance prior to surgery or under ultrasound guidance in the operating room. Radioactive seed localization is a more recently developed method and involves percutaneous placement of a radioactive seed by a radiologist at the site of the lesion. This is also performed under stereotactic or ultrasound guidance. A gamma detector probe is then used in the operating room to identify the site of the lesion for excision. While some early studies have suggested a decreased rate of positive margins using radioactive seed localization,[126] a more recently published randomized controlled trial comparing the two techniques showed no difference in the rate of positive margins between the two techniques.[127] One benefit of radioactive seed localization is more flexibility in surgical scheduling since seed placement can be performed several days or weeks prior to the planned operation.

8.3 Margin Status

Pathologic data from three large cohorts of patients has shown margin status to be a significant predictor of ipsilateral recurrence in patients with DCIS treated with lumpectomy.[18,105,128] There is a lot of variability in the literature as to the specific margin width deemed adequate for achieving local control. A retrospective review of 205 patients by Chan *et al.* found that margins at or less than 1 mm were associated with a significantly increased risk of ipsilateral recurrence, but no significant difference in recurrence among patients with various margin widths greater than 1 mm.[129] In 2001, Bijker *et al.* published data on pathologic review of specimens from patients in the EORTC 10583 trial, a large randomized controlled trial evaluating adjuvant radiation for patients with DCIS.[128] In this review of 775 cases, there was a significantly lower recurrence rate in patients with margins greater than 1 mm compared to those with margins classified as close (≤1 mm), involved or unspecified.[128] In a retrospective review of 469 patients by Silverstein *et al.*, recurrences were significantly higher in patients with margins less than 1 mm and very low when margins were 10 mm or more.[130] A metaanalysis published by Dunne *et al.* evaluating the effect of margin status on patients who underwent lumpectomy and adjuvant radiation for DCIS found that 2 mm margins were associated with significantly lower recurrences with no significant difference in recurrence for margins beyond 2 mm.[69] A more recent meta-analysis published in 2016 by Marinovich *et al.* came to a similar conclusion.[72] Another retrospective review of 2995 patients with DCIS found margins less than 3 mm to be associated with significantly higher risk of recurrence.[73] Alternatively, a third metaanalysis by Wang *et al.* suggests 10 mm margins as optimal given their data showing local recurrence was significantly lower when compared to patients with 2 mm margins.[131] Smaller, retrospective studies looking at recurrence among groups of patients with incrementally increasing margins show similar results, with the rate of recurrence decreasing significantly with incrementally wider margins.[18,132] To date, there have been no prospective randomized trials looking at specific margin widths and their effect on recurrence. Current guidelines in the United States for margin width in excision of DCIS are based on a consensus guideline produced by the Society of Surgical Oncology and the American Society for Radiation Oncology, which states that "no ink on tumor" is considered a negative margin in patients with invasive breast cancer undergoing breast conserving therapy.[125,133] The data supporting this guideline is contingent on the patients receiving adjuvant radiation therapy after lumpectomy. For the treatment of DCIS in patients not receiving adjuvant radiation, these guidelines recommend margin width of at least 2 mm, although the authors admit that data is limited.[125] Treatment of DCIS according to guidelines established for invasive cancer may be problematic, however, as DCIS has been shown to have discontinuous growth patterns in some cases. This could result in leaving disease behind if margins are too narrow.[134] A panel of international collaborators published a consensus statement on 2010 in *cancer* deeming 2 mm margins as adequate for excision of DCIS.[135]

8.4 Sentinel Lymph Node Biopsy

The concept of the sentinel lymph node, defined as the first lymph node draining a cancer, arose as early as the 1950s.[136] The use of intraoperative lymphatic mapping to identify sentinel nodes for excision was introduced in the early 1990s.[137] The use of sentinel lymph node biopsy (SLNB) to patients with invasive breast carcinoma allows a large proportion of patients to avoid the potential morbidity of an axillary dissection. The role of SLNB in patients with DCIS is varied. Because DCIS is considered preinvasive and therefore incapable of metastasis, logic would follow that axillary staging is not necessary. There are two scenarios in which SLNB is widely accepted as appropriate in the DCIS patient. This includes those patients undergoing mastectomy and those with lesions in the upper outer quadrant of the breast.[16,125] In both cases, the primary procedure compromises the ability to perform successful lymphatic mapping at a subsequent operation in the event that invasive disease is discovered on final pathology. Whether SLNB at the time of treatment for DCIS should be considered in other clinical scenarios is a matter of debate. Some argue that SLNB should not be considered given the low rate of node positivity in patients with DCIS and the little influence this intervention has on subsequent therapy. In a review of 134 patients with pure DCIS who underwent SLNB, 2% were found to have micrometastatic nodal disease.[138] Another retrospective study of 910 patients diagnosed with DCIS on core needle biopsy found that the rate of N1 disease in patients with pure DCIS on final pathology was 0.3%, while those upstaged to invasive disease on final pathology had a 10.7% rate of N1 disease ($P < 0.001$).[139] Similar rates of nodal disease have been reported in additional retrospective reviews.[140–143] Micrometastatic disease is defined as tumor metastases <0.2 cm, but this has not been shown to impact clinical outcomes.[144–147] As a result, no additional treatment of the axilla is necessary in the event of micrometastases according to published guidelines. In the small proportion of patients with pure DCIS and true N1 disease, a case can be made that this doesn't change treatment if these patients were to receive whole breast irradiation after lumpectomy for DCIS. The rate of upstaging from an initial diagnosis of DCIS after core needle biopsy to invasive carcinoma on final pathology is reported between 13%–30%.[142,148,149] A metaanalysis of 7350 cases of DCIS diagnosed on core needle biopsy found that characteristics associated with an increased risk of invasive carcinoma on final excision included lesions >2 cm, high histologic grade, palpable mass or mass present on mammography, and BIRADS 4 or 5.[150] Some groups advocate for use of SLNB in patients diagnosed with DCIS who are at risk of harboring invasive carcinoma as this avoids an additional procedure for axillary staging.[141,151]

9. ADJUVANT THERAPY

9.1 Role of Radiation

It is interesting to note that historically, the standardization of more conservative treatment for women diagnosed with DCIS lagged behind that of invasive breast cancer. A number of randomized controlled trials published in the 1980s and 90s[152–155] with subsequent follow-up studies[156–159] showed comparable overall and disease-free survival in patients with invasive cancer treated with mastectomy or lumpectomy with or without radiation. These trials also established the significant effect adjuvant radiation had on reducing local recurrence after lumpectomy, naturally paving the way for the application of these principles to the treatment of DCIS.

There have been five large randomized controlled trials evaluating the effect of radiation on outcomes in the treatment of DCIS, all of which demonstrated significant decrease in local recurrence with adjuvant radiation. The National Surgical Adjuvant Breast Project B-17 trial included a cohort of 818 women diagnosed with DCIS in the United States randomized to lumpectomy alone or lumpectomy plus radiation. The initial published results at 5-year follow-up showed the rate of ipsilateral recurrence in the lumpectomy alone group was 16.4%, compared to 7% in the lumpectomy plus radiation group ($P < 0.001$).[160] At 15 years, the overall frequency of ipsilateral invasive recurrence were 35 and 19.8% in the lumpectomy alone and lumpectomy plus radiation groups, respectively.[19] Another randomized controlled trial conducted by the United Kingdom Coordinating Committee on Cancer Research (UKCCCR), including DCIS patients from Australia, New Zealand, and the United Kingdom randomized patients to radiotherapy and/or tamoxifen or not after lumpectomy. Analysis at 5 years of the 1030 patients randomized to lumpectomy alone or lumpectomy with radiation found that radiation significantly reduced the rate of ipsilateral recurrence of both DCIS and invasive carcinoma (13.7% vs. 4.8%; $P < 0.0001$). This amounted to an absolute risk reduction of 8% for any ipsilateral recurrence (DCIS or invasive).[161] In 2000, the European Organization for Research and Treatment of Cancer (EORTC) published results of the effects of adjuvant radiation after lumpectomy on DCIS patients. At a median of 4-year follow-up, the rate of ipsilateral recurrence was 16% in those undergoing lumpectomy alone versus 9% in those who received adjuvant radiation (HR 0.62; 95% CI 0.44–0.87; $P = 0.005$).[162] A Swedish trial published in 2006 analyzed 1046 patients with DCIS who were randomized to lumpectomy alone or with radiation. At 5-year follow-up, the rate of ipsilateral recurrence was 22% in the lumpectomy only cohort and 7% in the cohort who received radiation ($P < 0.0001$).[163] Viani *et al.* conducted a metaanalysis of these four studies, with pooled results showing a significant reduction in recurrence of ipsilateral DCIS and invasive cancer with adjuvant radiation and no difference in overall survival.[164]

9.2 Rationale for Considering Not Undergoing Adjuvant Radiation

In the initial published results of the randomized controlled trials comparing lumpectomy with or without radiation, there is no stratification of patients according to grade of DCIS or margin status, which are two factors that have been shown in previous retrospective reviews to impact the rate of ipsilateral tumor recurrence.[18,100,106,111,165] While radiation was shown to decrease recurrence rate, there are differences in the degree of risk reduction based on the grade of the lesion and the width of the margin.[130] Many experts believe that some women with low-grade DCIS lesions could safely undergo excision alone with close follow up and avoid adjuvant radiation. In fact, in 2008, the updated guidelines of the National Comprehensive Cancer Network for the treatment of breast cancer modified its statement on treatment of DCIS, proposing that excision alone is a reasonable consideration for patients who have disease features that predict a low risk of recurrence.[125,166] In 1999, Fisher *et al.* published 8-year follow-up from the NSABP B17 trial, which analyzed the role of pathologic features on recurrence of ipsilateral breast tumors. This report identified moderate/marked necrosis as an independent predictor of ipsilateral recurrence. Although radiation still showed a significant benefit in ipsilateral recurrence among patients with slight/absent necrosis, the relative risk reduction was lower in this group (48%) compared to patients with moderate/marked necrosis who received radiation (70%).[167] Retrospective reviews have shown that patients with margins >10 mm treated with resection alone have acceptable recurrence rates at long-term follow-up.[70,130]

Several studies have analyzed outcomes with more specific selection of low-risk patients for treatment with excision alone. The Eastern Cooperative Oncology Group published a prospective style looking at the recurrence rate for patients with small, low-grade DCIS with margins 3 mm, or more treated by excision alone. At 6 years of follow-up, the rate of recurrence was 6.1% compared to 15.3% in patients with high grade DCIS treated by excision alone.[168] A recent retrospective study by Wehner *et al.* evaluated local recurrence rates in a cohort of 205 women treated by lumpectomy alone who were selected based on meeting NCCN criteria for low-risk. At a median of 51 months of follow up, the recurrence rate was 4.4%. Based on extrapolation of these results, the probability of recurrence at 12 years was 7.8%.[169] In a prospective randomized controlled trial comparing adjuvant radiation to excision alone in patients with DCIS, the recurrence rate was significantly higher in the excision alone group at 7-year follow-up (6.7% vs. 0.9%; $P < 0.001$).[169a] The cohort included only women with low/intermediate nuclear grade, tumors <2.5 cm, and margins of 3 mm or more. Although the difference was significant, the recurrence rate in the excision only cohort was reasonably low considering it is similar to local recurrence rates in the radiation cohorts from previous RCTs.

The Van Nuys Prognostic Index (VNPI) was developed as a method of classifying patients with DCIS based on risk of recurrence. The VNPI involves classification of a tumor according to histologic grade (according to the Van Nuys system), tumor size and margin width. This system is based on the logic that each of these factors had been previously shown to impact recurrence risk. Numbers of increasing value are assigned to each parameter, with higher risk features receiving high numbers, to achieve an overall score termed the VNPI. The first application of this involved a retrospective review of 333 patients with DCIS treated with either excision alone or excision with radiotherapy. After a median follow-up of 6.5 years, the local recurrence rate was not significantly different for excision alone versus excision with radiation in the cohort of patients with low VNPI scores (3 or 4). The benefit of radiation was significant in patients with intermediate and high VNPI scores.[132] In 2003, an updated version of the VNPI was published that incorporated age into the score and again found no significant effect of radiation on recurrence in a retrospective review of patients with a low VNPI.[170] Another retrospective review of 215 patients in the United Kingdom who underwent excision alone showed significantly increasing recurrence rates with increasing VNPI (0% vs. 21.5% vs. 32.1%; $P = 0.002$).[171] Retrospective review of 259 patients with DCIS found comparable recurrence rates between patients treated with excision alone and excision with radiation, with a majority of the patients in the excision alone cohort having low risk lesions according to the VPNI score.[172] The determination of the role of radiation based on histologic classification and patient factors is an important area of research, given more recent data showing that roughly 60% of patients do not receive radiation after lumpectomy for DCIS.[124]

Many studies have shown that in patients treated with breast conserving therapy, roughly half of recurrences are invasive carcinoma.[163,168,169] An updated review of the outcomes from the NSABP B-17 and B-24 trials demonstrated significantly increased mortality in patients who had recurrence of invasive disease (HR 1.75, $P < 0.001$).[19] In this respect, although adjuvant radiation after lumpectomy for DCIS had not yet been shown to affect overall survival, it has been shown to decrease the number of invasive recurrences, which is a predictor of mortality.

9.3 Chemotherapy

The estrogen receptor (ER) was the first molecular marker discovered to have a role in breast cancer.[173] This receptor became important clinically with the discovery of the therapeutic benefit of selective estrogen receptor modulators (SERMs), such as tamoxifen and raloxifene in reducing breast cancer recurrence in tumors with ER positivity.[174] In 1999, the National Surgical Adjuvant Breast and Bowel Project (NSABP) published the B-24 study. This was a randomized controlled trial of 1800 women with ER-positive DCIS randomized to receive adjuvant tamoxifen or placebo. At 5-year follow-up, the rate of recurrent breast cancer was significantly lower in the group receiving tamoxifen (8.2% vs. 13.4%; P = 0.009).[175] In a retrospective review of the results of this study, the benefit of tamoxifen was predicted by higher levels of ER expression.[176] The aromatase inhibitors (AIs) were introduced after the SERMs as another effective therapy for decreasing estrogen in postmenopausal women. These agents were initially shown to result in significantly higher disease-free survival compared to tamoxifen for adjuvant treatment of early stage breast cancer in postmenopausal women.[177] This led to two large randomized controlled trials looking at the effects of adjuvant therapy with aromatase inhibitors in postmenopausal patients with DCIS. In the NSABP B-35 study, adjuvant anastrozole treatment resulted in significant increase in disease-free survival compared to tamoxifen (HR 0.73; 95% CI 0.56–0.96; P = 0.023).[178] The International Breast Cancer Intervention Study II was a similar study performed in the United Kingdom. In this study, postmenopausal women were also randomized to adjuvant anastrozole or tamoxifen for DCIS, with results showing anastrozole to be noninferior to tamoxifen.[179] The side effect profiles of the two drugs differ as well, with tamoxifen being associated more frequently with deep venous thrombosis, embolism, and gynecologic malignancy while anastrozole was more frequently associated with arthralgia and myalgia.[178] As a result of these studies, clinical practice guidelines recommend the use of antiestrogen therapy for 5 years in patients with ER positive DCIS. Tamoxifen is acceptable for pre- or postmenopausal women, while aromatase inhibitors are recommended for postmenopausal women.[180]

10. SIGNIFICANCE OF BIOLOGIC MARKERS/MOLECULAR FEATURES—FUTURE DIRECTIONS IN POSSIBLE PROGNOSTIC INDICATORS

The underlying basis for the development of malignancy is a series of genetic mutations resulting in dysregulation of normal cellular replication resulting in the ability of uninhibited cell growth and tumor formation. In breast cancer, a number of important genetic mutations have been discovered that characterize tumor biology. This molecular characterization confers information about how aggressive a tumor is, which has important prognostic implications. Further, knowledge of the mechanisms that provide a survival advantage for the tumor has led to the development of agents targeted to these pathways resulting in tumor cell death. Most notable in the case of breast cancer are the development of drugs against the ER and

HER2 receptors, which are overexpressed in some tumors. Given the heterogeneity of DCIS, molecular characterization of tumors serves as a more objective method of determining how likely a particular lesion is to progress to malignancy or to recur after treatment. This has implications for the role of genetic and molecular characterization in guiding treatment decisions. Later we discuss some of the key molecular traits that distinguish low from high-risk DCIS lesions. While only a portion are used clinically in categorizing and guiding treatment of DCIS, recent work has focused on establishing genetic and molecular profiles for tumors.[181–183] This will likely lead to clinical implementation of more detailed molecular analysis of tumors in the future.

10.1 p53

P53 (or TP53), is a 54 kDa protein located on chromosome 17p13[184] The role of the p53 in tumorigenesis was first published in 1979,[185] with a number of studies demonstrating its interaction with oncogenic viral proteins.[186–189] While p53 itself was initially postulated to be an oncogene, it was eventually discovered that the wild type form was involved cell cycle arrest, apoptosis, and cell senescence in the presence of DNA damage.[190–192] As such, p53 is now well known as a tumor suppressor gene, earning the title "The Guardian of the Genome" by the scientific community.[193] Dysfunction of this gene due to binding and inhibition or mutation has been associated with a number of different malignancies, including approximately 25% of all breast cancers.[194,195] There are a multitude of different p53 somatic mutations that have been identified in patients with breast cancer, all of which have been shown to confer increased risk of death.[196] The most common mutations of p53 are missense mutations due to single amino acid changes in the protein and the majority of these are in the region encoding a DNA binding site.[194] These mutations result in stabilization and increased half life of the protein resulting in positive staining on immunohistochemistry (IHC).[197–199] The accessibility of IHC makes this a common detection method used in studies of breast cancer. Comparison of IHC with DNA sequencing for detection of p53 lesions has shown the sensitivity and specificity of IHC to be 72 and 92%, respectively.[200] False negative results are often due to mutations resulting in a truncated protein,[196] while false positives have been shown to occur in cases of overexpression of the wild-type protein.[201] Several studies have shown p53 expression to be associated with other high-risk tumor features in breast cancer. Overexpression of p53 protein is present in approximately 16%–25% of DCIS lesions.[202–204] In

DCIS, overexpression of p53 is associated with lesions possessing high-risk features including comedo histology, high nuclear grade, and poorer overall survival compared with tumors that lack p53 overexpression.[205,206] There is an inverse correlation between high levels of p53 and hormone receptor positivity, with a higher incidence of p53 staining in tumors that are hormone receptor negative and less-well differentiated.[202,205,207,208] Despite these findings, there is currently no clinical role for p53 mutation analysis in patients diagnosed with DCIS or invasive cancer, likely because this information does not impact clinical decisions regarding treatment. As more data is collected via genome sequencing and correlated to clinical outcomes, there may be a role for p53 mutation analysis in the future. Several publications have identified p53 mutations as a marker predicting poor response to current breast cancer therapies.[209,210] The resistance to chemotherapeutic agents related to p53 mutations is postulated to be due to the inability of mutated p53 to induce pathways for apoptosis in response to antitumor agents.[211,212] In the future, a more detailed understanding of the prognostic significance of certain p53 mutations and their potential role in cancer progression may result in a clinical role for p53 mutation analysis in patients with DCIS as an adjunct to guide therapeutic approach.

10.2 Hormone Receptors

The role of the estrogen in breast cancer was first reported upon in 1896 in a published report of breast cancer regression after oophorectomy.[173] Population studies have demonstrated increased risk of breast cancer in patients with higher endogenous estrogen levels or taking estrogen postmenopause.[213,214] Similarly, many of the risk factors associated with breast cancer share in common a higher rate of hormone stimulation throughout the life of the patient as a result of a higher number of ovulatory cycles. Examples include nulliparity, late parity, early menarche, and late menopause. Both estrogen and progesterone play an important role in breast development. The estrogen and progesterone receptors (ER and PR) are a part of a large family of nuclear receptors[215] that act as transcription factors and function in proliferation of breast epithelial cells when activated.[216] Two types of ERs have been identified in the breast, ERα, and ERβ. ERα has been shown to play a primary role in initiating breast development[217] and production of PR,[218,219] which plays an important role in maturation of the breast in pregnancy.[220] These hormone receptors are expressed in a high proportion of breast tumors and are implicated in tumorigenesis.[221–223] The role of estrogen in tumorigenesis is related to activation of the ERα receptor and subsequent downstream effects. There is also evidence that estrogen may have a direct effect on the genomic integrity of breast cancer cells aside from ER binding.[223] The downstream effects of ER binding are multiple and complex. The ER can bind DNA directly at sites termed estrogen response elements[224,225] and regulate gene transcription or bind and regulate the activity of other transcription factors.[226] Downstream effects of ER activation include activation of the MAPK and PI3K/AKT pathways, which results in increased DNA synthesis, increased expression of antiapoptotic proteins, suppression of proapoptotic proteins, and increased cyclin D1 expression facilitating cell cycle progression.[226] Although hormone receptor stimulation is a risk factor for breast cancer, in both invasive breast cancer and DCIS, ER/PR positivity is a marker of well differentiated, less aggressive tumors.[208,227] In DCIS, ER/PR positive lesions are more likely to be noncomedo with low nuclear grade.[207,228–230] The risk of recurrent DCIS or invasive cancer is lower in patients with ER/PR positive tumors.[231–233]

The role of hormone receptors in breast cancer lead to the discovery and development of therapies targeted at ER inhibition. Tamoxifen was the first antiestrogen therapy to be used for the treatment of breast cancer. An initial report in 1971 of treatment of 46 women with metastatic breast cancer with tamoxifen showed promising results.[234] Since then, the drug has made great strides in its application to the treatment of breast cancer. Several trials demonstrated its utility in reducing breast cancer recurrence and improving overall survival when used in the adjuvant setting in patients with early, ER positive tumors,[235] as well as prophylaxis in patients at increased risk of developing breast cancer.[236,237] The classification of tamoxifen as a SERM is a result of the discovery that the effects of receptor binding are dependent on the tissue, with inhibitor effects on ERs in the breast[238] and stimulatory effects in the bone[239,240] and endometrium.[241] This is a result of tissue-specific coregulators that bind the receptor.[242] This accounts in part, for the tissue-specific effects of ER binding. Raloxifene, a second-generation SERM initially used to treat osteoporosis postmenopause was shown to decrease the incidence of invasive breast carcinoma as well.[243] A randomized controlled trial comparing this to tamoxifen for reduction in breast cancer incidence in patients deemed high risk initially revealed similar rates of risk reduction between the two drugs[244] (STAR trial). Recent results from longer follow up this cohort reveal that tamoxifen is slightly more effective in reducing incidence, however, the more favorable side effect profile of raloxifene continues to make it an acceptable option for primary prevention.[245] The AIs are another class of drugs aimed at reducing ER activity. AIs inhibit production of estrogen from androgens, thus removing the ER stimulus. Current AIs used for breast cancer include anastrozole, exemestane, and letrozole. The Arimidex, Tamoxifen, Alone or in Combination (ATAC) trial was a randomized controlled study demonstrating a decreased risk of breast cancer recurrence in postmenopausal patients who received anastrazole versus tamoxifen.[246] Other studies have shown similar results for exemestane[247] and letrozole[248,249] in reducing risk of recurrence. Two randomized controlled trials comparing AI to placebo in high risk postmenopausal women showed a significantly lower incidence of breast cancer in patients taking AIs,[250,251] paving the way for use of these agents for primary prevention in postmenopausal patients at increased risk.[252]

10.3 EGFR

The EGFR (also called HER1 or ErbB1) is a tyrosine kinase that functions in stimulating DNA synthesis, cellular proliferation, and differentiation in the breast.[253,254] These effects are seen in breast tumor cells expressing this receptor as well.[255] In DCIS, EGFR expression has been shown to be associated with ER negative and poorly differentiated lesions.[256,257] Similar to antihormone receptor agents for ER positive tumors, discovery of the role of the EGFR in breast cancer led to the development of agents targeted to it. Current therapies targeting the EGFR include cetuximab, gefitinib, erlotinib, and lapatinib. Lapatinib is the only one FDA approved for treatment of breast cancer and this only includes HER2 positive metastatic breast cancer in which previous therapy has failed. Clinical trials are ongoing to determine the efficacy of these agents as adjuncts in the treatment of early breast cancer.[258] Research has demonstrated growth inhibition of EGFR positive DCIS with these targeted therapies as well,[259,260] paving the way for implementing these therapies in DCIS and chemoprevention in the future.

10.4 HER2

The human epidermal growth factor receptor 2 (HER2), also known as HER2/neu, EGFR 2, or c-erbB-2, has emerged as an important molecular marker in breast carcinoma. The gene is located on chromosome 17q21 and encodes a receptor tyrosine kinase that is a member of the EGFR family.[261,262] The HER2 protein has no known ligand, but forms heterodimers with other members of the EGFR family[263–265] resulting in an increased binding affinity for growth factors.[266] Receptor binding of EGFRs with ligands along with heterodimerization with the HER2 protein results in strong activation of the tyrosine kinase domains of each dimer.[267] Increased tyrosine kinase activity results in upregulation of cellular pathways, such as Ras-Raf-MAP kinase,[268,269] JNK,[266,270] and PI3-AKT[271,272] leading to expression of genes involved in cellular proliferation and inhibition of apoptosis. In the late 1980s, mutation of the HER2 gene leading to amplification and overexpression of the protein product was found in both invasive breast tumors, as well as DCIS.[273,274] Gene amplification of HER2 is correlated to overexpression of the protein as evidenced by positive membrane staining on immunohistochemical analysis.[273,275–277] Overexpression of the HER2 protein is associated with more aggressive tumor biology and has shown to be highly predictive of clinical outcome in breast cancer.[278] Studies examining DCIS in particular have shown HER2 positivity to be correlated other high risk features, including comedo histology,[274,279,280] high histologic grade,[281] a higher rate of cellular proliferation,[279] overexpression of p53,[208] and hormone receptor negativity.[208] The presence of HER2 positivity in both invasive and in situ components of the same tumor has led to the postulation that HER2 gene amplification is an early event in the pathogenesis of breast cancer.[282] The rate of HER2 positivity in DCIS had previously been reported between 50%–70%, which is higher than the 15%–20% reported for invasive carcinoma.[208,274,279,283,284] This observation could lend support to the postulation that HER2 amplification alone is insufficient for transformation to invasive carcinoma.[285] More recent data from large cohorts, however, has estimated the rate of overamplification in DCIS to be slightly lower than previously reported, ranging from 20%–30%.[284,286]

While HER2 overexpression previously portended a poor outcome, the introduction of trastuzumab to the drug market drastically changed this. Trastuzumab is a monoclonal antibody targeted to the extracellular domain of the HER2 protein.[287] Randomized clinical trials demonstrating the success of this agent on prolonging survival in patients with HER2 positive breast cancer[288–291] have made it a standard therapy in this patient population. The success of trastuzumab in the treatment of invasive breast cancer has lead to investigating the role of this drug in patients with HER2 positive DCIS. The NSABP B-43 trial is an ongoing randomized controlled study looking at the effects of adding trastuzumab to radiation after lumpectomy for DCIS.[286] If shown to have an impact on outcomes, this study could result in the routine assessment of HER2 receptor status in DCIS specimens as well.

10.5 Bcl-2

The Bcl-2 gene, located on chromosome 18q, was first discovered for its role in lymphoma as a result of constitutive activation.[292] The Bcl-2 protein is involved in the inhibition of apoptosis[293,294] and plays a role in fetal breast development.[295] Pathologic analysis shows that the expression of Bcl-2 is near 100% in normal breast tissue with loss of expression at increasing rates in DCIS and invasive disease.[296] This loss of bcl-2 expression as a cancer evolves has also been noted in colorectal cancer.[297] These data indicate that tumor aggressiveness and progression is favored by an increase in apoptosis in breast cancer.[298,299] Bcl-2 expression is associated with other tumor characteristics that confer less aggressive, prognostically favorable tumors. The loss of Bcl-2 is inversely correlated to p53 overexpression[296] and positively correlated with ER/PR positivity[300] and well-differentiated tumors.[301,302] In a study of over 11,000 patients with breast cancer, Bcl-2 positivity was shown to be significantly associated with a more favorable prognosis.[303] Although there is currently no role for Bcl-2 testing of DCIS lesions in clinical practice, it could play a role in molecular profiling of breast lesions in the future.

11. BREAST CANCER STEM CELLS

The cells making up solid tumors are well known to be a genetically heterogeneous population. The cancer stem cell theory postulates that there is a subset of tumor cells within this heterogeneous population that are capable of self-renewal and differentiation and are responsible for tumorigenesis. The first report on cancer stem cells was in 1997, when Bonnet and Dick demonstrated a group of highly tumorigenic, undifferentiated leukemia cells.[304] Cancer stem cells for a number of solid cancers,[305–307] including breast[308] have been discovered since that time. Cancer stem cells are potent initiators of tumor growth,[309] with the ability to generate tumors at a much lower inoculum than normal tumor cells.[308,310] They also harbor characteristics that make them resistant to standard chemotherapy drugs. The survival advantage in this small subset of tumor cells is thought to play a role in tumor recurrence after therapy. Several mechanisms have been described by which cancer stem cells can be identified. Cell surface marker expression patterns have been identified that are unique to cancer stem cells. Commonly reported phenotypic markers in breast cancer stem cells include $CD44^{+}/CD24^{-/low}$ $Lineage^{-}$,[308] ESA^{+} (epithelial specific antigen),[308,311] $CD133^{+}$,[312] increased ALDH1 (aldehyde dehydrogenase 1) activity,[310] ABCG2,[313] and CXCR4.[314] Some of these proteins are directly involved in cancer stem cell tumorigenesis and resistance. The ABCG2 protein, also termed breast cancer resistance protein (BCRP), is a member of the ATP binding cassette (ABC) transport protein family and confers drug resistance through efflux of drugs from the cell.[315] ALDH1 is involved in detoxification and may play a role in governing differentiation and drug resistance.[316,317] Its increased expression in breast cancer stem cells is associated with poor prognosis and early tumor metastasis.[310,318] CXCR4 is a chemokine receptor that has shown to confer metastatic potential and is particularly associated with metastasis to bone.[319,320]

In addition to protein products, a number of genetic signaling pathways have important roles in breast cancer stem cells. The Wnt, Notch, and Hedgehog pathways are involved in self-renewal of normal stem cells.[321,322] Not surprisingly, activation of these same pathways is also evident in cancer stem cells. The Wnt pathway governs cellular differentiation via activation of the transcription factor β-catenin and has been shown to be involved in tumorigenesis from stem cell progenitors.[323,324] The Notch signaling pathway is also implicated in breast tumorigenesis. The Notch family consists of four members of transmembrane receptors that, when bound by ligand, result in activation of a number different target genes. Activation of Notch-4 in breast cancer stem cells promotes tumorigenesis and is believed to be associated with tumor recurrence.[325,326] Activation of the Hedgehog pathway in breast cancer stem cells results in self-renewal via activation of GLI transcription factors and expression of the BMI1 transcriptional receptor important in maintenance of stem cells.[327,328] The JAK/STAT pathway is important in maintenance of stem cell populations[329] and has been shown to be constitutively active in breast cancer stem cells.[330] With the discovery of these various characteristics governing cancer stem cell behavior, there are a number of different studies investigating ways to interfere with these pathways and effector proteins that could lead to the development of more effective therapeutic agents in the future. Because cancer stem cells are present in preinvasive breast cancer as well,[331,332] successful implementation of targeted therapeutics in the future likely also has a role in treatment in patients with DCIS.

REFERENCES

1. Wellings SR, Jensen HM. On the origin and progression of ductal carcinoma in the human breast. *J Natl Cancer Inst.* 1973;50(5):1111–1118.
2. Rosner D, Bedwani RN, Vana J, Baker HW, Murphy GP. Noninvasive breast carcinoma: results of a national survey by the American College of Surgeons. *Ann Surg.* 1980;192(2):139–147.
3. Lagios MD. Duct carcinoma in situ. Pathology and treatment. *Surg Clin North Am.* 1990;70(4):853–871.
4. Margolin FR, Lagios MD. Mammographic detection of early breast cancer. Ten years' experience in a community hospital. *West J Med.* 1986;144(1):46–48.

5. Leonard GD, Swain SM. Ductal carcinoma in situ, complexities and challenges. *J Natl Cancer Inst.* 2004;96(12):906–920.
6. Jemal A, Siegel R, Ward E, Hao Y, Xu J, Murray T, et al. Cancer statistics, 2008. *CA CancerJ Clin.* 2008;58(2):71–96.
7. Siegel RL, Miller KD, Jemal A. Cancer statistics, 2017. *CA CancerJ Clin.* 2017;67 (1):7–30.
8. Kerlikowske K, Barclay J, Grady D, Sickles EA, Ernster V. Comparison of risk factors for ductal carcinoma in situ and invasive breast cancer. *J Natl Cancer Inst.* 1997;89 (1):76–82.
9. Weiss HA, Brinton LA, Brogan D, Coates RJ, Gammon MD, Malone KE, et al. Epidemiology of in situ and invasive breast cancer in women aged under 45. *Br J Cancer.* 1996;73(10):1298–1305.
10. Virnig BA, Wang SY, Shamilyan T, Kane RL, Tuttle TM. Ductal carcinoma in situ: risk factors and impact of screening. *J Natl Cancer Inst Monogr.* 2010;2010(41):113–116.
11. Betsill Jr WL, Rosen PP, Lieberman PH, Robbins GF. Intraductal carcinoma. Long-term follow-up after treatment by biopsy alone. *JAMA.* 1978;239(18):1863–1867.
12. McDivitt RW, Holleb AI, Foote Jr FW. Prior breast disease in patients treated for papillary carcinoma. *Arch Pathol.* 1968;85(2):117–124.
13. Black MM, Barclay TH, Cutler SJ, Hankey BF, Asire AJ. Association of atypical characteristics of benign breast lesions with subsequent risk of breast cancer. *Cancer.* 1972;29(2):338–343.
14. Sanders ME, Schuyler PA, Dupont WD, Page DL. The natural history of low-grade ductal carcinoma in situ of the breast in women treated by biopsy only revealed over 30 years of long-term follow-up. *Cancer.* 2005;103(12):2481–2484.
15. Page DL, Dupont WD, Rogers LW, Landenberger M. Intraductal carcinoma of the breast: follow-up after biopsy only. *Cancer.* 1982;49(4):751–758.
16. Silverstein MJ. Ductal Carcinoma in situ: treatment controversies and oncoplastic surgery. 2009 The Breast: Comprehansive Management of Benign and Malignant Disorders. Philadelphia, PA: Elvesier; 2017: pp. 927–952. Available from: https://www-clinicalkey-com.proxy.kumc.edu/-!/content/book/3-s2.0-B978141605221000053X
17. Narod SA, Iqbal J, Giannakeas V, Sopik V, Sun P. Breast cancer mortality after a diagnosis of ductal carcinoma in situ. *JAMA Oncol.* 2015;1(7):888–896.
18. Van Zee KJ, Subhedar P, Olcese C, Patil S, Morrow M. Relationship between margin width and recurrence of ductal carcinoma in situ: analysis of 2996 women treated with breast-conserving surgery for 30 years. *Ann Surg.* 2015;262(4):623–631.
19. Wapnir IL, Dignam JJ, Fisher B, Mamounas EP, Anderson SJ, Julian TB, et al. Long-term outcomes of invasive ipsilateral breast tumor recurrences after lumpectomy in NSABP B-17 and B-24 randomized clinical trials for DCIS. *J Natl Cancer Inst.* 2011;103 (6):478–488.
20. Tabar L, Fagerberg CJ, Gad A, Baldetorp L, Holmberg LH, Grontoft O, et al. Reduction in mortality from breast cancer after mass screening with mammography. Randomised trial from the Breast Cancer Screening Working Group of the Swedish National Board of Health and Welfare. *Lancet.* 1985;1(8433):829–832.
21. Bleyer A, Welch HG. Effect of three decades of screening mammography on breast-cancer incidence. *N Engl J Med.* 2012;367(21):1998–2005.
22. Ernster VL, Barclay J, Kerlikowske K, Grady D, Henderson C. Incidence of and treatment for ductal carcinoma in situ of the breast. *JAMA.* 1996;275(12):913–918.
23. Bassett LW. Breast imaging. In: Bland KI, ed. *The Breast: Comprehensive Management of Benign and Malignant Diseases. Philadelphia, PA: Saunders; 2009.*
24. Ernster VL, Ballard-Barbash R, Barlow WE, Zheng Y, Weaver DL, Cutter G, et al. Detection of ductal carcinoma in situ in women undergoing screening mammography. *J Natl Cancer Inst.* 2002;94(20):1546–1554.

25. Stomper PC, Connolly JL. Ductal carcinoma in situ of the breast: correlation between mammographic calcification and tumor subtype. *AJR Am J Roentgenol.* 1992;159 (3):483–485.
26. Holland R, Hendriks JH. Microcalcifications associated with ductal carcinoma in situ: mammographic-pathologic correlation. *Semin Diagn Pathol.* 1994;11(3):181–192.
27. Knutzen AM, Gisvold JJ. Likelihood of malignant disease for various categories of mammographically detected, nonpalpable breast lesions. *Mayo Clin Proc.* 1993;68 (5):454–460.
28. Tabar L, Dean PB. Basic principles of mammographic diagnosis. *Diagn Imaging Clin Med.* 1985;54(3–4):146–157.
29. Thomas J, Evans A, Macartney J, Pinder SE, Hanby A, Ellis I, et al. Radiological and pathological size estimations of pure ductal carcinoma in situ of the breast, specimen handling and the influence on the success of breast conservation surgery: a review of 2564 cases from the Sloane Project. *Br J Cancer.* 2010;102(2):285–293.
30. Holland R, Hendriks JH, Vebeek AL, Mravunac M, Schuurmans Stekhoven JH. Extent, distribution, and mammographic/histological correlations of breast ductal carcinoma in situ. *Lancet.* 1990;335(8688):519–522.
31. Wild JJ, Neal D. Use of high-frequency ultrasonic waves for detecting changes of texture in living tissues. *Lancet.* 1951;1(6656):655–657.
32. Kelly KM, Dean J, Lee SJ, Comulada WS. Breast cancer detection: radiologists' performance using mammography with and without automated whole-breast ultrasound. *Eur Radiol.* 2010;20(11):2557–2564.
33. Berg WA, Blume JD, Cormack JB, Mendelson EB, Lehrer D, Bohm-Velez M, et al. Combined screening with ultrasound and mammography vs mammography alone in women at elevated risk of breast cancer. *JAMA.* 2008;299(18):2151–2163.
34. Lee MH, Ko EY, Han BK, Shin JH, Ko ES, Hahn SY. Sonographic findings of pure ductal carcinoma in situ. *J Clin Ultrasound.* 2013;41(8):465–471.
35. Ross RJ, Thompson JS, Kim K, Bailey RA. Nuclear magnetic resonance imaging and evaluation of human breast tissue: preliminary clinical trials. *Radiology.* 1982;143 (1):195–205.
36. Heywang SH, Hahn D, Schmidt H, Krischke I, Eiermann W, Bassermann R, et al. MR imaging of the breast using gadolinium-DTPA. *J Comput Assist Tomogr.* 1986;10 (2):199–204.
37. Heywang SH, Wolf A, Pruss E, Hilbertz T, Eiermann W, Permanetter W. MR imaging of the breast with Gd-DTPA: use and limitations. *Radiology.* 1989;171(1):95–103.
38. Kaiser WA, Zeitler E. MR imaging of the breast: fast imaging sequences with and without Gd-DTPA. Preliminary observations. *Radiology.* 1989;170(3 Pt 1):681–686.
39. Orel SG, Schnall MD, LiVolsi VA, Troupin RH. Suspicious breast lesions: MR imaging with radiologic-pathologic correlation. *Radiology.* 1994;190(2):485–493.
40. Gilles R, Zafrani B, Guinebretiere JM, Meunier M, Lucidarme O, Tardivon AA, et al. Ductal carcinoma in situ: MR imaging-histopathologic correlation. *Radiology.* 1995;196(2):415–419.
41. Hansen NM, Growney A. Breast cancer: pre- and postoperative imaging for staging. *Surg Oncol Clin N Am.* 2007;16(2):447–463.
42. Matsubayashi R, Matsuo Y, Edakuni G, Satoh T, Tokunaga O, Kudo S. Breast masses with peripheral rim enhancement on dynamic contrast-enhanced MR images: correlation of MR findings with histologic features and expression of growth factors. *Radiology.* 2000;217(3):841–848.
43. Buadu LD, Murakami J, Murayama S, Hashiguchi N, Sakai S, Masuda K, et al. Breast lesions: correlation of contrast medium enhancement patterns on MR images with histopathologic findings and tumor angiogenesis. *Radiology.* 1996;200 (3):639–649.

44. Nunes LW, Schnall MD, Orel SG. Update of breast MR imaging architectural interpretation model. *Radiology.* 2001;219(2):484–494.
45. Sherif H, Mahfouz AE, Oellinger H, Hadijuana J, Blohmer JU, Taupitz M, et al. Peripheral washout sign on contrast-enhanced MR images of the breast. *Radiology.* 1997;205(1):209–213.
46. Newstead GM. MR imaging of ductal carcinoma in situ. *Magn Reson Imaging Clin N Am.* 2010;18(2):225–240.
47. Yamada T, Mori N, Watanabe M, Kimijima I, Okumoto T, Seiji K, et al. Radiologic-pathologic correlation of ductal carcinoma in situ. *Radiographics.* 2010;30(5): 1183–1198.
48. Soderstrom CE, Harms SE, Copit DS, Evans WP, Savino DA, Krakos PA, et al. Three-dimensional RODEO breast MR imaging of lesions containing ductal carcinoma in situ. *Radiology.* 1996;201(2):427–432.
49. Orel SG, Mendonca MH, Reynolds C, Schnall MD, Solin LJ, Sullivan DC. MR imaging of ductal carcinoma in situ. *Radiology.* 1997;202(2):413–420.
50. Peters NH, Borel Rinkes IH, Zuithoff NP, Mali WP, Moons KG, Peeters PH. Meta-analysis of MR imaging in the diagnosis of breast lesions. *Radiology.* 2008;246 (1):116–124.
51. Houssami N, Ciatto S, Macaskill P, Lord SJ, Warren RM, Dixon JM, et al. Accuracy and surgical impact of magnetic resonance imaging in breast cancer staging: systematic review and meta-analysis in detection of multifocal and multicentric cancer. *J Clin Oncol.* 2008;26(19):3248–3258.
52. Bone B, Aspelin P, Bronge L, Isberg B, Perbeck L, Veress B. Sensitivity and specificity of MR mammography with histopathological correlation in 250 breasts. *Acta Radiol.* 1996;37(2):208–213.
53. Gilles R, Guinebretiere JM, Lucidarme O, Cluzel P, Janaud G, Finet JF, et al. Nonpalpable breast tumors: diagnosis with contrast-enhanced subtraction dynamic MR imaging. *Radiology.* 1994;191(3):625–631.
54. Heywang-Koebrunner SH, Viehweg P. Sensitivity of contrast-enhanced MR imaging of the breast. *Magn Reson Imaging Clin N Am.* 1994;2(4):527–538.
55. Kuhl CK, Schrading S, Bieling HB, Wardelmann E, Leutner CC, Koenig R, et al. MRI for diagnosis of pure ductal carcinoma in situ: a prospective observational study. *Lancet.* 2007;370(9586):485–492.
56. Santamaria G, Velasco M, Farrus B, Zanon G, Fernandez PL. Preoperative MRI of pure intraductal breast carcinoma—a valuable adjunct to mammography in assessing cancer extent. *Breast.* 2008;17(2):186–194.
57. Allen LR, Lago-Toro CE, Hughes JH, Careaga E, Brown AT, Chernick M, et al. Is there a role for MRI in the preoperative assessment of patients with DCIS? *Ann Surg Oncol.* 2010;17(9):2395–2400.
58. Neubauer H, Li M, Kuehne-Heid R, Schneider A, Kaiser WA. High grade and non-high grade ductal carcinoma in situ on dynamic MR mammography: characteristic findings for signal increase and morphological pattern of enhancement. *Br J Radiol.* 2003;76(901):3–12.
59. Harms SE, Flamig DP, Hesley KL, Meiches MD, Jensen RA, Evans WP, et al. MR imaging of the breast with rotating delivery of excitation off resonance: clinical experience with pathologic correlation. *Radiology.* 1993;187(2):493–501.
60. Stomper PC, Herman S, Klippenstein DL, Winston JS, Edge SB, Arredondo MA, et al. Suspect breast lesions: findings at dynamic gadolinium-enhanced MR imaging correlated with mammographic and pathologic features. *Radiology.* 1995;197(2): 387–395.
61. Turner DA, Alcorn FS, Adler YT. Nuclear magnetic resonance in the diagnosis of breast cancer. *Radiol Clin North Am.* 1988;26(3):673–687.

62. Stack JP, Redmond OM, Codd MB, Dervan PA, Ennis JT. Breast disease: tissue characterization with Gd-DTPA enhancement profiles. *Radiology.* 1990;174(2):491–494.
63. Saslow D, Boetes C, Burke W, Harms S, Leach MO, Lehman CD, et al. American Cancer Society guidelines for breast screening with MRI as an adjunct to mammography. *CA Cancer J Clin.* 2007;57(2):75–89.
64. Lehman CD. Role of MRI in screening women at high risk for breast cancer. *J Magn Reson Imaging.* 2006;24(5):964–970.
65. Bleicher RJ, Ciocca RM, Egleston BL, Sesa L, Evers K, Sigurdson ER, et al. Association of routine pretreatment magnetic resonance imaging with time to surgery, mastectomy rate, and margin status. *J Am Coll Surg.* 2009;209(2):180–187.
66. Katipamula R, Degnim AC, Hoskin T, Boughey JC, Loprinzi C, Grant CS, et al. Trends in mastectomy rates at the Mayo Clinic Rochester: effect of surgical year and preoperative magnetic resonance imaging. *J Clin Oncol.* 2009;27(25):4082–4088.
67. Reddy DH, Mendelson EB. Incorporating new imaging models in breast cancer management. *Curr Treat Options Oncol.* 2005;6(2):135–145.
68. Pilewskie M, Kennedy C, Shappell C, Helenowski I, Scholtens D, Hansen N, et al. Effect of MRI on the management of ductal carcinoma in situ of the breast. *Ann Surg Oncol.* 2013;20(5):1522–1529.
69. Dunne C, Burke JP, Morrow M, Kell MR. Effect of margin status on local recurrence after breast conservation and radiation therapy for ductal carcinoma in situ. *J Clin Oncol.* 2009;27(10):1615–1620.
70. Macdonald HR, Silverstein MJ, Lee LA, Ye W, Sanghavi P, Holmes DR, et al. Margin width as the sole determinant of local recurrence after breast conservation in patients with ductal carcinoma in situ of the breast. *Am J Surg.* 2006;192(4):420–422.
71. Meijnen P, Gilhuijs KG, Rutgers EJ. The effect of margins on the clinical management of ductal carcinoma in situ of the breast. *J Surg Oncol.* 2008;98(8):579–584.
72. Marinovich ML, Azizi L, Macaskill P, Irwig L, Morrow M, Solin LJ, et al. The Association of Surgical Margins and Local Recurrence in Women with Ductal Carcinoma In Situ Treated with Breast-Conserving Therapy: a meta-analysis. *Ann Surg Oncol.* 2016;23(12):3811–3821.
73. Collins LC, Achacoso N, Haque R, Nekhlyudov L, Fletcher SW, Quesenberry Jr CP et al. Risk factors for non-invasive and invasive local recurrence in patients with ductal carcinoma in situ. *Breast Cancer Res Treat.* 2013;139(2):453–460.
74. Fancellu A, Turner RM, Dixon JM, Pinna A, Cottu P, Houssami N. Meta-analysis of the effect of preoperative breast MRI on the surgical management of ductal carcinoma in situ. *Br J Surg.* 2015;102(8):883–893.
75. Gonzalez V, Sandelin K, Karlsson A, Aberg W, Lofgren L, Iliescu G, et al. Preoperative MRI of the breast (POMB) influences primary treatment in breast cancer: a prospective, randomized, multicenter study. *World J Surg.* 2014;38(7):1685–1693.
76. Kropcho LC, Steen ST, Chung AP, Sim MS, Kirsch DL, Giuliano AE. Preoperative breast MRI in the surgical treatment of ductal carcinoma in situ. *Breast J.* 2012;18 (2):151–156.
77. Bedrosian I, Mick R, Orel SG, Schnall M, Reynolds C, Spitz FR, et al. Changes in the surgical management of patients with breast carcinoma based on preoperative magnetic resonance imaging. *Cancer.* 2003;98(3):468–473.
78. Miller BT, Abbott AM, Tuttle TM. The influence of preoperative MRI on breast cancer treatment. *Ann Surg Oncol.* 2012;19(2):536–540.
79. Tillman GF, Orel SG, Schnall MD, Schultz DJ, Tan JE, Solin LJ. Effect of breast magnetic resonance imaging on the clinical management of women with early-stage breast carcinoma. *J Clin Oncol.* 2002;20(16):3413–3423.
80. Kowalchik KV, Vallow LA, McDonough M, Thomas CS, Heckman MG, Peterson JL, et al. The role of preoperative bilateral breast magnetic resonance imaging in patient

selection for partial breast irradiation in ductal carcinoma in situ. *Int J Surg Oncol.* 2012;2012:206342.
81. Pilewskie M, Olcese C, Eaton A, Patil S, Morris E, Morrow M, et al. Perioperative breast MRI is not associated with lower locoregional recurrence rates in DCIS patients treated with or without radiation. *Ann Surg Oncol.* 2014;21(5):1552–1560.
82. Solin LJ, Orel SG, Hwang WT, Harris EE, Schnall MD. Relationship of breast magnetic resonance imaging to outcome after breast-conservation treatment with radiation for women with early-stage invasive breast carcinoma or ductal carcinoma in situ. *J Clin Oncol.* 2008;26(3):386–391.
83. Turnbull L, Brown S, Harvey I, Olivier C, Drew P, Napp V, et al. Comparative effectiveness of MRI in breast cancer (COMICE) trial: a randomised controlled trial. *Lancet.* 2010;375(9714):563–571.
84. Peters NH, van Esser S, van den Bosch MA, Storm RK, Plaisier PW, van Dalen T, et al. Preoperative MRI and surgical management in patients with nonpalpable breast cancer: the MONET—randomised controlled trial. *Eur J Cancer.* 2011;47(6):879–886.
85. Fischer U, Zachariae O, Baum F, von Heyden D, Funke M, Liersch T. The influence of preoperative MRI of the breasts on recurrence rate in patients with breast cancer. *Eur Radiol.* 2004;14(10):1725–1731.
86. Parker SH, Lovin JD, Jobe WE, Burke BJ, Hopper KD, Yakes WF. Nonpalpable breast lesions: stereotactic automated large-core biopsies. *Radiology.* 1991;180(2):403–407.
87. Fuhrman GM, Cederbom GJ, Bolton JS, King TA, Duncan JL, Champaign JL, et al. Image-guided core-needle breast biopsy is an accurate technique to evaluate patients with nonpalpable imaging abnormalities. *Ann Surg.* 1998;227(6):932–939.
88. O'Flynn EA, Wilson AR, Michell MJ. Image-guided breast biopsy: state-of-the-art. *Clin Radiol.* 2010;65(4):259–270.
89. Liberman L, Feng TL, Dershaw DD, Morris EA, Abramson AF. US-guided core breast biopsy: use and cost-effectiveness. *Radiology.* 1998;208(3):717–723.
90. Kettritz U, Rotter K, Schreer I, Murauer M, Schulz-Wendtland R, Peter D, et al. Stereotactic vacuum-assisted breast biopsy in 2874 patients: a multicenter study. *Cancer.* 2004;100(2):245–251.
91. Jackman RJ, Burbank F, Parker SH, Evans 3rd WP, Lechner MC, Richardson TR, et al. Stereotactic breast biopsy of nonpalpable lesions: determinants of ductal carcinoma in situ underestimation rates. *Radiology.* 2001;218(2):497–502.
92. Burbank F. Stereotactic breast biopsy of atypical ductal hyperplasia and ductal carcinoma in situ lesions: improved accuracy with directional, vacuum-assisted biopsy. *Radiology.* 1997;202(3):843–847.
93. Meyer JE, Smith DN, DiPiro PJ, Denison CM, Frenna TH, Harvey SC, et al. Stereotactic breast biopsy of clustered microcalcifications with a directional, vacuum-assisted device. *Radiology.* 1997;204(2):575–576.
94. Jaffer S, Bleiweiss IJ. Histologic classification of ductal carcinoma in situ. *Microsc Res Tech.* 2002;59(2):92–101.
95. Pinder A. Morphology of ductal carcinoma in situ. In: O'malley FP, ed. *Breast Pathology: A Volume in the Foundations in Diagnostic Pathology Series,* 2nd ed. Philadelphia, PA: Saunders; 2011.
96. Rakha E. Ductal carcinoma in situ. In: Dabbs D, ed. *Breast Pathology,* 2nd ed. Philadelphia, PA: Elsevier; 2017.
97. Hoda S. Ductal carcinoma in situ. In: Syed Hoda EB, Frederick Koerner, Paul Rosen, eds. *Rosens Breast Pathology,* 4th ed. Philadelphia, PA: Lippincott Williams and Wilkins; 2014.
98. Patchefsky AS, Schwartz GF, Finkelstein SD, Prestipino A, Sohn SE, Singer JS, et al. Heterogeneity of intraductal carcinoma of the breast. *Cancer.* 1989;63(4):731–741.
99. Lennington WJ, Jensen RA, Dalton LW, Page DL. Ductal carcinoma in situ of the breast. Heterogeneity of individual lesions. *Cancer.* 1994;73(1):118–124.

100. Bellamy CO, McDonald C, Salter DM, Chetty U, Anderson TJ. Noninvasive ductal carcinoma of the breast: the relevance of histologic categorization. *Hum Pathol.* 1993;24 (1):16–23.
101. Ottesen GL, Graversen HP, Blichert-Toft M, Zedeler K, Andersen JA. Ductal carcinoma in situ of the female breast. Short-term results of a prospective nationwide study. The Danish Breast Cancer Cooperative Group. *Am J Surg Pathol.* 1992;16(12): 1183–1196.
102. Sloane JP, Ellman R, Anderson TJ, Brown CL, Coyne J, Dallimore NS, et al. Consistency of histopathological reporting of breast lesions detected by screening: findings of the U.K. National External Quality Assessment (EQA) Scheme. U. K. National Coordinating Group for Breast Screening Pathology. *Eur J Cancer.* 1994;30A (10):1414–1419.
103. Badve S, A'Hern RP, Ward AM, Millis RR, Pinder SE, Ellis IO, et al. Prediction of local recurrence of ductal carcinoma in situ of the breast using five histological classifications: a comparative study with long follow-up. *Hum Pathol.* 1998;29(9):915–923.
104. Sneige N, McNeese MD, Atkinson EN, Ames FC, Kemp B, Sahin A, et al. Ductal carcinoma in situ treated with lumpectomy and irradiation: histopathological analysis of 49 specimens with emphasis on risk factors and long term results. *Hum Pathol.* 1995;26 (6):642–649.
105. Fisher ER, Costantino J, Fisher B, Palekar AS, Redmond C, Mamounas E. Pathologic findings from the National Surgical Adjuvant Breast Project (NSABP) Protocol B-17. Intraductal carcinoma (ductal carcinoma in situ). The National Surgical Adjuvant Breast and Bowel Project Collaborating Investigators. *Cancer.* 1995;75(6):1310–1319.
106. Lagios MD, Margolin FR, Westdahl PR, Rose MR. Mammographically detected duct carcinoma in situ. Frequency of local recurrence following tylectomy and prognostic effect of nuclear grade on local recurrence. *Cancer.* 1989;63(4):618–624.
107. Ross DS, Wen YH, Brogi E. Ductal carcinoma in situ: morphology-based knowledge and molecular advances. *Adv Anat Pathol.* 2013;20(4):205–216.
108. The Consensus Conference Committee. Consensus Conference of the Classification of Ductal Carcinoma in Situ. Cancer 1997;80:1798–1802
109. Harrison M, Coyne JD, Gorey T, Dervan PA. Comparison of cytomorphological and architectural heterogeneity in mammographically-detected ductal carcinoma in situ. *Histopathology.* 1996;28(5):445–450.
110. Killeen JL, Namiki H. DNA analysis of ductal carcinoma in situ of the breast. A comparison with histologic features. *Cancer.* 1991;68(12):2602–2607.
111. Silverstein MJ, Poller DN, Waisman JR, Colburn WJ, Barth A, Gierson ED, et al. Prognostic classification of breast ductal carcinoma-in-situ. *Lancet.* 1995;345(8958): 1154–1157.
112. Scott MA, Lagios MD, Axelsson K, Rogers LW, Anderson TJ, Page DL. Ductal carcinoma in situ of the breast: reproducibility of histological subtype analysis. *Hum Pathol.* 1997;28(8):967–973.
113. Holland R, Peterse JL, Millis RR, Eusebi V, Faverly D, van de Vijver MJ, et al. Ductal carcinoma in situ: a proposal for a new classification. *Semin Diagn Pathol.* 1994;11 (3):167–180.
114. Poller DN, Silverstein MJ, Galea M, Locker AP, Elston CW, Blamey RW, et al. Ideas in pathology. Ductal carcinoma in situ of the breast: a proposal for a new simplified histological classification association between cellular proliferation and c-erbB-2 protein expression. *Mod Pathol.* 1994;7(2):257–262.
115. Douglas-Jones AG, Gupta SK, Attanoos RL, Morgan JM, Mansel RE. A critical appraisal of six modern classifications of ductal carcinoma in situ of the breast (DCIS): correlation with grade of associated invasive carcinoma. *Histopathology.* 1996;29(5): 397–409.

116. Schuh F, Biazus JV, Resetkova E, Benfica CZ, Edelweiss MI. Reproducibility of three classification systems of ductal carcinoma in situ of the breast using a web-based survey. *Pathol Res Pract.* 2010;206(10):705–711.
117. Sloane JP, Amendoeira I, Apostolikas N, Bellocq JP, Bianchi S, Boecker W, et al. Consistency achieved by 23 European pathologists from 12 countries in diagnosing breast disease and reporting prognostic features of carcinomas. European Commission Working Group on Breast Screening Pathology. *Virchows Arch.* 1999;434(1):3–10.
118. Bethwaite P, Smith N, Delahunt B, Kenwright D. Reproducibility of new classification schemes for the pathology of ductal carcinoma in situ of the breast. *J Clin Pathol.* 1998;51 (6):450–454.
119. Tawfik O, Kimler BF, Davis M, Stasik C, Lai SM, Mayo MS, et al. Grading invasive ductal carcinoma of the breast: advantages of using automated proliferation index instead of mitotic count. *Virchows Arch.* 2007;450(6):627–636.
120. Stasik CJ, Davis M, Kimler BF, Fan F, Damjanov I, Thomas P, et al. Grading ductal carcinoma in situ of the breast using an automated proliferation index. *Ann Clin Lab Sci.* 2011;41(2):122–130.
121. Silverstein MJ, Barth A, Poller DN, Gierson ED, Colburn WJ, Waisman JR, et al. Ten-year results comparing mastectomy to excision and radiation therapy for ductal carcinoma in situ of the breast. *Eur J Cancer.* 1995;31A(9):1425–1427.
122. Worni M, Akushevich I, Greenup R, Sarma D, Ryser MD, Myers ER, et al. Trends in treatment patterns and outcomes for ductal carcinoma in situ. *J Natl Cancer Inst.* 2015;107(12).
123. Baxter NN, Virnig BA, Durham SB, Tuttle TM. Trends in the treatment of ductal carcinoma in situ of the breast. *J Natl Cancer Inst.* 2004;96(6):443–448.
124. Sumner 3rd WE, Koniaris LG, Snell SE, Spector S, Powell J, Avisar E, et al. Results of 23,810 cases of ductal carcinoma-in-situ. *Ann Surg Oncol.* 2007;14(5):1638–1643.
125. Network NCC. NCCN Clinical Practice Guidelines in Oncology: Breast Cancer Version 2.2017 2017. Available from: https://www.nccn.org/professionals/physician_gls/pdf/breast.pdf:%5B
126. Lovrics PJ, Cornacchi SD, Vora R, Goldsmith CH, Kahnamoui K. Systematic review of radioguided surgery for non-palpable breast cancer. *Eur J Surg Oncol.* 2011;37 (5):388–397.
127. Langhans L, Tvedskov TF, Klausen TL, Jensen MB, Talman ML, Vejborg I, et al. Radioactive seed localization or wire-guided localization of nonpalpable invasive and in Situ breast cancer: a randomized, multicenter, open-label trial. *Ann Surg.* 2017;266 (1):29–35.
128. Bijker N, Peterse JL, Duchateau L, Julien JP, Fentiman IS, Duval C, et al. Risk factors for recurrence and metastasis after breast-conserving therapy for ductal carcinoma-in-situ: analysis of European Organization for Research and Treatment of Cancer Trial 10853. *J Clin Oncol.* 2001;19(8):2263–2271.
129. Chan KC, Knox WF, Sinha G, Gandhi A, Barr L, Baildam AD, et al. Extent of excision margin width required in breast conserving surgery for ductal carcinoma in situ. *Cancer.* 2001;91(1):9–16.
130. Silverstein MJ, Lagios MD, Groshen S, Waisman JR, Lewinsky BS, Martino S, et al. The influence of margin width on local control of ductal carcinoma in situ of the breast. *N Engl J Med.* 1999;340(19):1455–1461.
131. Wang SY, Chu H, Shamliyan T, Jalal H, Kuntz KM, Kane RL, et al. Network meta-analysis of margin threshold for women with ductal carcinoma in situ. *J Natl Cancer Inst.* 2012;104(7):507–516.
132. Silverstein MJ, Lagios MD, Craig PH, Waisman JR, Lewinsky BS, Colburn WJ, et al. A prognostic index for ductal carcinoma in situ of the breast. *Cancer.* 1996;77(11): 2267–2274.

133. Moran MS, Schnitt SJ, Giuliano AE, Harris JR, Khan SA, Horton J, et al. Society of Surgical Oncology—American Society for Radiation Oncology consensus guideline on margins for breast-conserving surgery with whole-breast irradiation in stages I and II invasive breast cancer. *J Clin Oncol.* 2014;32(14):1507–1515.
134. Benson JR, Wishart GC. Predictors of recurrence for ductal carcinoma in situ after breast-conserving surgery. *Lancet Oncol.* 2013;14(9):e348–e357.
135. Kaufmann M, Morrow M, von Minckwitz G, Harris JR. Biedenkopf Expert Panel M. Locoregional treatment of primary breast cancer: consensus recommendations from an International Expert Panel. *Cancer.* 2010;116(5):1184–1191.
136. Gould EA, Winship T, Philbin PH, Kerr HH. Observations on a "sentinel node" in cancer of the parotid. *Cancer.* 1960;13:77–78.
137. Morton DL, Wen DR, Wong JH, Economou JS, Cagle LA, Storm FK, et al. Technical details of intraoperative lymphatic mapping for early stage melanoma. *Arch Surg.* 1992;127(4):392–399.
138. Kelly TA, Kim JA, Patrick R, Grundfest S, Crowe JP. Axillary lymph node metastases in patients with a final diagnosis of ductal carcinoma in situ. *Am J Surg.* 2003;186 (4):368–370.
139. van Roozendaal LM, Goorts B, Klinkert M, Keymeulen KB, De Vries B, Strobbe LJ, et al. Sentinel lymph node biopsy can be omitted in DCIS patients treated with breast conserving therapy. *Breast Cancer Res Treat.* 2016;156(3):517–525.
140. Intra M, Veronesi P, Mazzarol G, Galimberti V, Luini A, Sacchini V, et al. Axillary sentinel lymph node biopsy in patients with pure ductal carcinoma in situ of the breast. *Arch Surg.* 2003;138(3):309–313.
141. Veronesi P, Intra M, Vento AR, Naninato P, Caldarella P, Paganelli G, et al. Sentinel lymph node biopsy for localised ductal carcinoma in situ? *Breast.* 2005;14(6):520–522.
142. Francis AM, Haugen CE, Grimes LM, Crow JR, Yi M, Mittendorf EA, et al. Is sentinel lymph node dissection warranted for patients with a diagnosis of ductal carcinoma in situ? *Ann Surg Oncol.* 2015;22(13):4270–4279.
143. Silverstein MJ, Skinner KA, Lomis TJ. Predicting axillary nodal positivity in 2282 patients with breast carcinoma. *World J Surg.* 2001;25(6):767–772.
144. Millis RR, Springall R, Lee AH, Ryder K, Rytina ER, Fentiman IS. Occult axillary lymph node metastases are of no prognostic significance in breast cancer. *Br J Cancer.* 2002;86(3):396–401.
145. Wu Y, Mittendorf EA, Kelten C, Tucker SL, Wei W, Middleton LP, et al. Occult axillary lymph node metastases do not have prognostic significance in early stage breast cancer. *Cancer.* 2012;118(6):1507–1514.
146. Colpaert C, Vermeulen P, Jeuris W, van Beest P, Goovaerts G, Weyler J, et al. Early distant relapse in "node-negative" breast cancer patients is not predicted by occult axillary lymph node metastases, but by the features of the primary tumour. *J Pathol.* 2001;193(4):442–449.
147. Lara JF, Young SM, Velilla RE, Santoro EJ, Templeton SF. The relevance of occult axillary micrometastasis in ductal carcinoma in situ: a clinicopathologic study with long-term follow-up. *Cancer.* 2003;98(10):2105–2113.
148. Verkooijen HM, Peeters PH, Buskens E, Koot VC, Borel Rinkes IH, Mali WP, et al. Diagnostic accuracy of large-core needle biopsy for nonpalpable breast disease: a meta-analysis. *Br J Cancer.* 2000;82(5):1017–1021.
149. Renshaw AA. Predicting invasion in the excision specimen from breast core needle biopsy specimens with only ductal carcinoma in situ. *Arch Pathol Lab Med.* 2002;126 (1):39–41.
150. Brennan ME, Turner RM, Ciatto S, Marinovich ML, French JR, Macaskill P, et al. Ductal carcinoma in situ at core-needle biopsy: meta-analysis of underestimation and predictors of invasive breast cancer. *Radiology.* 2011;260(1):119–128.

151. Camp R, Feezor R, Kasraeian A, Cendan J, Schell S, Wilkinson E, et al. Sentinel lymph node biopsy for ductal carcinoma in situ: an evolving approach at the University of Florida. *Breast J.* 2005;11(6):394–397.
152. Fisher B, Bauer M, Margolese R, Poisson R, Pilch Y, Redmond C, et al. Five-year results of a randomized clinical trial comparing total mastectomy and segmental mastectomy with or without radiation in the treatment of breast cancer. *N Engl J Med.* 1985;312(11):665–673.
153. van Dongen JA, Bartelink H, Fentiman IS, Lerut T, Mignolet F, Olthuis G, et al. Randomized clinical trial to assess the value of breast-conserving therapy in stage I and II breast cancer, EORTC 10801 trial. *J Natl Cancer Inst Monogr.* 1992;(11):15–18.
154. Straus K, Lichter A, Lippman M, Danforth D, Swain S, Cowan K, et al. Results of the National Cancer Institute early breast cancer trial. *J Natl Cancer Inst Monogr.* 1992;(11):27–32.
155. Veronesi U, Saccozzi R, Del Vecchio M, Banfi A, Clemente C, De Lena M, et al. Comparing radical mastectomy with quadrantectomy, axillary dissection, and radiotherapy in patients with small cancers of the breast. *N Engl J Med.* 1981;305(1):6–11.
156. Fisher B, Anderson S, Redmond CK, Wolmark N, Wickerham DL, Cronin WM. Reanalysis and results after 12 years of follow-up in a randomized clinical trial comparing total mastectomy with lumpectomy with or without irradiation in the treatment of breast cancer. *N Engl J Med.* 1995;333(22):1456–1461.
157. Litiere S, Werutsky G, Fentiman IS, Rutgers E, Christiaens MR, Van Limbergen E, et al. Breast conserving therapy versus mastectomy for stage I-II breast cancer: 20 year follow-up of the EORTC 10801 phase 3 randomised trial. *Lancet Oncol.* 2012;13(4):412–419.
158. Poggi MM, Danforth DN, Sciuto LC, Smith SL, Steinberg SM, Liewehr DJ, et al. Eighteen-year results in the treatment of early breast carcinoma with mastectomy versus breast conservation therapy: the National Cancer Institute Randomized Trial. *Cancer.* 2003;98(4):697–702.
159. Veronesi U, Cascinelli N, Mariani L, Greco M, Saccozzi R, Luini A, et al. Twenty-year follow-up of a randomized study comparing breast-conserving surgery with radical mastectomy for early breast cancer. *N Engl J Med.* 2002;347(16):1227–1232.
160. Fisher B, Costantino J, Redmond C, Fisher E, Margolese R, Dimitrov N, et al. Lumpectomy compared with lumpectomy and radiation therapy for the treatment of intraductal breast cancer. *N Engl J Med.* 1993;328(22):1581–1586.
161. Houghton J, George WD, Cuzick J, Duggan C, Fentiman IS, Spittle M, et al. Radiotherapy and tamoxifen in women with completely excised ductal carcinoma in situ of the breast in the UK, Australia, and New Zealand: randomised controlled trial. *Lancet.* 2003;362(9378):95–102.
162. Julien JP, Bijker N, Fentiman IS, Peterse JL, Delledonne V, Rouanet P, et al. Radiotherapy in breast-conserving treatment for ductal carcinoma in situ: first results of the EORTC randomised phase III trial 10853. EORTC Breast Cancer Cooperative Group and EORTC Radiotherapy Group. *Lancet.* 2000;355(9203):528–533.
163. Emdin SO, Granstrand B, Ringberg A, Sandelin K, Arnesson LG, Nordgren H, et al. SweDCIS: radiotherapy after sector resection for ductal carcinoma in situ of the breast. Results of a randomised trial in a population offered mammography screening. *Acta Oncol.* 2006;45(5):536–543.
164. Viani GA, Stefano EJ, Afonso SL, De Fendi LI, Soares FV, Leon PG, et al. Breast-conserving surgery with or without radiotherapy in women with ductal carcinoma in situ: a meta-analysis of randomized trials. *Radiat Oncol.* 2007;2:28.
165. Fisher ER, Costantino J, Fisher B, Palekar AS, Paik SM, Suarez CM, et al. Pathologic findings from the National Surgical Adjuvant Breast Project (NSABP) Protocol B-17. Five-year observations concerning lobular carcinoma in situ. *Cancer.* 1996;78(7):1403–1416.

166. Carlson RW, Allred DC, Anderson BO, Burstein HJ, Carter WB, Edge SB, et al. Breast cancer. Clinical practice guidelines in oncology. *J Natl Compr Canc Netw.* 2009;7 (2):122–192.
167. Fisher ER, Dignam J, Tan-Chiu E, Costantino J, Fisher B, Paik S, et al. Pathologic findings from the National Surgical Adjuvant Breast Project (NSABP) eight-year update of Protocol B-17: intraductal carcinoma. *Cancer.* 1999;86(3):429–438.
168. Hughes LL, Wang M, Page DL, Gray R, Solin LJ, Davidson NE, et al. Local excision alone without irradiation for ductal carcinoma in situ of the breast: a trial of the Eastern Cooperative Oncology Group. *J Clin Oncol.* 2009;27(32):5319–5324.
169. Wehner P, Lagios MD, Silverstein MJ. DCIS treated with excision alone using the National Comprehensive Cancer Network (NCCN) guidelines. *Ann Surg Oncol.* 2013;20(10):3175–3179.
169a. McCormick B, Winter K, Hudis C, Kuerer HM, Rakovitch E, Smith BL, et al. RTOG 9804: a prospective randomized trial for good-risk ductal carcinoma in situ comparing radiotherapy with observation. *J Clin Oncol.* 2015;33(7):709–715.
170. Silverstein MJ. The University of Southern California/Van Nuys prognostic index for ductal carcinoma in situ of the breast. *Am J Surg.* 2003;186(4):337–343.
171. Gilleard O, Goodman A, Cooper M, Davies M, Dunn J. The significance of the Van Nuys prognostic index in the management of ductal carcinoma in situ. *World J Surg Oncol.* 2008;6:61.
172. Di Saverio S, Catena F, Santini D, Ansaloni L, Fogacci T, Mignani S, et al. 259 Patients with DCIS of the breast applying USC/Van Nuys prognostic index: a retrospective review with long term follow up. *Breast Cancer Res Treat.* 2008;109(3):405–416.
173. Beatson G. On the treatment of inoperable cases of carcinoma of the mamma: suggestions for a new method of treatment, with illustrative cases. *The Lancet.* 1896;148 (3803):162–165.
174. Fisher B, Costantino J, Redmond C, Poisson R, Bowman D, Couture J, et al. A randomized clinical trial evaluating tamoxifen in the treatment of patients with node-negative breast cancer who have estrogen-receptor-positive tumors. *N Engl J Med.* 1989;320(8):479–484.
175. Fisher B, Dignam J, Wolmark N, Wickerham DL, Fisher ER, Mamounas E, et al. Tamoxifen in treatment of intraductal breast cancer: National Surgical Adjuvant Breast and Bowel Project B-24 randomised controlled trial. *Lancet.* 1999;353(9169): 1993–2000.
176. Allred DC, Anderson SJ, Paik S, Wickerham DL, Nagtegaal ID, Swain SM, et al. Adjuvant tamoxifen reduces subsequent breast cancer in women with estrogen receptor-positive ductal carcinoma in situ: a study based on NSABP protocol B-24. *J Clin Oncol.* 2012;30(12):1268–1273.
177. Howell A, Cuzick J, Baum M, Buzdar A, Dowsett M, Forbes JF, et al. Results of the ATAC (Arimidex, tamoxifen, alone or in combination) trial after completion of 5 years' adjuvant treatment for breast cancer. *Lancet.* 2005;365(9453):60–62.
178. Margolese RG, Cecchini RS, Julian TB, Ganz PA, Costantino JP, Vallow LA, et al. Anastrozole versus tamoxifen in postmenopausal women with ductal carcinoma in situ undergoing lumpectomy plus radiotherapy (NSABP B-35): a randomised, double-blind, phase 3 clinical trial. *Lancet.* 2016;387(10021):849–856.
179. Forbes JF, Sestak I, Howell A, Bonanni B, Bundred N, Levy C, et al. Anastrozole versus tamoxifen for the prevention of locoregional and contralateral breast cancer in postmenopausal women with locally excised ductal carcinoma in situ (IBIS-II DCIS): a double-blind, randomised controlled trial. *Lancet.* 2016;387(10021): 866–873.
180. Carlson RW, Edge SB, Theriault RL. Panel NBCPG. NCCN: breast cancer. *Cancer Control.* 2001;8(6 suppl 2):54–61.

181. Parise CA, Bauer KR, Brown MM, Caggiano V. Breast cancer subtypes as defined by the estrogen receptor (ER), progesterone receptor (PR), and the human epidermal growth factor receptor 2 (HER2) among women with invasive breast cancer in California, 1999–2004. *Breast J.* 2009;15(6):593–602.
182. Hannemann J, Velds A, Halfwerk JB, Kreike B, Peterse JL, van de Vijver MJ. Classification of ductal carcinoma in situ by gene expression profiling. *Breast Cancer Res.* 2006;8(5):R61.
183. Tamimi RM, Baer HJ, Marotti J, Galan M, Galaburda L, Fu Y, et al. Comparison of molecular phenotypes of ductal carcinoma in situ and invasive breast cancer. *Breast Cancer Res.* 2008;10(4):R67.
184. Isobe M, Emanuel BS, Givol D, Oren M, Croce CM. Localization of gene for human p53 tumour antigen to band 17p13. *Nature.* 1986;320(6057):84–85.
185. Lane D, Levine A. p53 Research: the past thirty years and the next thirty years. *Cold Spring Harb Perspect Biol.* 2010;2(12):a000893.
186. Linzer DI, Levine AJ. Characterization of a 54K dalton cellular SV40 tumor antigen present in SV40-transformed cells and uninfected embryonal carcinoma cells. *Cell.* 1979;17(1):43–52.
187. Lane DP, Crawford LV. T antigen is bound to a host protein in SV40-transformed cells. *Nature.* 1979;278(5701):261–263.
188. Werness BA, Levine AJ, Howley PM. Association of human papillomavirus types 16 and 18 E6 proteins with p53. *Science.* 1990;248(4951):76–79.
189. Sarnow P, Ho YS, Williams J, Levine AJ. Adenovirus E1b-58kd tumor antigen and SV40 large tumor antigen are physically associated with the same 54 kd cellular protein in transformed cells. *Cell.* 1982;28(2):387–394.
190. Kastan MB, Onyekwere O, Sidransky D, Vogelstein B, Craig RW. Participation of p53 protein in the cellular response to DNA damage. *Cancer Res.* 1991;51(23 Pt 1):6304–6311.
191. Murray AW. Creative blocks: cell-cycle checkpoints and feedback controls. *Nature.* 1992;359(6396):599–604.
192. Maltzman W, Czyzyk L. UV irradiation stimulates levels of p53 cellular tumor antigen in nontransformed mouse cells. *Mol Cell Biol.* 1984;4(9):1689–1694.
193. Lane DP. Cancer. p53, guardian of the genome. *Nature.* 1992;358(6381):15–16.
194. Olivier M, Hollstein M, Hainaut P. TP53 mutations in human cancers: origins, consequences, and clinical use. *Cold Spring Harb Perspect Biol.* 2010;2(1):a001008.
195. Caleffi M, Teague MW, Jensen RA, Vnencak-Jones CL, Dupont WD, Parl FF. p53 gene mutations and steroid receptor status in breast cancer. Clinicopathologic correlations and prognostic assessment. *Cancer.* 1994;73(8):2147–2156.
196. Pharoah PD, Day NE, Caldas C. Somatic mutations in the p53 gene and prognosis in breast cancer: a meta-analysis. *Br J Cancer.* 1999;80(12):1968–1973.
197. Thomas R, Kaplan L, Reich N, Lane DP, Levine AJ. Characterization of human p53 antigens employing primate specific monoclonal antibodies. *Virology.* 1983;131(2):502–517.
198. Rotter V. p53, a transformation-related cellular-encoded protein, can be used as a biochemical marker for the detection of primary mouse tumor cells. *Proc Natl Acad Sci USA.* 1983;80(9):2613–2617.
199. Benchimol S, Pim D, Crawford L. Radioimmunoassay of the cellular protein p53 in mouse and human cell lines. *EMBO J.* 1982;1(9):1055–1062.
200. Norberg T, Lennerstrand J, Inganas M, Bergh J. Comparison between p53 protein measurements using the luminometric immunoassay and immunohistochemistry with detection of p53 gene mutations using cDNA sequencing in human breast tumors. *Int J Cancer.* 1998;79(4):376–383.
201. Sjogren S, Inganas M, Norberg T, Lindgren A, Nordgren H, Holmberg L, et al. The p53 gene in breast cancer: prognostic value of complementary DNA sequencing versus immunohistochemistry. *J Natl Cancer Inst.* 1996;88(3–4):173–182.

202. Poller DN, Roberts EC, Bell JA, Elston CW, Blamey RW, Ellis IO. p53 protein expression in mammary ductal carcinoma in situ: relationship to immunohistochemical expression of estrogen receptor and c-erbB-2 protein. *Hum Pathol.* 1993;24(5): 463–468.
203. Thor AD, Moore DH I, Edgerton SM, Kawasaki ES, Reihsaus E, Lynch HT, et al. Accumulation of p53 tumor suppressor gene protein: an independent marker of prognosis in breast cancers. *J Natl Cancer Inst.* 1992;84(11):845–855.
204. Sarode VR, Han JS, Morris DH, Peng Y, Rao R. A comparative analysis of biomarker expression and molecular subtypes of pure ductal carcinoma in situ and invasive breast carcinoma by image analysis: relationship of the subtypes with histologic grade, Ki67, p53 overexpression, and DNA ploidy. *Int J Breast Cancer.* 2011;2011:217060.
205. Walker RA, Dearing SJ, Lane DP, Varley JM. Expression of p53 protein in infiltrating and in-situ breast carcinomas. *J Pathol.* 1991;165(3):203–211.
206. Rajan PB, Scott DJ, Perry RH, Griffith CD. p53 protein expression in ductal carcinoma in situ (DCIS) of the breast. *Breast Cancer Res Treat.* 1997;42(3):283–290.
207. Baqai T, Shousha S. Oestrogen receptor negativity as a marker for high-grade ductal carcinoma in situ of the breast. *Histopathology.* 2003;42(5):440–447.
208. Bobrow LG, Happerfield LC, Gregory WM, Springall RD, Millis RR. The classification of ductal carcinoma in situ and its association with biological markers. *Semin Diagn Pathol.* 1994;11(3):199–207.
209. Garimella V, Hussain T, Agarwal V, Radhakrishna S, Fox JN, Kneeshaw PJ, et al. Clinical response to primary letrozole therapy in elderly patients with early breast cancer: possible role for p53 as a biomarker. *Int J Surg.* 2014;12(8):821–826.
210. Berns EM, Klijn JG, van Putten WL, de Witte HH, Look MP, Meijer-van Gelder ME, et al. p53 protein accumulation predicts poor response to tamoxifen therapy of patients with recurrent breast cancer. *J Clin Oncol.* 1998;16(1):121–127.
211. Muller M, Wilder S, Bannasch D, Israeli D, Lehlbach K, Li-Weber M, et al. p53 activates the CD95 (APO-1/Fas) gene in response to DNA damage by anticancer drugs. *J Exp Med.* 1998;188(11):2033–2045.
212. Nguyen DM, Spitz FR, Yen N, Cristiano RJ, Roth JA. Gene therapy for lung cancer: enhancement of tumor suppression by a combination of sequential systemic cisplatin and adenovirus-mediated p53 gene transfer. *J Thorac Cardiovasc Surg.* 1996;112(5): 1372–1376.
213. Key T, Appleby P, Barnes I, Reeves G, Endogenous H, Breast Cancer Collaborative G. Endogenous sex hormones and breast cancer in postmenopausal women: reanalysis of nine prospective studies. *J Natl Cancer Inst.* 2002;94(8):606–616.
214. Rossouw JE, Anderson GL, Prentice RL, LaCroix AZ, Kooperberg C, Stefanick ML, et al. Risks and benefits of estrogen plus progestin in healthy postmenopausal women: principal results from the women's health initiative randomized controlled trial. *JAMA.* 2002;288(3):321–333.
215. Mangelsdorf DJ, Thummel C, Beato M, Herrlich P, Schutz G, Umesono K, et al. The nuclear receptor superfamily: the second decade. *Cell.* 1995;83(6):835–839.
216. Ioan A, Klobocista M, Shirazi S, Press MF. Molecular Oncology of Breast Cancer. In: Bland KI, Copeland EM eds, Barsky SH, Gradishar WJ, Recht A, Urist MM associate eds. Breast: Comprehensive Management of Benign and Malignant Diseases. 4th ed. Philadelphia, PA: Elsevier; 2009. pp. 501–531
217. Bocchinfuso WP, Korach KS. Mammary gland development and tumorigenesis in estrogen receptor knockout mice. *J Mammary Gland Biol Neoplasia.* 1997;2(4): 323–334.
218. Petz LN, Ziegler YS, Schultz JR, Kim H, Kemper JK, Nardulli AM. Differential regulation of the human progesterone receptor gene through an estrogen response element half site and Sp1 sites. *J Steroid Biochem Mol Biol.* 2004;88(2):113–122.

219. Schultz JR, Petz LN, Nardulli AM. Estrogen receptor alpha and Sp1 regulate progesterone receptor gene expression. *Mol Cell Endocrinol.* 2003;201(1–2):165–175.
220. Lydon JP, DeMayo FJ, Funk CR, Mani SK, Hughes AR, Montgomery Jr CA et al. Mice lacking progesterone receptor exhibit pleiotropic reproductive abnormalities. *Genes Dev.* 1995;9(18):2266–2278.
221. Goepfert TM, McCarthy M, Kittrell FS, Stephens C, Ullrich RL, Brinkley BR, et al. Progesterone facilitates chromosome instability (aneuploidy) in p53 null normal mammary epithelial cells. *FASEB J.* 2000;14(14):2221–2229.
222. Brisken C, Hess K, Jeitziner R. Progesterone and overlooked endocrine pathways in breast cancer pathogenesis. *Endocrinology.* 2015;156(10):3442–3450.
223. Russo J, Russo IH. The role of estrogen in the initiation of breast cancer. *J Steroid Biochem Mol Biol.* 2006;102(1–5):89–96.
224. Wahli W, Martinez E, Corthesy B, Cardinaux JR. *Cis*- and *trans*-acting elements of the estrogen-regulated vitellogenin gene B1 of *Xenopus laevis*. *J Steroid Biochem.* 1989;34 (1–6):17–32.
225. Klinge CM. Estrogen receptor interaction with estrogen response elements. *Nucleic Acids Res.* 2001;29(14):2905–2919.
226. Marino M, Galluzzo P, Ascenzi P. Estrogen signaling multiple pathways to impact gene transcription. *Curr Genomics.* 2006;7(8):497–508.
227. Walker KJ, Bouzubar N, Robertson J, Ellis IO, Elston CW, Blamey RW, et al. Immunocytochemical localization of estrogen receptor in human breast tissue. *Cancer Res.* 1988;48(22):6517–6522.
228. Giri DD, Dundas SA, Nottingham JF, Underwood JC. Oestrogen receptors in benign epithelial lesions and intraduct carcinomas of the breast: an immunohistological study. *Histopathology.* 1989;15(6):575–584.
229. Poller DN, Snead DR, Roberts EC, Galea M, Bell JA, Gilmour A, et al. Oestrogen receptor expression in ductal carcinoma in situ of the breast: relationship to flow cytometric analysis of DNA and expression of the c-erbB-2 oncoprotein. *Br J Cancer.* 1993;68(1):156–161.
230. Bur M, Shiraki W, Masood S. Estrogen and progesterone receptor detection in neoplastic and non-neoplastic thyroid tissues. *Mod Pathol.* 1993;6(4):469–472.
231. Williams KE, Barnes NL, Cramer A, Johnson R, Cheema K, Morris J, et al. Molecular phenotypes of DCIS predict overall and invasive recurrence. *Ann Oncol.* 2015;26 (5):1019–1025.
232. Provenzano E, Hopper JL, Giles GG, Marr G, Venter DJ, Armes JE. Biological markers that predict clinical recurrence in ductal carcinoma in situ of the breast. *Eur J Cancer.* 2003;39(5):622–630.
233. Roka S, Rudas M, Taucher S, Dubsky P, Bachleitner-Hofmann T, Kandioler D, et al. High nuclear grade and negative estrogen receptor are significant risk factors for recurrence in DCIS. *Eur J Surg Oncol.* 2004;30(3):243–247.
234. Cole MP, Jones CT, Todd ID. A new anti-oestrogenic agent in late breast cancer. An early clinical appraisal of ICI46474. *Br J Cancer.* 1971;25(2):270–275.
235. Tamoxifen for early breast cancer: an overview of the randomised trials. Early Breast Cancer Trialists' Collaborative Group. *Lancet.* 1998;351(9114):1451–1467.
236. Fisher B, Costantino JP, Wickerham DL, Redmond CK, Kavanah M, Cronin WM, et al. Tamoxifen for prevention of breast cancer: report of the National Surgical Adjuvant Breast and Bowel Project P-1 Study. *J Natl Cancer Inst.* 1998;90(18):1371–1388.
237. Cuzick J, Forbes JF, Sestak I, Cawthorn S, Hamed H, Holli K, et al. Long-term results of tamoxifen prophylaxis for breast cancer—96-month follow-up of the randomized IBIS-I trial. *J Natl Cancer Inst.* 2007;99(4):272–282.
238. Shang Y, Brown M. Molecular determinants for the tissue specificity of SERMs. *Science.* 2002;295(5564):2465–2468.

239. Jordan VC, Phelps E, Lindgren JU. Effects of anti-estrogens on bone in castrated and intact female rats. *Breast Cancer Res Treat.* 1987;10(1):31–35.
240. Love RR, Mazess RB, Barden HS, Epstein S, Newcomb PA, Jordan VC, et al. Effects of tamoxifen on bone mineral density in postmenopausal women with breast cancer. *N Engl J Med.* 1992;326(13):852–856.
241. Gottardis MM, Robinson SP, Satyaswaroop PG, Jordan VC. Contrasting actions of tamoxifen on endometrial and breast tumor growth in the athymic mouse. *Cancer Res.* 1988;48(4):812–815.
242. Feng Q, O'Malley BW. Nuclear receptor modulation—role of coregulators in selective estrogen receptor modulator (SERM) actions. *Steroids.* 2014;90:39–43.
243. Cummings SR, Eckert S, Krueger KA, Grady D, Powles TJ, Cauley JA, et al. The effect of raloxifene on risk of breast cancer in postmenopausal women: results from the MORE randomized trial. Multiple outcomes of raloxifene evaluation. *JAMA.* 1999;281(23):2189–2197.
244. Vogel VG, Costantino JP, Wickerham DL, Cronin WM, Cecchini RS, Atkins JN, et al. Effects of tamoxifen vs raloxifene on the risk of developing invasive breast cancer and other disease outcomes: the NSABP study of tamoxifen and raloxifene (STAR) P-2 trial. *JAMA.* 2006;295(23):2727–2741.
245. Vogel VG, Costantino JP, Wickerham DL, Cronin WM, Cecchini RS, Atkins JN, et al. Update of the National Surgical Adjuvant Breast and Bowel Project Study of Tamoxifen and Raloxifene (STAR) P-2 Trial: preventing breast cancer. *Cancer Prev Res.* 2010;3(6):696–706.
246. Baum M, Budzar AU, Cuzick J, Forbes J, Houghton JH, Klijn JG, et al. Anastrozole alone or in combination with tamoxifen versus tamoxifen alone for adjuvant treatment of postmenopausal women with early breast cancer: first results of the ATAC randomised trial. *Lancet.* 2002;359(9324):2131–2139.
247. Coombes RC, Hall E, Gibson LJ, Paridaens R, Jassem J, Delozier T, et al. A randomized trial of exemestane after two to three years of tamoxifen therapy in postmenopausal women with primary breast cancer. *N Engl J Med.* 2004;350(11):1081–1092.
248. Goss PE, Ingle JN, Martino S, Robert NJ, Muss HB, Piccart MJ, et al. A randomized trial of letrozole in postmenopausal women after five years of tamoxifen therapy for early-stage breast cancer. *N Engl J Med.* 2003;349(19):1793–1802.
249. Thurlimann B, Keshaviah A, Coates AS, Mouridsen H, Mauriac L, et al. A comparison of letrozole and tamoxifen in postmenopausal women with early breast cancer. *N Engl J Med.* 2005;353(26):2747–2757.
250. Cuzick J, Sestak I, Forbes JF, Dowsett M, Knox J, Cawthorn S, et al. Anastrozole for prevention of breast cancer in high-risk postmenopausal women (IBIS-II): an international, double-blind, randomised placebo-controlled trial. *Lancet.* 2014;383(9922):1041–1048.
251. Goss PE, Ingle JN, Ales-Martinez JE, Cheung AM, Chlebowski RT, Wactawski-Wende J, et al. Exemestane for breast-cancer prevention in postmenopausal women. *N Engl J Med.* 2011;364(25):2381–2391.
252. Bevers TB, Ward JH, Arun BK, Colditz GA, Cowan KH, Daly MB, et al. Breast cancer risk reduction, version 2.2015. *J Natl Compr Canc Netw.* 2015;13(7):880–915.
253. Taketani Y, Oka T. Epidermal growth factor stimulates cell proliferation and inhibits functional differentiation of mouse mammary epithelial cells in culture. *Endocrinology.* 1983;113(3):871–877.
254. Imai Y, Leung CK, Friesen HG, Shiu RP. Epidermal growth factor receptors and effect of epidermal growth factor on growth of human breast cancer cells in long-term tissue culture. *Cancer Res.* 1982;42(11):4394–4398.
255. Osborne CK, Hamilton B, Titus G, Livingston RB. Epidermal growth factor stimulation of human breast cancer cells in culture. *Cancer Res.* 1980;40(7):2361–2366.

256. Lebeau A, Unholzer A, Amann G, Kronawitter M, Bauerfeind I, Sendelhofert A, et al. EGFR, HER-2/neu, cyclin D1, p21 and p53 in correlation to cell proliferation and steroid hormone receptor status in ductal carcinoma in situ of the breast. *Breast Cancer ResTreat.* 2003;79(2):187–198.
257. Livasy CA, Perou CM, Karaca G, Cowan DW, Maia D, Jackson S, et al. Identification of a basal-like subtype of breast ductal carcinoma in situ. *HumPathol.* 2007;38(2):197–204.
258. Saxena R, Dwivedi A. ErbB family receptor inhibitors as therapeutic agents in breast cancer: current status and future clinical perspective. *MedResRev.* 2012;32(1):166–215.
259. Chan KC, Knox WF, Gee JM, Morris J, Nicholson RI, Potten CS, et al. Effect of epidermal growth factor receptor tyrosine kinase inhibition on epithelial proliferation in normal and premalignant breast. *CancerRes.* 2002;62(1):122–128.
260. Chan AK, Kalmes A, Hawkins S, Daum G, Clowes AW. Blockade of the epidermal growth factor receptor decreases intimal hyperplasia in balloon-injured rat carotid artery. *J Vasc Surg.* 2003;37(3):644–649.
261. Yamamoto T, Ikawa S, Akiyama T, Semba K, Nomura N, Miyajima N, et al. Similarity of protein encoded by the human c-erb-B-2 gene to epidermal growth factor receptor. *Nature.* 1986;319(6050):230–234.
262. Akiyama T, Sudo C, Ogawara H, Toyoshima K, Yamamoto T. The product of the human c-erbB-2 gene: a 185-kilodalton glycoprotein with tyrosine kinase activity. *Science.* 1986;232(4758):1644–1646.
263. Goldman R, Levy RB, Peles E, Yarden Y. Heterodimerization of the erbB-1 and erbB-2 receptors in human breast carcinoma cells: a mechanism for receptor transregulation. *Biochemistry.* 1990;29(50):11024–11028.
264. Wada T, Qian XL, Greene MI. Intermolecular association of the p185neu protein and EGF receptor modulates EGF receptor function. *Cell.* 1990;61(7):1339–1347.
265. Carraway 3rd KL, Cantley LC. A neu acquaintance for erbB3 and erbB4: a role for receptor heterodimerization in growth signaling. *Cell.* 1994;78(1):5–8.
266. Karunagaran D, Tzahar E, Beerli RR, Chen X, Graus-Porta D, Ratzkin BJ, et al. ErbB-2 is a common auxiliary subunit of NDF and EGF receptors: implications for breast cancer. *EMBOJ.* 1996;15(2):254–264.
267. Rubin I, Yarden Y. The basic biology of HER2. *Ann Oncol.* 2001;12(suppl 1):S3–S8.
268. Graus-Porta D, Beerli RR, Daly JM, Hynes NE. ErbB-2, the preferred heterodimerization partner of all ErbB receptors, is a mediator of lateral signaling. *EMBOJ.* 1997;16 (7):1647–1655.
269. Ben-Levy R, Paterson HF, Marshall CJ, Yarden Y. A single autophosphorylation site confers oncogenicity to the Neu/ErbB-2 receptor and enables coupling to the MAP kinase pathway. *EMBOJ.* 1994;13(14):3302–3311.
270. Yarden Y, Sliwkowski MX. Untangling the ErbB signalling network. *Nat Rev Mol Cell Biol.* 2001;2(2):127–137.
271. Wallasch C, Weiss FU, Niederfellner G, Jallal B, Issing W, Ullrich A. Heregulin-dependent regulation of HER2/neu oncogenic signaling by heterodimerization with HER3. *EMBOJ.* 1995;14(17):4267–4275.
272. Soltoff SP, Carraway 3rd KL, Prigent SA, Gullick WG, Cantley LC. ErbB3 is involved in activation of phosphatidylinositol 3-kinase by epidermal growth factor. *MolCellBiol.* 1994;14(6):3550–3558.
273. Slamon DJ, Godolphin W, Jones LA, Holt JA, Wong SG, Keith DE, et al. Studies of the HER-2/neu proto-oncogene in human breast and ovarian cancer. *Science.* 1989;244 (4905):707–712.
274. van de Vijver MJ, Peterse JL, Mooi WJ, Wisman P, Lomans J, Dalesio O, et al. Neu-protein overexpression in breast cancer. Association with comedo-type ductal carcinoma in situ and limited prognostic value in stage II breast cancer. *N Engl J Med.* 1988;319(19):1239–1245.

275. Venter DJ, Tuzi NL, Kumar S, Gullick WJ. Overexpression of the c-erbB-2 oncoprotein in human breast carcinomas: immunohistological assessment correlates with gene amplification. *Lancet.* 1987;2(8550):69–72.
276. van de Vijver MJ, Mooi WJ, Wisman P, Peterse JL, Nusse R. Immunohistochemical detection of the neu protein in tissue sections of human breast tumors with amplified neu DNA. *Oncogene.* 1988;2(2):175–178.
277. Gusterson BA, Gullick WJ, Venter DJ, Powles TJ, Elliott C, Ashley S, et al. Immunohistochemical localization of c-erbB-2 in human breast carcinomas. *Mol Cell Probes.* 1987;1(4):383–391.
278. Ross JS, Fletcher JA. The HER-2/neu oncogene in breast cancer: prognostic factor, predictive factor, and target for therapy. *Oncologist.* 1998;3(4):237–252.
279. Barnes DM, Meyer JS, Gonzalez JG, Gullick WJ, Millis RR. Relationship between c-erbB-2 immunoreactivity and thymidine labelling index in breast carcinoma in situ. *Breast Cancer Res Treat.* 1991;18(1):11–17.
280. Brower ST, Ahmed S, Tartter PI, Bleiweiss I, Amberson JB. Prognostic variables in invasive breast cancer: contribution of comedo versus noncomedo in situ component. *Ann Surg Oncol.* 1995;2(5):440–444.
281. Leal CB, Schmitt FC, Bento MJ, Maia NC, Lopes CS. Ductal carcinoma in situ of the breast. Histologic categorization and its relationship to ploidy and immunohistochemical expression of hormone receptors, p53, and c-erbB-2 protein. *Cancer.* 1995;75 (8):2123–2131.
282. Barnes DM, Lammie GA, Millis RR, Gullick WL, Allen DS, Altman DG. An immunohistochemical evaluation of c-erbB-2 expression in human breast carcinoma. *Br J Cancer.* 1988;58(4):448–452.
283. Bartkova J, Barnes DM, Millis RR, Gullick WJ. Immunohistochemical demonstration of c-erbB-2 protein in mammary ductal carcinoma in situ. *Hum Pathol.* 1990;21 (11):1164–1167.
284. Zhou W, Jirstrom K, Amini RM, Fjallskog ML, Sollie T, Lindman H, et al. Molecular subtypes in ductal carcinoma in situ of the breast and their relation to prognosis: a population-based cohort study. *BMC Cancer.* 2013;13:512.
285. Ramachandra S, Machin L, Ashley S, Monaghan P, Gusterson BA. Immunohistochemical distribution of c-erbB-2 in in situ breast carcinoma—a detailed morphological analysis. *J Pathol.* 1990;161(1):7–14.
286. Siziopikou KP, Anderson SJ, Cobleigh MA, Julian TB, Arthur DW, Zheng P, et al. Preliminary results of centralized HER2 testing in ductal carcinoma in situ (DCIS): NSABP B-43. *Breast Cancer Res Treat.* 2013;142(2):415–421.
287. Menard S, Pupa SM, Campiglio M, Tagliabue E. Biologic and therapeutic role of HER2 in cancer. *Oncogene.* 2003;22(42):6570–6578.
288. Cobleigh MA, Vogel CL, Tripathy D, Robert NJ, Scholl S, Fehrenbacher L, et al. Multinational study of the efficacy and safety of humanized anti-HER2 monoclonal antibody in women who have HER2-overexpressing metastatic breast cancer that has progressed after chemotherapy for metastatic disease. *J Clin Oncol.* 1999;17 (9):2639–2648.
289. Baselga J. Clinical trials of Herceptin(R) (trastuzumab). *Eur J Cancer.* 2001;37(suppl 1): 18–24.
290. Slamon DJ, Leyland-Jones B, Shak S, Fuchs H, Paton V, Bajamonde A, et al. Use of chemotherapy plus a monoclonal antibody against HER2 for metastatic breast cancer that overexpresses HER2. *N Engl J Med.* 2001;344(11):783–792.
291. Vogel CL, Cobleigh MA, Tripathy D, Gutheil JC, Harris LN, Fehrenbacher L, et al. First-line Herceptin monotherapy in metastatic breast cancer. *Oncology.* 2001;61(suppl 2):37–42.
292. Fukuhara S, Rowley JD, Variakojis D, Golomb HM. Chromosome abnormalities in poorly differentiated lymphocytic lymphoma. *Cancer Res.* 1979;39(8):3119–3128.

293. Nunez G, London L, Hockenbery D, Alexander M, McKearn JP, Korsmeyer SJ. Deregulated Bcl-2 gene expression selectively prolongs survival of growth factor-deprived hemopoietic cell lines. *J Immunol.* 1990;144(9):3602–3610.
294. Hockenbery D, Nunez G, Milliman C, Schreiber RD, Korsmeyer SJ. Bcl-2 is an inner mitochondrial membrane protein that blocks programmed cell death. *Nature.* 1990;348 (6299):334–336.
295. Nathan B, Anbazhagan R, Clarkson P, Bartkova J, Gusterson B. Expression of BCL-2 in the developing human fetal and infant breast. *Histopathology.* 1994;24(1):73–76.
296. Zhang GJ, Kimijima I, Abe R, Kanno M, Katagata N, Hara K, et al. Correlation between the expression of apoptosis-related bcl-2 and p53 oncoproteins and the carcinogenesis and progression of breast carcinomas. *Clin Cancer Res.* 1997;3(12 Pt 1):2329–2335.
297. Watson AJ, Merritt AJ, Jones LS, Askew JN, Anderson E, Becciolini A, et al. Evidence of reciprocity of bcl-2 and p53 expression in human colorectal adenomas and carcinomas. *Br J Cancer.* 1996;73(8):889–895.
298. Mustonen M, Raunio H, Paakko P, Soini Y. The extent of apoptosis is inversely associated with bcl-2 expression in premalignant and malignant breast lesions. *Histopathology.* 1997;31(4):347–354.
299. Bai M, Agnantis NJ, Kamina S, Demou A, Zagorianakou P, Katsaraki A, et al. In vivo cell kinetics in breast carcinogenesis. *Breast Cancer Res.* 2001;3(4):276–283.
300. Clark SE, Warwick J, Carpenter R, Bowen RL, Duffy SW, Jones JL. Molecular subtyping of DCIS: heterogeneity of breast cancer reflected in pre-invasive disease. *Br J Cancer.* 2011;104(1):120–127.
301. Siziopikou KP, Prioleau JE, Harris JR, Schnitt SJ. bcl-2 expression in the spectrum of preinvasive breast lesions. *Cancer.* 1996;77(3):499–506.
302. Kapucuoglu N, Losi L, Eusebi V. Immunohistochemical localization of Bcl-2 and Bax proteins in in situ and invasive duct breast carcinomas. *Virchows Arch.* 1997;430(1):17–22.
303. Dawson SJ, Makretsov N, Blows FM, Driver KE, Provenzano E, Le Quesne J, et al. BCL2 in breast cancer: a favourable prognostic marker across molecular subtypes and independent of adjuvant therapy received. *Br J Cancer.* 2010;103(5):668–675.
304. Bonnet D, Dick JE. Human acute myeloid leukemia is organized as a hierarchy that originates from a primitive hematopoietic cell. *Nat Med.* 1997;3(7):730–737.
305. Collins AT, Berry PA, Hyde C, Stower MJ, Maitland NJ. Prospective identification of tumorigenic prostate cancer stem cells. *Cancer Res.* 2005;65(23):10946–10951.
306. Prince ME, Sivanandan R, Kaczorowski A, Wolf GT, Kaplan MJ, Dalerba P, et al. Identification of a subpopulation of cells with cancer stem cell properties in head and neck squamous cell carcinoma. *Proc Natl Acad Sci USA.* 2007;104(3):973–978.
307. Li C, Lee CJ, Simeone DM. Identification of human pancreatic cancer stem cells. *Methods Mol Biol.* 2009;568:161–173.
308. Al-Hajj M, Wicha MS, Benito-Hernandez A, Morrison SJ, Clarke MF. Prospective identification of tumorigenic breast cancer cells. *Proc Natl Acad Sci USA.* 2003;100 (7):3983–3988.
309. Ichim CV, Wells RA. First among equals: the cancer cell hierarchy. *Leuk Lymphoma.* 2006;47(10):2017–2027.
310. Ginestier C, Hur MH, Charafe-Jauffret E, Monville F, Dutcher J, Brown M, et al. ALDH1 is a marker of normal and malignant human mammary stem cells and a predictor of poor clinical outcome. *Cell Stem Cell.* 2007;1(5):555–567.
311. Dontu G, Abdallah WM, Foley JM, Jackson KW, Clarke MF, Kawamura MJ, et al. In vitro propagation and transcriptional profiling of human mammary stem/progenitor cells. *Genes Dev.* 2003;17(10):1253–1270.
312. Wright MH, Calcagno AM, Salcido CD, Carlson MD, Ambudkar SV, Varticovski L. Brca1 breast tumors contain distinct CD44+/CD24- and CD133+ cells with cancer stem cell characteristics. *Breast Cancer Res.* 2008;10(1):R10.

313. Zhou S, Schuetz JD, Bunting KD, Colapietro AM, Sampath J, Morris JJ, et al. The ABC transporter Bcrp1/ABCG2 is expressed in a wide variety of stem cells and is a molecular determinant of the side-population phenotype. *Nat Med.* 2001;7(9):1028–1034.
314. Krohn A, Song YH, Muehlberg F, Droll L, Beckmann C, Alt E. CXCR4 receptor positive spheroid forming cells are responsible for tumor invasion in vitro. *Cancer Lett.* 2009;280(1):65–71.
315. Doyle L, Ross DD. Multidrug resistance mediated by the breast cancer resistance protein BCRP (ABCG2). *Oncogene.* 2003;22(47):7340–7358.
316. Marchitti SA, Brocker C, Stagos D, Vasiliou V. Non-P450 aldehyde oxidizing enzymes: the aldehyde dehydrogenase superfamily. *Expert Opin Drug Metab Toxicol.* 2008;4(6):697–720.
317. Sladek NE. Aldehyde dehydrogenase-mediated cellular relative insensitivity to the oxazaphosphorines. *Curr Pharm Des.* 1999;5(8):607–625.
318. Charafe-Jauffret E, Ginestier C, Iovino F, Tarpin C, Diebel M, Esterni B, et al. Aldehyde dehydrogenase 1-positive cancer stem cells mediate metastasis and poor clinical outcome in inflammatory breast cancer. *Clin Cancer Res.* 2010;16(1):45–55.
319. Mukherjee S, Manna A, Bhattacharjee P, Mazumdar M, Saha S, Chakraborty S, et al. Non-migratory tumorigenic intrinsic cancer stem cells ensure breast cancer metastasis by generation of CXCR4(+) migrating cancer stem cells. *Oncogene.* 2016;35(37):4937–4948.
320. Muller A, Homey B, Soto H, Ge N, Catron D, Buchanan ME, et al. Involvement of chemokine receptors in breast cancer metastasis. *Nature.* 2001;410(6824):50–56.
321. Beachy PA, Karhadkar SS, Berman DM. Tissue repair and stem cell renewal in carcinogenesis. *Nature.* 2004;432(7015):324–331.
322. Liu J, Sato C, Cerletti M, Wagers A. Notch signaling in the regulation of stem cell self-renewal and differentiation. *Curr Top Dev Biol.* 2010;92:367–409.
323. Lindvall C, Bu W, Williams BO, Li Y. Wnt signaling, stem cells, and the cellular origin of breast cancer. *Stem Cell Rev.* 2007;3(2):157–168.
324. Li Y, Welm B, Podsypanina K, Huang S, Chamorro M, Zhang X, et al. Evidence that transgenes encoding components of the Wnt signaling pathway preferentially induce mammary cancers from progenitor cells. *Proc Natl Acad Sci USA.* 2003;100(26):15853–15858.
325. Harrison H, Farnie G, Howell SJ, Rock RE, Stylianou S, Brennan KR, et al. Regulation of breast cancer stem cell activity by signaling through the Notch4 receptor. *Cancer Res.* 2010;70(2):709–718.
326. Callahan R, Raafat A. Notch signaling in mammary gland tumorigenesis. *J Mammary Gland Biol Neoplasia.* 2001;6(1):23–36.
327. Kasper M, Regl G, Frischauf AM, Aberger F. GLI transcription factors: mediators of oncogenic hedgehog signalling. *Eur J Cancer.* 2006;42(4):437–445.
328. Liu S, Dontu G, Mantle ID, Patel S, Ahn NS, Jackson KW, et al. Hedgehog signaling and Bmi-1 regulate self-renewal of normal and malignant human mammary stem cells. *Cancer Res.* 2006;66(12):6063–6071.
329. Kiger AA, Jones DL, Schulz C, Rogers MB, Fuller MT. Stem cell self-renewal specified by JAK-STAT activation in response to a support cell cue. *Science.* 2001;294(5551):2542–2545.
330. Hernandez-Vargas H, Ouzounova M, Le Calvez-Kelm F, Lambert MP, McKay-Chopin S, Tavtigian SV, et al. Methylome analysis reveals Jak-STAT pathway deregulation in putative breast cancer stem cells. *Epigenetics.* 2011;6(4):428–439.
331. Li Q, Yao Y, Eades G, Liu Z, Zhang Y, Zhou Q. Downregulation of miR-140 promotes cancer stem cell formation in basal-like early stage breast cancer. *Oncogene.* 2014;33(20):2589–2600.
332. Tsukabe M, Shimazu K, Morimoto K, Naoi Y, Kagara N, Shimoda M, et al. Clinicopathological analysis of breast ductal carcinoma in situ with ALDH1-positive cancer stem cells. *Oncology.* 2013;85(4):248–256.

CHAPTER THREE

Pregnancy and Breast Cancer

Ramadevi Subramani*, Rajkumar Lakshmanaswamy†,1

* Center of Emphasis in Cancer Research, Paul L. Foster School of Medicine, Texas Tech University Health Sciences Center El Paso, El Paso, TX, United States

† Center of Emphasis in Cancer Research, Paul L. Foster School of Medicine, Graduate School of Biomedical Sciences, Texas Tech University Health Sciences Center El Paso, El Paso, TX, United States

1 Corresponding author. E-mail address: rajkumar.lakshmanaswamy@ttuhsc.edu

Contents

Abstract

Breast cancer is the most commonly diagnosed type of cancer among women worldwide. The majority of breast cancers are sporadic and the etiology is not well

Progress in Molecular Biology and Translational Science, Volume 151
ISSN 1877-1173
http://dx.doi.org/10.1016/bs.pmbts.2017.07.006

understood. Several factors have been attributed to altering the risk of breast cancer. A full-term pregnancy is a crucial factor in altering the risk. Early full-term pregnancy has been shown to reduce the lifetime risk of breast cancer, while a later first full-term pregnancy increases breast cancer risk. Epidemiological and experimental data demonstrate that spontaneous or induced abortions do not significantly alter the risk of breast cancer. In this study, we briefly discuss the different types and stages of breast cancer, various risk factors, and potential mechanisms involved in early full-term pregnancy–induced protection against breast cancer. Understanding how early full-term pregnancy induces protection against breast cancer will help design innovative preventive and therapeutic strategies. This understanding can also help in the development of molecular biomarkers that can be of tremendous help in predicting the risk of breast cancer in the general population.

1. INTRODUCTION

Breast cancer is the most predominant cancer in women, currently affecting over 1.6 million women worldwide. Despite improved detection and treatment strategies, breast cancer incidence rates continue to increase.[1,2] Moreover, metastatic breast cancer quickly acquires resistance to therapies and there is currently no effective cure.[3–6] Current therapies for preventing and treating breast cancer also have many undesirable side effects.[7–11] Given these factors, there has been a concerted effort to develop better strategies for the prevention and treatment of breast cancer. In this regard, one primary area of study is parity-induced protection against breast cancer. It is well known that early pregnancy (before age 20) reduces breast cancer risk by as much as 50%.[12–32] Multiple pregnancies further reduce a woman's risk of developing breast cancer by an another 7% for each additional pregnancy.[33] In contrast, nulliparity or a first pregnancy later in life (after age 35) increases breast cancer risk.[27,30,34–36]

A complete understanding of the natural and universal phenomenon of decreased breast cancer risk in early parous women will allow for the identification of biomarkers for predicting the risks of breast cancer development and disease progression. This would provide enormous benefit to both the patients and physicians, as caregivers can focus on women who are truly at a high risk of disease development and progression. In addition, more effective prevention and treatment strategies are expected to emerge from this line of investigation.

The phenomenon of parity-induced protection against breast cancer is also observed in rodent models, where others and we have shown that both pregnancy and short-term treatment with pregnancy-level estrogen/progesterone protects against carcinogen-induced mammary cancer.[37–41] Epidemiologic studies and in vivo rodent studies have revealed notable changes in the systemic hormone and growth factor profiles of parous females.[42–47] These persistent systemic alterations have been shown to contribute significantly to parity-induced protection against breast cancer. However, many questions still remain regarding systemic versus local tissue-level changes that mediate the protective effects of early parity.

Ongoing investigations suggest that changes in mammary gland responsiveness to hormones and growth factors may play an important role in parity-mediated protection.[38,48–53] Indeed, substantial evidence from both epidemiologic studies and animal models indicate that the responsiveness of mammary epithelial cells to hormones and growth factors can change based on differences in receptor subtypes, coactivators, and hormone levels.[54–56] Studies to date reveal that estrogen receptor alpha (ERα) and epidermal growth factor receptor (EGFR) expressions decrease in parous mammary tissue, while estrogen receptor-beta (ERβ) expression increases.[30,40,57–62] These observations are consistent with decreased estrogen sensitivity and increased cell maturation. Unfortunately, the precise molecular circuitry involved in these changes and the significance of these observations in protection against mammary carcinogenesis are still not well defined. To understand these factors in depth, it is important to first understand growth of normal breast tissue, the types and stages of breast cancer, and the risk factors of the disease.

2. DEVELOPMENT OF THE MAMMARY GLAND

The mammary gland is a unique organ that is not completely developed at birth. The mammary gland is mainly composed of epithelium, stroma, and fat tissue. The ductal and alveolar epithelium is made up of luminal epithelial cells and basal myoepithelial cells. Luminal epithelial cells line the ducts and alveoli and the myoepithelial cells form the basal component. The myoepithelial cells contract in response to oxytocin to force milk stored in the lobules to flow through the ducts to the nipple. These cells also contain the mammary stem cells, which maintain the epithelium that undergoes cyclic changes based on the menstrual cycles, pregnancy, and lactation.[63–67]

Postnatal development of mammary gland occurs in two stages.[68–73] The first stage of development occurs during puberty, when the mammary duct elongates under the influence of estrogen, an ovarian steroid hormone.[74–77]

Progesterone, another ovarian steroid, induces side branching of the ducts.[78–81] These structures form the terminal ductal lobular unit (TDLU).[82,83] The second stage of mammary gland development occurs during pregnancy under the influence of several hormones, such as estrogen, progesterone, growth hormone, and prolactin.[84–86] Initially, alveolar buds sprout and they finally form lobules of alveoli where milk is produced and stored. Upon nipple stimulation by the suckling action of the offspring, milk stored in the lobules is secreted to feed the newborn. Once the offspring is weaned, the mammary gland undergoes a major regression through a process called involution. After involution, the mammary gland resembles, to a large extent, the nonpregnant mammary gland but with a more complex ductal network.

The endocrine system, including ovarian hormones estrogen and progesterone, plays a major role in development of the mammary gland. In addition to the ovarian hormones, pituitary hormones, such as growth hormone and prolactin, also play a key role in influencing the growth of the mammary gland and the process of lactation. Receptors for ovarian steroid hormones are expressed at very early stages of mammary gland development. However, these receptors are not responsive to hormones early in life, but become responsive before puberty.[87–89] The ER is detected in the nuclei of luminal cells of the ducts. ERα-positive cells are present in all ductal areas and at budding points,[90–92] while ERβ is present throughout the breast.[93–97] ERβ has not been as extensively studied as ERα. The progesterone receptor (PR) is used as an indicator of estrogen action. Progesterone receptor A (PRA) is expressed at different stages of mammary gland development and progesterone receptor B (PRB) is expressed predominately in alveolar structures and during pregnancy.[98–100]

3. TYPES OF BREAST CANCER

Breast cancer is not a single disease, but rather a group of heterogeneous diseases.[66] Breast cancers have been classified into various subtypes using histological and molecular characteristics. These classifications continue to improve with advances in cancer research, leading to better treatment and prognosis.

The majority of breast cancers are carcinomas, particularly adenocarcinomas that originate in the ducts or lobules. Other types of breast cancer, such as sarcomas, originate in the cells of the muscle, fat, or connective tissue.

Breast cancers are classified as either invasive or noninvasive. Ductal carcinoma in situ (DCIS) and lobular carcinoma in situ (LCIS) are noninvasive premalignant breast lesions and are confined to the ducts and lobules, respectively. DCIS has been further subclassified into five different subtypes: comedo, cribriform, micropapillary, papillary, and solid.[101–103]

Invasive (or infiltrating) ductal carcinoma (IDC) is the most commonly diagnosed type of breast cancer. These cancer cells break through the ductal wall and invade the surrounding fatty tissue. Invasive lobular carcinoma (ILC) can spread to other parts of the body as well. There are several subtypes of invasive carcinomas, such as adenocystic carcinomas, medullary carcinomas, metaplastic carcinomas, mucinous carcinomas, papillary carcinomas, and tubular carcinomas. In addition, there are various other types of breast cancer that are less common, such as inflammatory breast cancer, Paget disease of the nipple, phyllodes tumor, and angiosarcoma. In some situations, a single breast cancer diagnosis can contain a combination of different types of carcinomas.

This histological classification scheme has been a valuable tool that has evolved over decades, but the scheme relies solely on histology. To overcome this limitation, various classification systems that incorporate molecular markers, such as ER, PR, human epidermal growth factor receptor 2 (ErbB2 or HER2/neu), and tumor protein p53 have been generated.[104–107] In general, breast cancers are classified as hormone receptor positive or hormone receptor negative.

Hormone receptor–positive breast cancer cells express ERs and/or PRs. The cancers that possess these cells are called hormone-positive breast cancers. The presence of these receptors increases the likelihood of developing targeted therapies, which can be effective in treating breast cancer patients. Currently these cancers are treated by lowering the levels of estrogen using aromatase inhibitors or selective ER modulators, such as tamoxifen. Hormone receptor–positive breast cancers are more commonly diagnosed in postmenopausal women. Hormone receptor–responsive breast cancers grow slower and present a better prognosis for patients than hormone receptor–negative breast cancers.

Hormone receptor–negative, or hormone-negative, breast cancers do not express ER or PR. For this reason, hormonal therapy is not feasible. These cancers are generally diagnosed in premenopausal women and tend to grow fast. The HER2 protein is overexpressed in a subset of breast cancers, which usually are hormone receptor–negative cancers. Herceptin is a drug that has been widely used to treat the HER2-positive breast cancers.

There are some breast cancers that are positive for ERs, PRs, and HER2. These cancers are known as triple-positive breast cancers. Another type of breast cancer is the triple-negative breast cancer, which is more common in African–American and Hispanic women. These cancers do not express ERs, PRs, or HER2. Triple-negative breast cancer is very aggressive and is generally diagnosed in younger women. As there are no identified receptors or proteins, these cancers do not have targeted therapies and the patients mainly receive chemotherapy.

The use of ER, PR, and HER2/neu as molecular biomarkers is important in making clinical decisions regarding treatment options.[108–110] While the current model for breast cancer classification has prognostic value, a lack of adding molecular components to the scheme limits the ability to predict response to newer targeted therapies. Utilizing the recent advances in molecular markers along with the histological classification should improve treatment and prognosis.

More recently, molecular subtyping of breast cancer has enhanced prediction of breast cancer response to and outcomes of treatments. Currently, at least five molecular subtypes have been identified: normal breast like, luminal subtype A, luminal subtype B, $ErbB^{2+}$, and basal like.[111–114] A new subtype classified as "claudin low" has also been identified.[115,116] It is clear that utilizing a combination of clinical, histological, and molecular data will help design effective treatments and predict responses to them.[117–121]

4. STAGES OF BREAST CANCER

Breast cancer stage is determined based on four major characteristics: (1) the size of the cancer, (2) if the cancer is invasive or noninvasive, (3) if the cancer is detected in the lymph nodes, and (4) if the cancer has metastasized to other parts of the body beyond the breast. During staging of breast cancer it is also further classified as local, regional, or distant. "Local" refers to cancer that is confined within the breast, "regional" refers to cancer detected in the lymph nodes of the armpit, and "distant" refers to breast cancer detected in other parts of the body.

Breast cancer stages are described on a number scale ranging from O to IV. Stage O represents breast cancers that are noninvasive. Stage I breast cancers are invasive and are further classified as Stage IA and IB. In Stage IA, the cancer measures up to 2 cm with no involvement of lymph nodes, while in Stage IB the cancer can be found in lymph nodes. Stage II is also classified into two subcategories: Stage IIA and Stage IIB. Stage IIA describes invasive breast cancers that measure up to 2 cm with involvement of one to three axillary lymph nodes. Stage IIB breast cancers measure between 2 cm and 5 cm with involvement of one to three axillary lymph nodes. Stage III is divided into three subcategories. Stage IIIA describes invasive breast cancers that may be of any size with involvement of four to nine axillary lymph nodes. Stage IIIB breast cancers may be of any size and have spread to the chest wall, skin of the breast, and up to nine axillary lymph nodes. Stage IIIC breast cancers have spread to 10 or more axillary lymph nodes. Stage IV describes invasive breast cancer that has spread to other organs of the body, such as the lungs, bones, liver, or brain. Identifying the stage of the breast cancer can help patients and doctors understand prognosis, pathology results, and the best treatment options.

TNM is another cancer staging system that provides information about how the cancer looks and behaves; T stands for tumor size, N stands for lymph node involvement, and M stands for metastasis. The T number is based on the size of the tumor and the extent of its growth, and a higher number indicates a larger tumor. The N number is based on the number of lymph nodes involved and how much cancer is found in them. The higher the N number, the greater the extent of the lymph node involvement. The M category describes if the cancer has spread to other parts of the body.

5. RISK FACTORS

The etiology of breast cancer is not well understood because a vast majority of breast cancers develop sporadically. There are several factors that have been associated with increasing or decreasing the risk of breast cancer.

5.1 Age and Gender

Many types of cancers are considered as a disease of aging. The incidence of breast cancer increases with age, doubling about every 10 years.[122] Almost 80% of breast cancers are diagnosed in women over the age of 50. Although men are also diagnosed with breast cancer, the incidence is relatively very low compared to women.[123]

5.2 Geographical Location

Age-adjusted breast cancer incidence and mortality rates differ significantly between countries. Women in Western countries tend to have a higher incidence of breast cancer compared to women from Far Eastern countries. Interestingly, immigrants have similar rates of breast cancer as that of the host country they immigrated to within one or two generations. This emphasizes the importance of environmental factors in breast cancer development.[122,124]

5.3 Personal and Family History

A woman diagnosed with cancer in one of her breasts has 3–4 times the risk of developing a new breast cancer, unrelated to the first one, in the contralateral breast or in another part of the same breast. If a woman's mother, sister, or daughter has been diagnosed with breast cancer, then the woman is at a higher risk for developing the disease.[125,126]

5.4 Heredity

Roughly 5%–10% of breast cancer cases are thought to be of hereditary origin. Women who have mutated BRCA1 or BRCA2 genes are at a particularly high risk. Carriers of BRCA1 mutations have a 55%–65% chance of developing breast cancer in their lifetime and women who carry BRCA2 mutations have a 45% chance of developing the disease. Ashkenazi Jewish women tend to have a high incidence of BRCA mutations.[127,128]

5.5 Breast Lesions

With the advent of digital mammography and vacuum-assisted biopsy systems, breast lesions are being detected at a high frequency. A previous breast biopsy result of atypical hyperplasia (lobular or ductal) or lobular carcinoma in situ increases a woman's breast cancer risk by 4–5 times. Lesions of uncertain malignant potential within the breast are associated with possible development of invasive breast cancer in the future. A heterogeneous group of lesions, which although benign on biopsy, have associations with current or future invasive breast malignancy.

5.6 Race

Caucasian women have a higher risk of developing breast cancer compared to African–American, Asian, Hispanic, and Native American women. This difference in risk is slowly disappearing as more epidemiological and clinical data addressing breast cancer in the various races are being collected and reported.[129,130]

5.7 Weight

The risk of breast cancer in obese premenopausal women is believed to be low, but is not fully established.[131–134] On the other hand, obese postmenopausal women have an increased risk of developing breast cancer. Truncal obesity in postmenopausal women has, in particular, been associated with increased risk. A metaanalysis of four independent studies observed that obese postmenopausal women had a 70% greater risk of developing ER-positive breast cancer.[135,136] An unusual gain of body weight during adulthood significantly increases the risk of breast cancer before and after menopause.[137–139]

5.8 Alcohol

Consumption of alcohol is correlated with an increased breast cancer risk. Women who consume one alcoholic drink a day have a slightly increased risk of breast cancer compared to nondrinkers. Teenaged girls who drink three to five drinks a week develop breast lumps more frequently than nondrinkers of the same age. A metaanalysis of data from 53 studies found one alcoholic drink consumed per day increased relative risk of breast cancer by about 7%.[140] Women who consumed two to three alcoholic drinks per day had a 20% increased risk.[140,141]

5.9 Radiation

Women exposed to excessive radiation are at a higher risk as well. The younger a women is at exposure to excessive radiation, the higher is her lifetime risk of breast cancer.[142–146] Women with Hodgkin's disease given radiation therapy at a young age have approximately 7 times the risk of breast cancer compared to women with Hodgkin's disease who did receive radiation therapy.[142]

5.10 Hormone Replacement Therapy

Many epidemiological studies have reported that hormone replacement therapy (HRT), often used by postmenopausal women to enhance quality of life, is associated with increased breast cancer risk. Data from the Million Women Study, Women's Health Initiative (WHI), and Heart and Estrogen/Progestin Replacement Study demonstrated that combined estrogen and progestin replacement therapy significantly impacted incidence of breast cancer.[147–152] The risk of breast cancer is higher in women who use estrogen and progestin combination replacement therapy compared with women who take estrogen-only replacement therapy. Estrogen-only replacement therapy is only given to hysterectomized women and data indicate that these women have a reduced risk of breast cancer.[153–155]

5.11 Reproductive History

Women who have never undergone a full-term pregnancy and women who had their first full-term pregnancy after the age of 35 have a higher risk of breast cancer. Women who had their first full-term pregnancy before the age of 20 have a 50% reduced risk of developing breast cancer.[156,157]

5.12 Breastfeeding

Breastfeeding has been identified as a modifiable risk factor of breast cancer. Epidemiological data indicate that breastfeeding decreases the risk of breast cancer.[158] A large metaanalysis showed that for every 12 months of breastfeeding, the risk of breast cancer is reduced by 4.3%.[33] However, the risk reduction is very minor and it is mainly associated with premenopausal women.[159] The reason for lactation-induced risk reduction is not understood.

6. EARLY FULL-TERM PREGNANCY AND BREAST CANCER

In the 1700s, Bernardino Ramazzini associated the childlessness of nuns with their increased breast cancer risk.[160,161] Later, in 1842, Rigoni Stern observed and reported a similar phenomenon.[160] MacMahon *et al.* did a comprehensive epidemiological study in Brazil, Greece, Japan, Taiwan, USA, Wales, and Yugoslavia in 1970 and reported that women who had their first birth under 20 years of age had about one-third the risk of developing breast cancer compared to women who had their first birth at the age of 35 years or older.[12] This phenomenon of early parity–induced protection against breast cancer has been reported by several groups.[26,157,162–165] Studies report that multiple full-term pregnancies further decrease the lifetime risk of breast cancer. Each additional pregnancy after the first child birth is expected to reduce the risk of breast cancer by another 7%.[33] However, it is important to note that the protective effect of early full-time pregnancy is not immediate. Actually, parous women have an increased risk of breast cancer immediately after delivery. This increased risk declines over a period of 5–10 years.[28,34] Women who have their first full-term pregnancy after 30 years of age have the highest transient risk of breast cancer after delivery.[166–169]

There is a lot debate on why a full-term pregnancy is required to confer protection against breast cancer and how a miscarriage or an abortion impacts the risk. Experimental models have clearly demonstrated that a full-term pregnancy is required for protection against breast cancer.[170,171] If pregnancy is interrupted prior to full-term, then there is only partial protection.

For instance, the gestational period of rats is around 21–22 days. When pregnancy is interrupted at days 5, 10, or 15 in rats exposed to dimethylbenz(a)anthracene, approximately 45%–50% cancer protection is observed compared to rats that underwent full-term pregnancy. Several studies have demonstrated that full-term parous rats have 75%–95% protection against chemical carcinogen–induced mammary carcinogenesis.[37,38,40,172,173] Therefore, early interruption of pregnancy results in reduced protection against mammary carcinogenesis compared to full-term parous animals.

In humans, an interrupted pregnancy has not been shown to provide partial protection against breast cancer, suggesting events that take place during late pregnancy and parturition are significant in this protection. Studies conducted with small groups of women could not clearly establish the impact of miscarriage or abortion on breast cancer risk; some report no increase in risk, while others do cite elevated risk levels.[174–176] Later, a prospective analysis of data collected from the medical records of a large number of women reliably demonstrated that abortions did not increase the risk of breast cancer.[177–182]

Similar to early full-term pregnancy, short-term treatment with estrogen and progesterone has been shown to confer significant protection against chemical carcinogen–induced mammary carcinogenesis, as well as in transgenic animals predisposed to developing spontaneous mammary cancers.[41,183–188] It was demonstrated that exogenous treatment with estrogen for 1–3 weeks, which resulted in pregnancy levels of estrogen, was highly effective in inhibiting mammary carcinogenesis.[186] Administration of human chorionic gonadotropin (hCG) has also been shown to reduce the incidence of mammary cancers.[189–193]

7. POTENTIAL MECHANISMS OF EARLY FULL-TERM PREGNANCY–INDUCED PROTECTION AGAINST BREAST CANCER

Though it has been known for centuries that early pregnancy significantly reduces the risk of developing breast cancer, the mechanism behind this protection is still not clear. There are several theories that have been proposed on the possible mechanism for this early full-term pregnancy–induced protection against breast cancer.

Early pregnancy–induced alterations in the systemic hormonal environment has been demonstrated to be closely linked with the reduction in risk. Alterations in growth hormone and prolactin have been primarily implicated in early pregnancy–induced protection against breast cancer.

Terminal differentiation of the mammary epithelium has been attributed as a possible reason for this protection. Terminal end buds in rodents and terminal ducal lobular units in women undergo major differentiation during pregnancy and it is believed that this differentiated state is maintained even after involution of the breast that occurs after cessation of lactation. This differentiated state of the mammary gland is thought to make it less proliferative, thus making it refractory to mammary carcinogenesis. During involution, a high degree of apoptosis occurs, during which it is suggested that the potential target cells for carcinogenesis are lost, leading to reduced risk of breast cancer. Genome-wide gene expression studies have provided evidence of this differentiated state of the parous breast.[52,194] On the contrary, it has been shown that complete differentiation of the mammary gland by administration of differentiating agents did not reduce the incidence of breast cancer, indicating that differentiation alone cannot be the cause for early full-term pregnancy–induced protection against breast cancer.[185]

Early full-term pregnancy induces persistent alterations in the epithelial–stromal interactions in the breast. The importance epithelial–stromal interaction in the development of the mammary gland is well established.[195] Further, it has been demonstrated that there is a significant alteration in epithelial–stromal interaction during breast cancer development and metastasis.[196–199] The altered epithelial–stromal dialogue results in a microenvironment that favors cancer growth, invasion, and metastasis by remodeling of the extracellular matrix.[200,201] The fibroblasts adjacent to the tumor and the tumor cells secrete soluble factors that influence each other.[197,198,200,202–207] Several growth factors, such as the transforming growth factor-β (TGF-β),[208] fibroblast growth factors,[209] insulin-like growth factors,[210] and hepatocyte growth factor (HGF)[207] are released. These growth factors are vital for mammary carcinogenesis. This interaction between the cancer and the surrounding stroma plays a key role in metastasis,[211–213] influencing the genes involved in cell proliferation and adhesion.[214] Though literature demonstrates the significance of the epithelial–stromal interaction in breast cancer development, it is still not clear if and how early full-term pregnancy impacts this interaction to result in reduced risk of breast cancer.

Another aspect of early full-term pregnancy–induced protection is altered stem cell activity. Stem cells are thought to be the targets for mammary carcinogenesis. Early full-term pregnancy has been shown to influence the stemness and number of stem cells. There are contradicting reports regarding the change in number of mammary stem cells in response to early full-term pregnancy. Although the findings are inconclusive, it is important to note that early full-term pregnancy could impact breast cancer risk by altering mammary stem cells.

Early full-term pregnancy–induced protection against breast cancer is believed to be only against ER-positive breast cancer. Estrogen-sensing cells are less sensitive and fewer in number after early first full-term pregnancy. This lowered sensitivity and number of estrogen-responsive cells have been attributed as potential contributors to protection against breast cancer. It has also been shown that short-term treatment with pregnancy levels of ovarian steroids, estrogen and progesterone, can be very effective in reducing the incidence of ER-negative mammary cancers.[187] Overall, there are several possible mechanisms by which early full-term pregnancy could confer protection against breast cancer.

7.1 Altered Systemic Hormonal Environment and Hormone Responsiveness

Hormones play a vital role in growth of normal breast tissues and breast cancer. During pregnancy, the body is exposed quite high levels of progesterone, placental lactogen, prolactin, estrogens, and glucocorticoids. This results in a unique hormonal profile in parous women compared to nulliparous women that is not well understood. Several research groups have shown that there is a persistent alteration in the systemic hormonal environments of parous subjects compared to nulliparous subjects. In the early to mid-1980s and in the late 2000s, different groups reported that circulating levels of prolactin were lower in parous women.[43,44,215] Musey *et al.* demonstrated that parous women had a reduced level of circulating prolactin in comparison with nulliparous women. They followed these women for 150 months and also demonstrated that parous women responded differently to treatment with perphenazine than did nulliparous women. Perphenazine is a dopamine antagonist.[43] Treatment with this drug results in copious prolactin secretion. When women were treated with perphenazine, the level of prolactin in parous women was significantly lower than in nulliparous women. In mouse models, prolactin treatment significantly increased mammary carcinogenesis, while prolactin-suppressing drugs drastically reduced mammary carcinogenesis.[216] Based on these findings, it was speculated that there is a persistent change in the systemic hormonal environment.

Chemical carcinogenesis experiments in animal models have demonstrated the significance of altered systemic hormonal environment in early pregnancy–induced protection against mammary carcinogenesis. Using classic mammary fat pad transplantation studies, it was shown that a systemic hormonal environment is critical in parity-induced protection against mammary cancer. In this study, mammary epithelial cells were isolated from virgin mice who were exposed to dimethylbenz(a)anthracene and transplanted to mammary gland-free fat pads of parous and virgin mice. Carcinogen-exposed mammary epithelial cells formed tumors in virgin hosts, while tumor formation was inhibited in parous hosts.[38] These findings indicate that the systemic hormonal environment is important for the development of mammary cancer.

The circulating level of growth hormone is also decreased in parous rats compared to nulliparous rats.[37,217,218] Further, suppression of growth hormone secretion results in regression of mammary tumors that have been induced by a chemical carcinogen.[219] In addition, growth hormone–deficient rats are refractory to chemical carcinogen–induced mammary cancers.[39,220] Supplementation of exogenous growth hormone to these carcinogen-exposed, growth hormone–deficient rats resulted in a very high incidence of mammary tumors.[39]

Some studies reported that E_2 levels were decreased in parous subjects,[47,221] while other studies did not observe a significant change.[43] There could be several reasons for these contrasting findings, such as the age of the women, stage of menstrual cycle, day and time of sampling, etc. Further, it is vital to understand the hormone sensitivity of mammary gland in parous subjects compared to nulliparous subjects. In this regard, estrogen has been studied quite extensively. There is some evidence that mammary glands from parous women have altered estrogen responses. Levels of ERα were lower in parous rats compared with age-matched nulliparous rats.[37,222] ERα signaling is known to induce PR. Approximately 75% of breast cancers express ERα and a very high percentage of these breast cancer also express PRs.[223] Breast cancers that are ER and PR positive are more responsive to endocrine therapies and have a good prognosis.[224–231] Clinical data have implicated that expression of PRs is helpful in defining the use of aromatase inhibitors over tamoxifen as the treatment of choice.

Mammary glands of parous females that have decreased levels of ERα also have reduced levels of PRs.[37,232] Alternatively, expression levels of ERβ is increased in mammary glands of parous females compared to a nulliparous group. This is prominent in the mammary glands of multiparous animals.[233]

In response to these changes in the ER levels, the downstream signaling is altered significantly. Several mediators of the classical and nonclassical estrogen signaling pathways, such as epithelial growth factor receptor, GPR30, JunB, p300/CBP, and TGF-β1 are believed to play a major role in early pregnancy–induced protection against mammary carcinogenesis.[48,234–238]

The data are convincing that early pregnancy and short-term hormone treatments alter the systemic hormonal environment and confer protection against breast cancer. Still, it is difficult to conclude that this is the cause of protection, as there are several variations in the experimental design, end points, timelines, the type and length of treatment, etc. Hence, the molecular and physiological bases of this protection are not well understood.

7.2 Terminal Differentiation of the Breast

Pregnancy-induced terminal differentiation of the breast has been considered a factor for protection from breast cancer. One group of scientists has classified the lobules in the breast as types 1, 2, and 3 based on the increasing number of clusters of ductules per lobule. Type 1 and 2 lobules are predominant in the nulliparous breast, while type 3 lobules are seen during pregnancy.[239,240] To design a large-scale study in which breast tissue can be obtained before and after pregnancy is difficult, and without strong data from such a study, it is difficult to conclude that there is a discrete and persistent morphological difference between parous and nulliparous breast tissue. So while these observations are intriguing, it is difficult to make a decisive statement.

Molecular analysis of mammary gland tissue from parous and nulliparous subjects revealed differences in genes associated with differentiation and proliferation. Using gene expression microarray analysis, it was shown that multiple growth factor genes were downregulated, while genes encoding growth inhibitory molecules were upregulated in parous mammary glands compared to nulliparous mammary glands.[48] Further, parous mammary glands had higher levels of whey acidic proteins, caseins, and adipocyte differentiation–related proteins, indicating a differentiated state. Another study also reported that an increased expression of genes encoded for milk proteins. Genes involved in immunity were upregulated, while genes involved in growth factor signaling were downregulated in parous mammary glands.[235,241] Short-term estrogen and progesterone treatment resulted in differential gene expression as assessed by suppression subtractive hybridization. Genes involved in differentiation, cell–cell contact, metabolism, and cell homeostasis were differentially expressed in hormone-treated mammary glands compared to the control.[242] All these studies imply that the mammary glands of parous subjects are more differentiated than nulliparous subjects, which could be a factor for parity-induced protection against mammary cancer.

Prolactin plays a major role in differentiation of breast tissue. Synthesis and secretion of prolactin is controlled by dopamine secreted from the hypothalamus. Perphenazine is a dopamine receptor antagonist and administration of perphenazine results in copious amounts of prolactin secretion. This high level of prolactin causes a high degree of differentiation of the mammary gland. Rats that were exposed to chemical carcinogen *N*-methyl-*N*-nitrosourea developed a high incidence of mammary cancers. Another group of rats were exposed to the carcinogen treatment and then received a perphenazine treatment. The mammary glands of these rats treated were fully differentiated, but still had a high incidence of mammary cancer.[185] These findings indicate the differentiation alone is not sufficient to confer protection against mammary cancer. In addition, it has also been shown that carcinogen-exposed rats treated with short-term estrogen expressed a high degree of protection against mammary carcinogenesis.[186] This short-term estrogen treatment did not induce high degree of differentiation. Moreover, protective estrogen plus progesterone treatment also did not induce full lobulo–alveolar differentiation, but was effective in inhibiting mammary carcinogenesis.[238] All of these studies suggest that terminal differentiation alone is unlikely to explain the protective effect of early full-term pregnancy.

7.3 Effect of Early Full-Term Pregnancy on Mammary Stem Cells

Stem cells give rise to various kinds of cells in the body. Totipotent stem cells are embryonic stem cells that are present during the first few cell divisions postfertilization and can form any of the different types of cells in the body. Multipotent stem cells are adult stem cells that can form other cell types, but have limited potency. All of the body's organs have stem cells that are vital for maintenance and repair of organs.

Stem cells undergo both symmetric and asymmetric cell divisions. Symmetric division results in the production of two identical daughter stem cells.[243–245] Mutations that occur during symmetric cell divisions of stem cells can be repaired and retained for long periods of time, unlike terminally differentiated cells where mutations are usually short lived. In asymmetric stem cell division, one of the two daughter is identical to the parental stem cell, while the other daughter becomes a committed cell and eventually differentiates into cell types of the specific tissue/organ. The potency of stem cells are thought to decrease with aging.

Breast stem cells possess a self-renewal capacity, which is a characteristic of cancer cells, and are thought to be the target cells for breast cancer development. Stem cells have a long cell life that can result in an accumulation of mutations that could favor cancer development. The stem cell theory of cancer suggests that a small group of cancer cells act as cancer stem cells that reproduce themselves, producing the bulk of cancer cells. Cancer stem cells are implicated in the development of metastasis and relapse after different cancer treatments. Epidemiological studies suggest that there is a significant increase in breast cancer stem cells during puberty. This expansion of stem cells is required for the rapid growth of the breast that occurs during this time. Any mutations during puberty are expected to have a profound impact on risk of breast cancer. This finding is supported by the Hiroshima and Nagasaki atomic bomb–related radiation and cancer data. These data indicated that women in those areas during the attack under the age of 20 had a 13-fold increased risk of developing breast cancer.[246,247] Male breast cancer incidence and mortality risk among the Japanese atomic bomb survivors is very high.[248]

Breast tissue is primarily made up of luminal epithelial, myoepithelial, stromal, and fat cells. The breast is partly made up of stem cells that have the ability to self-renew and are responsible for the generation of all cell types specific to the gland. It is proposed that breast cancer arises primarily from the stem cells and that pregnancy may lead to protective changes in the stem cell population of the mammary gland. However, it remains debatable whether the mammary epithelial stem cell population is a primary contributing factor to the phenomenon of parity-induced protection.[23,30,249–253] A study has suggested that there is a persistent decrease in the number of mammary-repopulating units (mammary epithelial stem cells) after parity.[249] On the other hand, there are reports that parity-induced protection is not due to changes in the number of mammary stem cells.[254,255]

Early parity induces reduction in ER- and PR-positive cells. It also reduces Wnt signaling and decreases cell proliferation. Alternatively, late pregnancy does not significantly reduce the number of hormone receptor–positive cells and does not inhibit Wnt signaling.[165,256] Breast tissues of parous women have reduced frequency of progenitor cells. Thus, Wnt signaling seems to play a vital role in early parity–induced protection against breast cancer. It is evident, to a certain extent, given the experimental and epidemiological data that breast stem cells can be altered in response to early parity, thereby rendering the breast resistant to cancer development. Further in-depth research is required to thoroughly understand the role of stem cells and their influence on breast cancer risk.

Early full-term pregnancy reduces the lifetime risk of breast cancer, but these women have a slightly increased risk of breast cancer for roughly 5–10 years after first birth. The younger a women is when she gives birth to her first child, the sooner her risk is reduced. An earlier study investigated the difference in expression of genes between parous and nulliparous women as a function of time after the last full-term pregnancy. Genes controlling immune cells, such as dendritic cells and T killer cells, are not activated during the first 5 years after pregnancy. This could explain the slight increase in risk of breast cancer following birth. Furthermore, some genes that play a vital role in several key signaling pathways are activated during the first 5 years after birth, and this increased signaling might also contribute to the increased breast cancer risk.[169,257,258]

8. CONCLUSIONS

Early full-term pregnancy is the only known physiological factor that drastically reduces the risk of breast cancer. The protection against breast cancer conferred by early full-term pregnancy could have evolutionary significance. Some of the main factors in this protection are: (1) altered systemic hormonal environment, (2) terminal differentiation of the mammary gland, (3) altered stem cell activity, (4) altered epithelial-stromal interactions, and (5) altered hormonal sensitivity. Understanding the mechanisms involved in lifetime reduction in risk of breast cancer due to early full-term pregnancy will lead to better preventive and therapeutic strategies. Short-term protective hormone treatments have been developed and demonstrated to be highly effective against mammary carcinogenesis in rodent models. Further investigation to translate this short-term protective hormone treatment to humans could help in preventing breast cancer globally. Finally, identified molecular alterations can serve as biomarkers that help predict the risk of breast cancer in the general population.

ACKNOWLEDGMENTS

This article was supported by USAMRMC proposal BC160999.

REFERENCES

1. DeSantis C, Ma J, Bryan L, Jemal A. Breast cancer statistics, 2013. *CA Cancer J Clin.* 2014;64(1):52–62.
2. Fedewa SA, de Moor JS, Ward EM, et al. Mammography use and physician recommendation after the 2009 U.S. preventive services task force breast cancer screening recommendations. *Am J Prev Med.* 2016;50(5):e123–e131.
3. de la Mare JA, Sterrenberg JN, Sukhthankar MG, et al. Assessment of potential anticancer stem cell activity of marine algal compounds using an in vitro mammosphere assay. *Cancer Cell Int.* 2013;13(1):39.
4. Parvani JG, Davuluri G, Wendt MK, et al. Deptor enhances triple-negative breast cancer metastasis and chemoresistance through coupling to survivin expression. *Neoplasia.* 2015;17(3):317–328.
5. Mateo F, Arenas EJ, Aguilar H, et al. Stem cell-like transcriptional reprogramming mediates metastatic resistance to mTOR inhibition. *Oncogene.* 2017;36(19):2737–2749.
6. Gomez-Miragaya J, Palafox M, Pare L, et al. Resistance to taxanes in triple-negative breast cancer associates with the dynamics of a CD49f+ tumor-initiating population. *Stem Cell Rep.* 2017;8(5):1392–1407.
7. Malfatti CR, Dos Santos FS, Wouk J, et al. Intracerebroventricular administration of the (1→6)-beta-D-glucan (lasiodiplodan) in male rats prevents D-penicillamine-induced behavioral alterations and lipoperoxidation in the cortex. *Pharm Biol.* 2017;55(1): 1289–1294.
8. Kim GM, Kim S, Park HS, et al. Chemotherapy-induced irreversible alopecia in early breast cancer patients. *Breast Cancer Res Treat.* 2017;163(3):527–533.
9. Ansari L, Shiehzadeh F, Taherzadeh Z, et al. The most prevalent side effects of pegylated liposomal doxorubicin monotherapy in women with metastatic breast cancer: a systematic review of clinical trials. *Cancer Gene Ther.* 2017;24(5):189–193.
10. Meattini I, Curigliano G, Terziani F, et al. SAFE trial: an ongoing randomized clinical study to assess the role of cardiotoxicity prevention in breast cancer patients treated with anthracyclines with or without trastuzumab. *Med Oncol.* 2017;34(5):75.
11. Li H, Sun X, Miller E, et al. BMI, reproductive factors, and breast cancer molecular subtypes: a case-control study and meta-analysis. *J Epidemiol.* 2017;27(4):143–151.
12. MacMahon B, Cole P, Lin TM, et al. Age at first birth and breast cancer risk. *Bull World Health Organ.* 1970;43(2):209–221.
13. Sartwell PE, Arthes FG, Tonascia JA. Exogenous hormones, reproductive history, and breast cancer. *J Natl Cancer Inst.* 1977;59(6):1589–1592.
14. Woods KL, Smith SR, Morrison JM. Parity and breast cancer: evidence of a dual effect. *Br Med J.* 1980;281(6237):419–421.
15. Trapido E. Parity and breast cancer. *Br Med J.* 1980;281(6247):1072.
16. Macaluso M, Tamburini M, Massara G, Bertario L, Di Pietro S. Parity and breast cancer: confirmed evidence of an effect on age at diagnosis. *Breast Cancer Res Treat.* 1982;2 (3):257–260.
17. Trichopoulos D, Hsieh CC, MacMahon B, et al. Age at any birth and breast cancer risk. *Int J Cancer.* 1983;31(6):701–704.
18. Trapido EJ. Age at first birth, parity, and breast cancer risk. *Cancer.* 1983;51(5):946–948.
19. Pathak DR, Speizer FE, Willett WC, Rosner B, Lipnick RJ. Parity and breast cancer risk: possible effect on age at diagnosis. *Int J Cancer.* 1986;37(1):21–25.
20. Kvale G, Heuch I. A prospective study of reproductive factors and breast cancer. II. Age at first and last birth. *Am J Epidemiol.* 1987;126(5):842–850.
21. Kvale G, Heuch I, Eide GE. A prospective study of reproductive factors and breast cancer. I. Parity. *Am J Epidemiol.* 1987;126(5):831–841.
22. Layde PM, Webster LA, Baughman AL, Wingo PA, Rubin GL, Ory HW. The independent associations of parity, age at first full term pregnancy, and duration of

breastfeeding with the risk of breast cancer. Cancer and steroid hormone study group. *J Clin Epidemiol.* 1989;42(10):963–973.
23. Dall G, Risbridger G, Britt K. Mammary stem cells and parity-induced breast cancer protection—new insights. *J Steroid Biochem Mol Biol.* 2017;170:54–60.
24. Negri E, La Vecchia C, Duffy SW, Bruzzi P, Parazzini F, Day NE. Age at first and second births and breast cancer risk in biparous women. *Int J Cancer.* 1990;45(3):428–430.
25. Thompson WD, Janerich DT. Maternal age at birth and risk of breast cancer in daughters. *Epidemiology.* 1990;1(2):101–106.
26. Kelsey JL, Gammon MD, John EM. Reproductive factors and breast cancer. *Epidemiol Rev.* 1993;15(1):36–47.
27. Hsieh C, Pavia M, Lambe M, et al. Dual effect of parity on breast cancer risk. *Eur J Cancer.* 1994;30A(7):969–973.
28. Rosner B, Colditz GA, Willett WC. Reproductive risk factors in a prospective study of breast cancer: the nurses' health study. *Am J Epidemiol.* 1994;139(8):819–835.
29. Kroman N, Wohlfahrt J, Andersen KW, Mouridsen HT, Westergaard T, Melbye M. Parity, age at first childbirth and the prognosis of primary breast cancer. *Br J Cancer.* 1998;78(11):1529–1533.
30. Britt K, Ashworth A, Smalley M. Pregnancy and the risk of breast cancer. *Endocr Relat Cancer.* 2007;14(4):907–933.
31. Horn J, Vatten LJ. Reproductive and hormonal risk factors of breast cancer: a historical perspective. *Int J Womens Health.* 2017;9:265–272.
32. von Au A, Klotzbuecher M, Uhlmann L, et al. Impact of reproductive factors on breast cancer subtypes in postmenopausal women: a retrospective single-center study. *Arch Gynecol Obstet.* 2017;295(4):971–978.
33. Collaborative Group on Hormonal Factors in Breast Cancer.Breast cancer and breast-feeding: collaborative reanalysis of individual data from 47 epidemiological studies in 30 countries, including 50302 women with breast cancer and 96973 women without the disease. *Lancet.* 2002;360(9328):187–195.
34. Albrektsen G, Heuch I, Hansen S, Kvale G. Breast cancer risk by age at birth, time since birth and time intervals between births: exploring interaction effects. *Br J Cancer.* 2005;92(1):167–175.
35. Newcomb PA, Trentham-Dietz A, Hampton JM, et al. Late age at first full term birth is strongly associated with lobular breast cancer. *Cancer.* 2011;117(9):1946–1956.
36. Chakravarthi BV, Varambally S. Targeting the link between late pregnancy and breast cancer. *Elife.* 2013;2:e01926.
37. Thordarson G, Jin E, Guzman RC, Swanson SM, Nandi S, Talamantes F. Refractoriness to mammary tumorigenesis in parous rats: is it caused by persistent changes in the hormonal environment or permanent biochemical alterations in the mammary epithelia? *Carcinogenesis.* 1995;16(11):2847–2853.
38. Abrams TJ, Guzman RC, Swanson SM, Thordarson G, Talamantes F, Nandi S. Changes in the parous rat mammary gland environment are involved in parity-associated protection against mammary carcinogenesis. *Anticancer Res.* 1998;18(6A):4115–4121.
39. Thordarson G, Semaan S, Low C, et al. Mammary tumorigenesis in growth hormone deficient spontaneous dwarf rats; effects of hormonal treatments. *Breast Cancer Res Treat.* 2004;87(3):277–290.
40. Thordarson G, Slusher N, Leong H, et al. Insulin-like growth factor (IGF)-I obliterates the pregnancy-associated protection against mammary carcinogenesis in rats: evidence that IGF-I enhances cancer progression through estrogen receptor-alpha activation via the mitogen-activated protein kinase pathway. *Breast Cancer Res.* 2004;6(4):R423–R436.
41. Rajkumar L, Guzman RC, Yang J, Thordarson G, Talamantes F, Nandi S. Prevention of mammary carcinogenesis by short-term estrogen and progestin treatments. *Breast Cancer Res.* 2004;6(1):R31–R37.

42. Musey VC, Collins DC, Musey PI, Martino-Saltzman D, Preedy JR. Age-related changes in the female hormonal environment during reproductive life. *Am J Obstet Gynecol.* 1987;157(2):312–317.
43. Musey VC, Collins DC, Musey PI, Martino-Saltzman D, Preedy JR. Long-term effect of a first pregnancy on the secretion of prolactin. *N Engl J Med.* 1987;316(5): 229–234.
44. Kwa HG, Cleton F, Bulbrook RD, Wang DY, Hayward JL. Plasma prolactin levels and breast cancer: relation to parity, weight and height, and age at first birth. *Int J Cancer.* 1981;28(1):31–34.
45. Yu MC, Gerkins VR, Henderson BE, Brown JB, Pike MC. Elevated levels of prolactin in nulliparous women. *Br J Cancer.* 1981;43(6):826–831.
46. Xu B, Lipworth L, Wide L, et al. Maternal and gestational correlates of pregnancy prolactin and growth hormone in USA and china. *Eur J Cancer Prev.* 2003;12(1):35–42.
47. Bernstein L, Pike MC, Ross RK, Judd HL, Brown JB, Henderson BE. Estrogen and sex hormone-binding globulin levels in nulliparous and parous women. *J Natl Cancer Inst.* 1985;74(4):741–745.
48. D'Cruz CM, Moody SE, Master SR, et al. Persistent parity-induced changes in growth factors, TGF-beta3, and differentiation in the rodent mammary gland. *Mol Endocrinol.* 2002;16(9):2034–2051.
49. Verlinden I, Gungor N, Wouters K, Janssens J, Raus J, Michiels L. Parity-induced changes in global gene expression in the human mammary gland. *Eur J Cancer Prev.* 2005;14(2):129–137.
50. Uehara N, Unami A, Kiyozuka Y, Shikata N, Oishi Y, Tsubura A. Parous mammary glands exhibit distinct alterations in gene expression and proliferation responsiveness to carcinogenic stimuli in Lewis rats. *Oncol Rep.* 2006;15(4):903–911.
51. Asztalos S, Gann PH, Hayes MK, et al. Gene expression patterns in the human breast after pregnancy. *Cancer Prev Res.* 2010;3(3):301–311.
52. Peri S, de Cicco RL, Santucci-Pereira J, et al. Defining the genomic signature of the parous breast. *BMC Med Genomics.* 2012;5:46.
53. Russo J, Santucci-Pereira J, de Cicco RL, et al. Pregnancy-induced chromatin remodeling in the breast of postmenopausal women. *Int J Cancer.* 2012;131(5):1059–1070.
54. Brisken C, O'Malley B. Hormone action in the mammary gland. *Cold Spring Harb Perspect Biol.* 2010;2(12):a003178.
55. Klinge CM, Jernigan SC, Mattingly KA, Risinger KE, Zhang J. Estrogen response element-dependent regulation of transcriptional activation of estrogen receptors alpha and beta by coactivators and corepressors. *J Mol Endocrinol.* 2004;33(2):387–410.
56. Eng FC, Barsalou A, Akutsu N, et al. Different classes of coactivators recognize distinct but overlapping binding sites on the estrogen receptor ligand binding domain. *J Biol Chem.* 1998;273(43):28371–28377.
57. Arumugam A, Subramani R, Nandy S, Lopez R, Boopalan T, Lakshmanaswamy R. Parity and short-term estradiol treatment utilizes similar cellular mechanisms to confer protection against breast cancer. *Cell Physiol Biochem.* 2014;34(2):491–505.
58. Chang TH, Kunasegaran K, Tarulli GA, De Silva D, Voorhoeve PM, Pietersen AM. New insights into lineage restriction of mammary gland epithelium using parity-identified mammary epithelial cells. *Breast Cancer Res.* 2014;16(1):R1.
59. Huh JI, Qiu TH, Chandramouli GV, et al. 2-Methoxyestradiol induces mammary gland differentiation through amphiregulin-epithelial growth factor receptor-mediated signaling: molecular distinctions from the mammary gland of pregnant mice. *Endocrinology.* 2007;148(3):1266–1277.
60. Snedeker SM, Brown CF, DiAugustine RP. Expression and functional properties of transforming growth factor alpha and epidermal growth factor during mouse mammary gland ductal morphogenesis. *Proc Natl Acad Sci USA.* 1991;88(1):276–280.

61. Forster C, Makela S, Warri A, et al. Involvement of estrogen receptor beta in terminal differentiation of mammary gland epithelium. *Proc Natl Acad Sci USA.* 2002;99(24):15578–15583.
62. Saji S, Sakaguchi H, Andersson S, Warner M, Gustafsson J. Quantitative analysis of estrogen receptor proteins in rat mammary gland. *Endocrinology.* 2001;142(7):3177–3186.
63. Silva HG, Tortora RP, Farias ML. Increased bone turnover during the third trimester of pregnancy and decreased bone mineral density after parturition in adolescents as compared to age-matched control patients. *Gynecol Endocrinol.* 2005;21(3):174–179.
64. Shackleton M, Vaillant F, Simpson KJ, et al. Generation of a functional mammary gland from a single stem cell. *Nature.* 2006;439(7072):84–88.
65. Jay SM, Dawson D, Lamond N. Train drivers' sleep quality and quantity during extended relay operations. *Chronobiol Int.* 2006;23(6):1241–1252.
66. Stingl J, Caldas C. Molecular heterogeneity of breast carcinomas and the cancer stem cell hypothesis. *Nat Rev Cancer.* 2007;7(10):791–799.
67. Shipitsin M, Campbell LL, Argani P, et al. Molecular definition of breast tumor heterogeneity. *Cancer Cell.* 2007;11(3):259–273.
68. Chatterton Jr RT. Mammary gland: development and secretion. *Obstet Gynecol Annu.* 1978;7:303–324.
69. Ceriani RL. Proceedings: hormones and other factors controlling growth in the mammary gland: a review. *J Invest Dermatol.* 1974;63(1):93–108.
70. Silberstein GB. Postnatal mammary gland morphogenesis. *Microsc Res Tech.* 2001;52(2):155–162.
71. Hovey RC, Trott JF, Vonderhaar BK. Establishing a framework for the functional mammary gland: from endocrinology to morphology. *J Mammary Gland Biol Neoplasia.* 2002;7(1):17–38.
72. Berryhill GE, Trott JF, Hovey RC. Mammary gland development—it's not just about estrogen. *J Dairy Sci.* 2016;99(1):875–883.
73. McBryan J, Howlin J. Pubertal mammary gland development: elucidation of in vivo morphogenesis using murine models. *Methods Mol Biol.* 2017;1501:77–114.
74. Pike MC, Spicer DV, Dahmoush L, Press MF. Estrogens, progestogens, normal breast cell proliferation, and breast cancer risk. *Epidemiol Rev.* 1993;15(1):17–35.
75. Nandi S. Endocrine control of mammary gland development and function in the C3H/he crgl mouse. *J Natl Cancer Inst.* 1958;21(6):1039–1063.
76. Deroo BJ, Hewitt SC, Collins JB, Grissom SF, Hamilton KJ, Korach KS. Profile of estrogen-responsive genes in an estrogen-specific mammary gland outgrowth model. *Mol Reprod Dev.* 2009;76(8):733–750.
77. Feng Y, Manka D, Wagner KU, Khan SA. Estrogen receptor-alpha expression in the mammary epithelium is required for ductal and alveolar morphogenesis in mice. *Proc Natl Acad Sci USA.* 2007;104(37):14718–14723.
78. Robinson GW, Hennighausen L, Johnson PF. Side-branching in the mammary gland: the progesterone-wnt connection. *Genes Dev.* 2000;14(8):889–894.
79. Atwood CS, Hovey RC, Glover JP, et al. Progesterone induces side-branching of the ductal epithelium in the mammary glands of peripubertal mice. *J Endocrinol.* 2000;167(1):39–52.
80. Ismail PM, Amato P, Soyal SM, et al. Progesterone involvement in breast development and tumorigenesis—as revealed by progesterone receptor "knockout" and "knockin" mouse models. *Steroids.* 2003;68(10–13):779–787.
81. Arendt LM, Kuperwasser C. Form and function: how estrogen and progesterone regulate the mammary epithelial hierarchy. *J Mammary Gland Biol Neoplasia.* 2015;20(1–2):9–25.
82. Cardiff RD, Wellings SR. The comparative pathology of human and mouse mammary glands. *J Mammary Gland Biol Neoplasia.* 1999;4(1):105–122.

83. Smalley M, Ashworth A. Stem cells and breast cancer: a field in transit. *Nat Rev Cancer.* 2003;3(11):832–844.
84. Fendrick JL, Raafat AM, Haslam SZ. Mammary gland growth and development from the postnatal period to postmenopause: ovarian steroid receptor ontogeny and regulation in the mouse. *J Mammary Gland Biol Neoplasia.* 1998;3(1):7–22.
85. Russo IH, Russo J. Role of hormones in mammary cancer initiation and progression. *J Mammary Gland Biol Neoplasia.* 1998;3(1):49–61.
86. Russo J, Russo IH. Breast development, hormones and cancer. *Adv Exp Med Biol.* 2008;630:52–56.
87. Haslam SZ, Shyamala G. Progesterone receptors in normal mammary gland: receptor modulations in relation to differentiation. *J Cell Biol.* 1980;86(3):730–737.
88. Haslam SZ. The ontogeny of mouse mammary gland responsiveness to ovarian steroid hormones. *Endocrinology.* 1989;125(5):2766–2772.
89. Haslam SZ, Nummy KA. The ontogeny and cellular distribution of estrogen receptors in normal mouse mammary gland. *J Steroid Biochem Mol Biol.* 1992;42(6):589–595.
90. Dickson RB, Stancel GM. Estrogen receptor-mediated processes in normal and cancer cells. *J Natl Cancer Inst Monogr.* 2000;(27):135–145.
91. Ferguson AT, Davidson NE. Regulation of estrogen receptor alpha function in breast cancer. *Crit Rev Oncog.* 1997;8(1):29–46.
92. Russo J, Ao X, Grill C, Russo IH. Pattern of distribution of cells positive for estrogen receptor alpha and progesterone receptor in relation to proliferating cells in the mammary gland. *Breast Cancer Res Treat.* 1999;53(3):217–227.
93. Warner M, Saji S, Gustafsson JA. The normal and malignant mammary gland: a fresh look with ER beta onboard. *J Mammary Gland Biol Neoplasia.* 2000;5(3):289–294.
94. Lazennec G, Bresson D, Lucas A, Chauveau C, Vignon F. ER beta inhibits proliferation and invasion of breast cancer cells. *Endocrinology.* 2001;142(9):4120–4130.
95. Warner M, Huang B, Gustafsson JA. Estrogen receptor beta as a pharmaceutical target. *Trends Pharmacol Sci.* 2017;38(1):92–99.
96. Heldring N, Pike A, Andersson S, et al. Estrogen receptors: how do they signal and what are their targets. *Physiol Rev.* 2007;87(3):905–931.
97. Palmieri C, Saji S, Sakaguchi H, et al. The expression of oestrogen receptor (ER)-beta and its variants, but not ERalpha, in adult human mammary fibroblasts. *J Mol Endocrinol.* 2004;33(1):35–50.
98. Mote PA, Johnston JF, Manninen T, Tuohimaa P, Clarke CL. Detection of progesterone receptor forms A and B by immunohistochemical analysis. *J Clin Pathol.* 2001;54 (8):624–630.
99. Anderson E. The role of oestrogen and progesterone receptors in human mammary development and tumorigenesis. *Breast Cancer Res.* 2002;4(5):197–201.
100. Clarke RB. Ovarian steroids and the human breast: regulation of stem cells and cell proliferation. *Maturitas.* 2006;54(4):327–334.
101. Recommendations for the reporting of breast carcinoma. Association of directors of anatomic and surgical pathology. *Hum Pathol.* 1996;27(3):220–224.
102. Recommendations for the reporting of breast, carcinoma, association of directors of anatomic and surgical, pathology. *Mod Pathol.* 1996;9(1):77–81.
103. Malhotra GK, Zhao X, Band H, Band V, Histological. molecular and functional subtypes of breast cancers. *Cancer Biol Ther.* 2010;10(10):955–960.
104. Lagios MD, Margolin FR, Westdahl PR, Rose MR. Mammographically detected duct carcinoma in situ. Frequency of local recurrence following tylectomy and prognostic effect of nuclear grade on local recurrence. *Cancer.* 1989;63(4):618–624.
105. Poller DN, Silverstein MJ, Galea M, et al. Ideas in pathology. Ductal carcinoma in situ of the breast: a proposal for a new simplified histological classification association between cellular proliferation and c-erbB-2 protein expression. *Mod Pathol.* 1994;7(2):257–262.

106. Holland R, Peterse JL, Millis RR, et al. Ductal carcinoma in situ: a proposal for a new classification. *Semin Diagn Pathol.* 1994;11(3):167–180.
107. Silverstein MJ, Poller DN, Waisman JR, et al. Prognostic classification of breast ductal carcinoma-in-situ. *Lancet.* 1995;345(8958):1154–1157.
108. Maughan KL, Lutterbie MA, Ham PS. Treatment of breast cancer. *Am Fam Physician.* 2010;81(11):1339–1346.
109. Rakha EA, Reis-Filho JS, Ellis IO. Combinatorial biomarker expression in breast cancer. *Breast Cancer Res Treat.* 2010;120(2):293–308.
110. Payne SJ, Bowen RL, Jones JL, Wells CA. Predictive markers in breast cancer—the present. *Histopathology.* 2008;52(1):82–90.
111. Perou CM, Sorlie T, Eisen MB, et al. Molecular portraits of human breast tumours. *Nature.* 2000;406(6797):747–752.
112. Brenton JD, Aparicio SA, Caldas C. Molecular profiling of breast cancer: portraits but not physiognomy. *Breast Cancer Res.* 2001;3(2):77–80.
113. Sorlie T, Perou CM, Tibshirani R, et al. Gene expression patterns of breast carcinomas distinguish tumor subclasses with clinical implications. *Proc Natl Acad Sci USA.* 2001;98 (19):10869–10874.
114. Sorlie T, Tibshirani R, Parker J, et al. Repeated observation of breast tumor subtypes in independent gene expression data sets. *Proc Natl Acad Sci USA.* 2003;100(14): 8418–8423.
115. Herschkowitz JI, Simin K, Weigman VJ, et al. Identification of conserved gene expression features between murine mammary carcinoma models and human breast tumors. *Genome Biol.* 2007;8(5):R76.
116. Prat A, Parker JS, Karginova O, et al. Phenotypic and molecular characterization of the claudin-low intrinsic subtype of breast cancer. *Breast Cancer Res.* 2010;12(5):R68.
117. Parker JS, Mullins M, Cheang MC, et al. Supervised risk predictor of breast cancer based on intrinsic subtypes. *J Clin Oncol.* 2009;27(8):1160–1167.
118. Sontrop HM, Reinders MJ, Moerland PD. Breast cancer subtype predictors revisited: from consensus to concordance? *BMC Med Genomics.* 2016;9(1):26.
119. Ross JS, Hatzis C, Symmans WF, Pusztai L, Hortobagyi GN. Commercialized multigene predictors of clinical outcome for breast cancer. *Oncologist.* 2008;13(5): 477–493.
120. Ross JS. Multigene classifiers, prognostic factors, and predictors of breast cancer clinical outcome. *Adv Anat Pathol.* 2009;16(4):204–215.
121. Yau C, Esserman L, Moore DH, Waldman F, Sninsky J, Benz CC. A multigene predictor of metastatic outcome in early stage hormone receptor-negative and triple-negative breast cancer. *Breast Cancer Res.* 2010;12(5):R85.
122. McPherson K, Steel CM, Dixon JM. ABC of breast diseases. Breast cancer-epidemiology, risk factors, and genetics. *BMJ.* 2000;321(7261):624–628.
123. Losurdo A, Rota S, Gullo G, et al. Controversies in clinicopathological characteristics and treatment strategies of male breast cancer: a review of the literature. *Crit Rev Oncol Hematol.* 2017;113:283–291.
124. Andreeva VA, Unger JB, Pentz MA. Breast cancer among immigrants: a systematic review and new research directions. *J Immigr Minor Health.* 2007;9(4):307–322.
125. Viehweg P, Bernerth T, Heinig A, et al. MR-guided intervention in women at high hereditary risk of breast cancer due to both family and personal history of breast cancer. *Breast J.* 2006;12(6):549–558.
126. Schwab FD, Kilic N, Huang DJ, et al. Personal or first-degree family breast cancer history: which has higher impact on tumor detection and tumor size in breast cancer. *Arch Gynecol Obstet.* 2015;291(6):1387–1394.
127. Wellings E, Vassiliades L, Abdalla R. Breast cancer screening for high-risk patients of different ages and risk—which modality is most effective? *Cureus.* 2016;8(12):e945.

128. Skol AD, Sasaki MM, Onel K. The genetics of breast cancer risk in the post-genome era: thoughts on study design to move past BRCA and towards clinical relevance. *Breast Cancer Res.* 2016;18(1):99.
129. Brennan M. Breast cancer in ethnic minority groups in developed nations: case studies of the United Kingdom and Australia. *Maturitas.* 2017;99:16–19.
130. Bandera EV, Maskarinec G, Romieu I, John EM. Racial and ethnic disparities in the impact of obesity on breast cancer risk and survival: a global perspective. *Adv Nutr.* 2015;6(6):803–819.
131. Nelson HD, Zakher B, Cantor A, et al. Risk factors for breast cancer for women aged 40 to 49 years: a systematic review and meta-analysis. *Ann Intern Med.* 2012;156 (9):635–648.
132. Huang Z, Hankinson SE, Colditz GA, et al. Dual effects of weight and weight gain on breast cancer risk. *JAMA.* 1997;278(17):1407–1411.
133. van den Brandt PA, Spiegelman D, Yaun SS, et al. Pooled analysis of prospective cohort studies on height, weight, and breast cancer risk. *Am J Epidemiol.* 2000;152(6):514–527.
134. Reeves GK, Pirie K, Beral V, et al. Cancer incidence and mortality in relation to body mass index in the million women study: cohort study. *BMJ.* 2007;335(7630):1134.
135. Suzuki R, Orsini N, Saji S, Key TJ, Wolk A. Body weight and incidence of breast cancer defined by estrogen and progesterone receptor status—a meta-analysis. *Int J Cancer.* 2009;124(3):698–712.
136. Vrieling A, Buck K, Kaaks R, Chang-Claude J. Adult weight gain in relation to breast cancer risk by estrogen and progesterone receptor status: a meta-analysis. *Breast Cancer Res Treat.* 2010;123(3):641–649.
137. Rosner B, Eliassen AH, Toriola AT, et al. Weight and weight changes in early adulthood and later breast cancer risk. *Int J Cancer.* 2017;140(9):2003–2014.
138. Ahn J, Schatzkin A, Lacey Jr JV et al. Adiposity, adult weight change, and postmenopausal breast cancer risk. *Arch Intern Med.* 2007;167(19):2091–2102.
139. Eliassen AH, Colditz GA, Rosner B, Willett WC, Hankinson SE. Adult weight change and risk of postmenopausal breast cancer. *JAMA.* 2006;296(2):193–201.
140. Hamajima N, Hirose K, Tajima K, et al. Alcohol, tobacco and breast cancer—collaborative reanalysis of individual data from 53 epidemiological studies, including 58,515 women with breast cancer and 95,067 women without the disease. *Br J Cancer.* 2002;87 (11):1234–1245.
141. Allen NE, Beral V, Casabonne D, et al. Moderate alcohol intake and cancer incidence in women. *J Natl Cancer Inst.* 2009;101(5):296–305.
142. Ibrahim EM, Abouelkhair KM, Kazkaz GA, Elmasri OA, Al-Foheidi M. Risk of second breast cancer in female Hodgkin's lymphoma survivors: a meta-analysis. *BMC Cancer.* 2012;12:197.
143. Koo E, Henderson MA, Dwyer M, Skandarajah AR. Management and prevention of breast cancer after radiation to the chest for childhood, adolescent, and young adulthood malignancy. *Ann Surg Oncol.* 2015;22(suppl 3):S545–S551.
144. Moskowitz CS, Chou JF, Wolden SL, et al. Breast cancer after chest radiation therapy for childhood cancer. *J Clin Oncol.* 2014;32(21):2217–2223.
145. Guibout C, Adjadj E, Rubino C, et al. Malignant breast tumors after radiotherapy for a first cancer during childhood. *J Clin Oncol.* 2005;23(1):197–204.
146. Inskip PD, Sigurdson AJ, Veiga L, et al. Radiation-related new primary solid cancers in the childhood cancer survivor study: comparative radiation dose response and modification of treatment effects. *Int J Radiat Oncol Biol Phys.* 2016;94(4):800–807.
147. Collaborative Group on Hormonal Factors in Breast Cancer.Breast cancer and hormone replacement therapy: collaborative reanalysis of data from 51 epidemiological studies of 52,705 women with breast cancer and 108,411 women without breast cancer. *Lancet.* 1997;350(9084):1047–1059.

148. Banks E, Beral V, Reeves G. Million Women Study Collaborators.Published results on breast cancer and hormone replacement therapy in the million women study are correct. *Climacteric.* 2004;7(4):415–416.
149. Beral V. Million Women Study Collaborators.Breast cancer and hormone-replacement therapy in the million women study. *Lancet.* 2003;362(9382):419–427.
150. Colditz GA. Decline in breast cancer incidence due to removal of promoter: combination estrogen plus progestin. *Breast Cancer Res.* 2007;9(4):108.
151. Chlebowski RT, Hendrix SL, Langer RD, et al. Influence of estrogen plus progestin on breast cancer and mammography in healthy postmenopausal women: the Women's Health Initiative randomized trial. *JAMA.* 2003;289(24):3243–3253.
152. Collaborative Group on Hormonal Factors in Breast Cancer.Familial breast cancer: collaborative reanalysis of individual data from 52 epidemiological studies including 58,209 women with breast cancer and 101,986 women without the disease. *Lancet.* 2001;358(9291):1389–1399.
153. Hormone replacement therapy and cancer. *Gynecol Endocrinol.* 2001;15(6):453–465.
154. Narod SA. Hormone replacement therapy and the risk of breast cancer. *Nat Rev Clin Oncol.* 2011;8(11):669–676.
155. Chlebowski RT, Aragaki AK, Anderson GL. Menopausal hormone therapy influence on breast cancer outcomes in the Women's Health Initiative. *J Natl Compr Canc Netw.* 2015;13(7):917–924.
156. Lambertini M, Santoro L, Del Mastro L, et al. Reproductive behaviors and risk of developing breast cancer according to tumor subtype: a systematic review and meta-analysis of epidemiological studies. *CancerTreat Rev.* 2016;49:65–76.
157. Kobayashi S, Sugiura H, Ando Y, et al. Reproductive history and breast cancer risk. *Breast Cancer.* 2012;19(4):302–308.
158. Lipworth L, Bailey LR, Trichopoulos D. History of breast-feeding in relation to breast cancer risk: a review of the epidemiologic literature. *J Natl Cancer Inst.* 2000;92 (4):302–312.
159. Newcomb PA, Storer BE, Longnecker MP, et al. Lactation and a reduced risk of premenopausal breast cancer. *N Engl J Med.* 1994;330(2):81–87.
160. Mustacchi P. Ramazzini and Rigoni-Stern on parity and breast cancer. Clinical impression and statistical corroboration. *Arch Intern Med.* 1961;108:639–642.
161. Franco G. Bernardino Ramazzini and women workers' health in the second half of the XVIIth century. *J Public Health.* 2012;34(2):305–308.
162. Lilienfeld AM, Coombs J, Bross ID, Chamberlain A. Marital and reproductive experience in a community-wide epidemiological study of breast cancer. *Johns Hopkins Med J.* 1975;136(4):157–162.
163. Beral V. Parity and susceptibility to cancer. *Ciba Found Symp.* 1983;96:182–203.
164. Enger SM, Ross RK, Henderson B, Bernstein L. Breastfeeding history, pregnancy experience and risk of breast cancer. *Br J Cancer.* 1997;76(1):118–123.
165. Meier-Abt F, Bentires-Alj M. How pregnancy at early age protects against breast cancer. *Trends Mol Med.* 2014;20(3):143–153.
166. Janerich DT, Hoff MB. Evidence for a crossover in breast cancer risk factors. *Am J Epidemiol.* 1982;116(5):737–742.
167. Lambe M, Hsieh C, Trichopoulos D, Ekbom A, Pavia M, Adami HO. Transient increase in the risk of breast cancer after giving birth. *N Engl J Med.* 1994;331(1):5–9.
168. Lambe M, Hsieh CC, Tsaih SW, Ekbom A, Trichopoulos D, Adami HO. Parity, age at first birth and the risk of carcinoma in situ of the breast. *Int J Cancer.* 1998;77(3):330–332.
169. Schedin P. Pregnancy-associated breast cancer and metastasis. *Nat Rev Cancer.* 2006;6 (4):281–291.
170. Sinha DK, Pazik JE, Dao TL. Prevention of mammary carcinogenesis in rats by pregnancy: effect of full-term and interrupted pregnancy. *Br J Cancer.* 1988;57(4):390–394.

171. Russo J, Rivera R, Russo IH. Influence of age and parity on the development of the human breast. *Breast Cancer Res Treat.* 1992;23(3):211–218.
172. McCormick GM, Moon RC. Effect of pregnancy and lactation on growth of mammary tumours induced by 7,12-dimethlbenz(a) anthracene (DMBA). *Br J Cancer.* 1965;19: 160–166.
173. Swanson SM, Guzman RC, Collins G, et al. Refractoriness to mammary carcinogenesis in the parous mouse is reversible by hormonal stimulation induced by pituitary isografts. *Cancer Lett.* 1995;90(2):171–181.
174. Pike MC, Henderson BE, Casagrande JT, Rosario I, Gray GE. Oral contraceptive use and early abortion as risk factors for breast cancer in young women. *Br J Cancer.* 1981;43 (1):72–76.
175. Daling JR, Malone KE, Voigt LF, White E, Weiss NS. Risk of breast cancer among young women: relationship to induced abortion. *J Natl Cancer Inst.* 1994;86(21): 1584–1592.
176. Daling JR, Brinton LA, Voigt LF, et al. Risk of breast cancer among white women following induced abortion. *Am J Epidemiol.* 1996;144(4):373–380.
177. Michels KB, Willett WC. Does induced or spontaneous abortion affect the risk of breast cancer? *Epidemiology.* 1996;7(5):521–528.
178. Melbye M, Wohlfahrt J, Olsen JH, et al. Induced abortion and the risk of breast cancer. *N Engl J Med.* 1997;336(2):81–85.
179. Tang MT, Weiss NS, Malone KE. Induced abortion in relation to breast cancer among parous women: a birth certificate registry study. *Epidemiology.* 2000;11(2):177–180.
180. Goldacre MJ, Kurina LM, Seagroatt V, Yeates D. Abortion and breast cancer: a case-control record linkage study. *J Epidemiol Community Health.* 2001;55(5):336–337.
181. Erlandsson G, Montgomery SM, Cnattingius S, Ekbom A. Abortions and breast cancer: record-based case-control study. *Int J Cancer.* 2003;103(5):676–679.
182. Michels KB, Xue F, Colditz GA, Willett WC. Induced and spontaneous abortion and incidence of breast cancer among young women: a prospective cohort study. *Arch Intern Med.* 2007;167(8):814–820.
183. Huggins C, Moon RC, Morii S. Extinction of experimental mammary cancer. I. Estradiol-17beta and progesterone. *Proc Natl Acad Sci USA.* 1962;48:379–386.
184. Grubbs CJ, Peckham JC, McDonough KD. Effect of ovarian hormones on the induction of 1-methyl-1-nitrosourea-induced mammary cancer. *Carcinogenesis.* 1983;4(4): 495–497.
185. Guzman RC, Yang J, Rajkumar L, Thordarson G, Chen X, Nandi S. Hormonal prevention of breast cancer: mimicking the protective effect of pregnancy. *Proc Natl Acad Sci USA.* 1999;96(5):2520–2525.
186. Rajkumar L, Guzman RC, Yang J, Thordarson G, Talamantes F, Nandi S. Short-term exposure to pregnancy levels of estrogen prevents mammary carcinogenesis. *Proc Natl Acad Sci USA.* 2001;98(20):11755–11759.
187. Rajkumar L, Kittrell FS, Guzman RC, Brown PH, Nandi S, Medina D. Hormone-induced protection of mammary tumorigenesis in genetically engineered mouse models. *Breast Cancer Res.* 2007;9(1):R12.
188. Rajkumar L, Canada A, Esparza D, Collins K, Moreno E, Duong H. Decreasing hormonal promotion is key to breast cancer prevention. *Endocrine.* 2009;35(2): 220–226.
189. Russo IH, Koszalka M, Russo J. Human chorionic gonadotropin and rat mammary cancer prevention. *J Natl Cancer Inst.* 1990;82(15):1286–1289.
190. Russo IH, Koszalka M, Gimotty PA, Russo J. Protective effect of chorionic gonadotropin on DMBA-induced mammary carcinogenesis. *Br J Cancer.* 1990;62(2):243–247.
191. Russo IH, Koszalka M, Russo J. Effect of human chorionic gonadotropin on mammary gland differentiation and carcinogenesis. *Carcinogenesis.* 1990;11(10):1849–1855.

192. Russo I, Russo J. Role of HCG and inhibin in breast-cancer (review). *Int J Oncol.* 1994;4 (2):297–306.
193. Srivastava P, Russo J, Russo IH. Chorionic gonadotropin inhibits rat mammary carcinogenesis through activation of programmed cell death. *Carcinogenesis.* 1997;18(9): 1799–1808.
194. Belitskaya-Levy I, Zeleniuch-Jacquotte A, Russo J, et al. Characterization of a genomic signature of pregnancy identified in the breast. *Cancer Prev Res.* 2011;4(9):1457–1464.
195. Angelucci C, Maulucci G, Lama G, et al. Epithelial-stromal interactions in human breast cancer: effects on adhesion, plasma membrane fluidity and migration speed and directness. *PLoS One.* 2012;7(12):e50804.
196. Shekhar MP, Werdell J, Santner SJ, Pauley RJ, Tait L. Breast stroma plays a dominant regulatory role in breast epithelial growth and differentiation: implications for tumor development and progression. *Cancer Res.* 2001;61(4):1320–1326.
197. Wiseman BS, Werb Z. Stromal effects on mammary gland development and breast cancer. *Science.* 2002;296(5570):1046–1049.
198. Polyak K, Kalluri R. The role of the microenvironment in mammary gland development and cancer. *Cold Spring Harb Perspect Biol.* 2010;2(11):a003244.
199. Elenbaas B, Weinberg RA. Heterotypic signaling between epithelial tumor cells and fibroblasts in carcinoma formation. *Exp Cell Res.* 2001;264(1):169–184.
200. Wang TN, Albo D, Tuszynski GP. Fibroblasts promote breast cancer cell invasion by upregulating tumor matrix metalloproteinase-9 production. *Surgery.* 2002;132(2): 220–225.
201. Casey T, Bond J, Tighe S, et al. Molecular signatures suggest a major role for stromal cells in development of invasive breast cancer. *Breast Cancer Res Treat.* 2009;114(1): 47–62.
202. Hu M, Yao J, Carroll DK, et al. Regulation of in situ to invasive breast carcinoma transition. *Cancer Cell.* 2008;13(5):394–406.
203. Orimo A, Gupta PB, Sgroi DC, et al. Stromal fibroblasts present in invasive human breast carcinomas promote tumor growth and angiogenesis through elevated SDF-1/CXCL12 secretion. *Cell.* 2005;121(3):335–348.
204. Kunz-Schughart LA, Knuechel R. Tumor-associated fibroblasts (part II): functional impact on tumor tissue. *Histol Histopathol.* 2002;17(2):623–637.
205. Lebret SC, Newgreen DF, Thompson EW, Ackland ML. Induction of epithelial to mesenchymal transition in PMC42-LA human breast carcinoma cells by carcinoma-associated fibroblast secreted factors. *Breast Cancer Res.* 2007;9(1):R19.
206. Kleer CG, Bloushtain-Qimron N, Chen YH, et al. Epithelial and stromal cathepsin K and CXCL14 expression in breast tumor progression. *Clin Cancer Res.* 2008;14 (17):5357–5367.
207. Tyan SW, Kuo WH, Huang CK, et al. Breast cancer cells induce cancer-associated fibroblasts to secrete hepatocyte growth factor to enhance breast tumorigenesis. *PLoS One.* 2011;6(1):e15313.
208. Akhurst RJ, Balmain A. Genetic events and the role of TGF beta in epithelial tumour progression. *J Pathol.* 1999;187(1):82–90.
209. Giulianelli S, Cerliani JP, Lamb CA, et al. Carcinoma-associated fibroblasts activate progesterone receptors and induce hormone independent mammary tumor growth: a role for the FGF-2/FGFR-2 axis. *Int J Cancer.* 2008;123(11):2518–2531.
210. Singer C, Rasmussen A, Smith HS, Lippman ME, Lynch HT, Cullen KJ. Malignant breast epithelium selects for insulin-like growth factor II expression in breast stroma: evidence for paracrine function. *Cancer Res.* 1995;55(11):2448–2454.
211. LeBedis C, Chen K, Fallavollita L, Boutros T, Brodt P. Peripheral lymph node stromal cells can promote growth and tumorigenicity of breast carcinoma cells through the release of IGF-I and EGF. *Int J Cancer.* 2002;100(1):2–8.

212. Montel V, Mose ES, Tarin D. Tumor-stromal interactions reciprocally modulate gene expression patterns during carcinogenesis and metastasis. *Int J Cancer.* 2006;119 (2):251–263.
213. Santos RP, Benvenuti TT, Honda ST, et al. Influence of the interaction between nodal fibroblast and breast cancer cells on gene expression. *Tumour Biol.* 2011;32(1):145–157.
214. Buess M, Nuyten DS, Hastie T, Nielsen T, Pesich R, Brown PO. Characterization of heterotypic interaction effects in vitro to deconvolute global gene expression profiles in cancer. *Genome Biol.* 2007;8(9):R191.
215. Eliassen AH, Tworoger SS, Hankinson SE. Reproductive factors and family history of breast cancer in relation to plasma prolactin levels in premenopausal and postmenopausal women. *Int J Cancer.* 2007;120(7):1536–1541.
216. Welsch CW, Nagasawa H. Prolactin and murine mammary tumorigenesis: a review. *Cancer Res.* 1977;37(4):951–963.
217. Bridges RS, Hammer Jr RP. Parity-associated alterations of medial preoptic opiate receptors in female rats. *Brain Res.* 1992;578(1–2):269–274.
218. Mann PE, Bridges RS. Neural and endocrine sensitivities to opioids decline as a function of multiparity in the rat. *Brain Res.* 1992;580(1–2):241–248.
219. Rose DP, Gottardis M, Noonan JJ. Rat mammary carcinoma regressions during suppression of serum growth hormone and prolactin. *Anticancer Res.* 1983;3(5): 323–325.
220. Swanson SM, Unterman TG. The growth hormone-deficient spontaneous dwarf rat is resistant to chemically induced mammary carcinogenesis. *Carcinogenesis.* 2002;23 (6):977–982.
221. Dorgan JF, Reichman ME, Judd JT, et al. Relationships of age and reproductive characteristics with plasma estrogens and androgens in premenopausal women. *Cancer Epidemiol Biomarkers Prev.* 1995;4(4):381–386.
222. Yang J, Yoshizawa K, Nandi S, Tsubura A. Protective effects of pregnancy and lactation against *N*-methyl-*N*-nitrosourea-induced mammary carcinomas in female Lewis rats. *Carcinogenesis.* 1999;20(4):623–628.
223. McGuire WL. Steroid receptors in human breast cancer. *Cancer Res.* 1978;38(11 pt 2): 4289–4291.
224. Leavitt WW, Chen TJ, Allen TC. Regulation of progesterone receptor formation by estrogen action. *Ann NY Acad Sci.* 1977;286:210–225.
225. Horwitz KB, McGuire WL. Predicting response to endocrine therapy in human breast cancer: a hypothesis. *Science.* 1975;189(4204):726–727.
226. Osborne CK, Yochmowitz MG, Knight 3rd WA, McGuire WL. The value of estrogen and progesterone receptors in the treatment of breast cancer. *Cancer.* 1980;46 (12 suppl):2884–2888.
227. Osborne CK, Knight 3rd WA, Yochmowitz MG, McGuire WL. Modern approaches to the treatment of breast cancer. *Blood.* 1980;56(5):745–752.
228. Gross GE, Clark GM, Chamness GC, McGuire WL. Multiple progesterone receptor assays in human breast cancer. *Cancer Res.* 1984;44(2):836–840.
229. Ravdin PM, Green S, Dorr TM, et al. Prognostic significance of progesterone receptor levels in estrogen receptor-positive patients with metastatic breast cancer treated with tamoxifen: results of a prospective southwest oncology group study. *J Clin Oncol.* 1992;10(8):1284–1291.
230. Elledge RM, Green S, Pugh R, et al. Estrogen receptor (ER) and progesterone receptor (PgR), by ligand-binding assay compared with ER, PgR and pS2, by immunohistochemistry in predicting response to tamoxifen in metastatic breast cancer: a Southwest Oncology Group study. *Int J Cancer.* 2000;89(2):111–117.
231. Rugo HS, Vidula N, Ma C. Improving response to hormone therapy in breast cancer: new targets, new therapeutic options. *Am Soc Clin Oncol Educ Book.* 2016;35:e40–e54.

232. Aupperlee MD, Smith KT, Kariagina A, Haslam SZ. Progesterone receptor isoforms A and B: temporal and spatial differences in expression during murine mammary gland development. *Endocrinology*. 2005;146(8):3577–3588.
233. Kass L, Durando M, Ramos JG, et al. Association of increased estrogen receptor beta2 expression with parity-induced alterations in the rat mammary gland. *J Steroid Biochem Mol Biol*. 2004;91(1–2):29–39.
234. Balogh GA, Heulings R, Mailo DA, et al. Genomic signature induced by pregnancy in the human breast. *Int J Oncol*. 2006;28(2):399–410.
235. Blakely CM, Stoddard AJ, Belka GK, et al. Hormone-induced protection against mammary tumorigenesis is conserved in multiple rat strains and identifies a core gene expression signature induced by pregnancy. *Cancer Res*. 2006;66(12):6421–6431.
236. Ewan KB, Oketch-Rabah HA, Ravani SA, Shyamala G, Moses HL, Barcellos-Hoff MH. Proliferation of estrogen receptor-alpha-positive mammary epithelial cells is restrained by transforming growth factor-beta1 in adult mice. *Am J Pathol*. 2005;167 (2):409–417.
237. Russo J, Russo IH. Cellular basis of breast cancer susceptibility. *Oncol Res*. 1999;11 (4):169–178.
238. Sivaraman L, Stephens LC, Markaverich BM, et al. Hormone-induced refractoriness to mammary carcinogenesis in Wistar-Furth rats. *Carcinogenesis*. 1998;19(9): 1573–1581.
239. Russo J, Moral R, Balogh GA, Mailo D, Russo IH. The protective role of pregnancy in breast cancer. *Breast Cancer Res*. 2005;7(3):131–142.
240. Russo J, Lynch H, Russo IH. Mammary gland architecture as a determining factor in the susceptibility of the human breast to cancer. *Breast J*. 2001;7(5):278–291.
241. Gillgrass A, Ashkar A. Stimulating natural killer cells to protect against cancer: recent developments. *Expert Rev Clin Immunol*. 2011;7(3):367–382.
242. Ginger MR, Gonzalez-Rimbau MF, Gay JP, Rosen JM. Persistent changes in gene expression induced by estrogen and progesterone in the rat mammary gland. *Mol Endocrinol*. 2001;15(11):1993–2009.
243. Kimble JE. Strategies for control of pattern formation in *Caenorhabditis elegans*. *Philos Trans R Soc Lond B*. 1981;295(1078):539–551.
244. Morrison SJ, Hemmati HD, Wandycz AM, Weissman IL. The purification and characterization of fetal liver hematopoietic stem cells. *Proc Natl Acad Sci USA*. 1995;92 (22):10302–10306.
245. Morrison M. A good collaboration is based on unique contributions from each side: assessing the dynamics of collaboration in stem cell science. *Life Sci Soc Policy*. 2017;13(1):7.
246. Tokunaga M, Land CE, Tokuoka S, Nishimori I, Soda M, Akiba S. Incidence of female breast cancer among atomic bomb survivors, 1950-1985. *Radiat Res*. 1994;138 (2):209–223.
247. Land CE, Tokunaga M, Koyama K, et al. Incidence of female breast cancer among atomic bomb survivors, Hiroshima and Nagasaki, 1950-1990. *Radiat Res*. 2003;160 (6):707–717.
248. Little MP, McElvenny DM. Male breast cancer incidence and mortality risk in the Japanese atomic bomb survivors—differences in excess relative and absolute risk from female breast cancer. *Environ Health Perspect*. 2017;125(2):223–229.
249. Siwko SK, Dong J, Lewis MT, Liu H, Hilsenbeck SG, Li Y. Evidence that an early pregnancy causes a persistent decrease in the number of functional mammary epithelial stem cells—implications for pregnancy-induced protection against breast cancer. *Stem Cells*. 2008;26(12):3205–3209.
250. Russo J, Balogh GA, Heulings R, et al. Molecular basis of pregnancy-induced breast cancer protection. *Eur J Cancer Prev*. 2006;15(4):306–342.

251. Trichopoulos D, Lagiou P, Adami HO. Towards an integrated model for breast cancer etiology: the crucial role of the number of mammary tissue-specific stem cells. *Breast Cancer Res.* 2005;7(1):13–17.
252. Wagner KU, Smith GH. Pregnancy and stem cell behavior. *J Mammary Gland Biol Neoplasia.* 2005;10(1):25–36.
253. Matulka LA, Triplett AA, Wagner KU. Parity-induced mammary epithelial cells are multipotent and express cell surface markers associated with stem cells. *Dev Biol.* 2007;303(1):29–44.
254. Smith GH, Medina D. A morphologically distinct candidate for an epithelial stem cell in mouse mammary gland. *J Cell Sci.* 1988;90(Pt 1):173–183.
255. Nandy SB, Subramani R, Rajamanickam V, et al. microRNA alterations in ALDH positive mammary epithelial cells: a crucial contributing factor towards breast cancer risk reduction in case of early pregnancy. *BMC Cancer.* 2014;14:644.
256. Jarde T, Dale T. Wnt signalling in murine postnatal mammary gland development. *Acta Physiol.* 2012;204(1):118–127.
257. Lyons TR, Schedin PJ, Borges VF. Pregnancy and breast cancer: when they collide. *J Mammary Gland Biol Neoplasia.* 2009;14(2):87–98.
258. Faupel-Badger JM, Arcaro KF, Balkam JJ, et al. Postpartum remodeling, lactation, and breast cancer risk: summary of a national cancer institute-sponsored workshop. *J Natl Cancer Inst.* 2013;105(3):166–174.

CHAPTER FOUR

Association of Cytokines and Chemokines in Pathogenesis of Breast Cancer

Jeronay King, Hina Mir, Shailesh Singh[1]
Morehouse School of Medicine, Atlanta, GA, United States
[1] Corresponding author. E-mail address: shsingh@msm.edu

Contents

Abstract

Breast cancer touches women's life worldwide. Expected outcome is not achieved due to molecular heterogeneity and complex biology despite substantial advancement in diagnosis, prevention and treatment of breast cancer. Patients with estrogen receptor, progesterone receptor, and human epidermal growth factor receptor 2 (Her2) positive tumors receive hormone ablation and Her2 directed therapy. While patients diagnosed with triple-negative breast cancer receive chemotherapy in both the early and advanced stages. However, chemotherapeutic efficacies are not the same in every patient, which has fostered a major effort to identify new targets to treat breast cancer. Positive therapeutic outcome after immune checkpoint inhibitors emphasizes the significance of the host immune system in breast cancer. Cancer develops in immune competent host wherein cytokines, while shaping the immune system, also serve as growth signals for cancer cells. The dynamics of cross talk between immune system

Progress in Molecular Biology and Translational Science, Volume 151
ISSN 1877-1173
http://dx.doi.org/10.1016/bs.pmbts.2017.07.003

and cancer cells mediated by cytokines and chemokines changes during cancer initiation, progression, and therapeutic interventions. Hence, better understanding of molecular footprint of cancer cells, as well as crosstalk between cancer cells and host immune system is needed to develop patient specific treatment and management of breast cancer.

1. INTRODUCTION

Breast cancer (BrCa) affects about one in eight women in their lifetime.[1] Etiology of BrCa is yet to be defined, however, lifestyle, genetic and environmental factors are often associated with this multifactorial disease.[2] BrCa is primarily classified by pathologists based on estrogen (ER+), progesterone (PR+), and human epidermal growth factor receptor 2 (HER 2) status. Moreover, the five molecular BrCa subtypes identified by gene expression profiling and immunohistochemistry include: (1) Luminal A representing ER+/PR+, of low-grade and low-Ki67 index (a proliferative marker); (2) Luminal B, ER+/PR+ of higher grade and proliferative index; (3) HER2+ with or without ER; and (4) the "basal-like," or triple negative that do not express any of the three receptors.[3] These hormone receptors offer therapeutic targets for decreasing tumor growth by inhibiting downstream survival pathways. However, owing to the heterogeneity of the disease treatment results in differential responses among different patients.[4] Other treatment options include: chemotherapy, immunotherapy, and radiation. Most commonly a combination of two or three drugs is used to effectively treat BrCa. However, in many cases even after greatly reducing the tumor these regimens fail due to resistance leading to recurrence underscored by epithelial to mesenchymal transition (EMT) and metastasis.[5,6] This happens because BrCa cells often adopt mechanisms to dodge the cytotoxic effects of multimodal regime, as well as the immune surveillance, consequentially allowing them to not only survive but also proliferate and metastasize. Cancer cells develop immune tolerance by leaning immunity toward TH2 phenotype, and suppressing innate and adaptive immune system by supporting expansion and function of suppressor phenotype. On a systemic scale, neoplastic cells avoid immune recognition by reducing antigen presentation and secreting immune suppressive cytokines and more.[7] Cancer cells that have successfully evaded the immune system continue to grow, essentially resulting in clonal expansion of resistant variants.[8] These are common characteristics of cancer development. It is crucial to delineate the dialogue between cancer and the host immune system to offer new treatment modality.

2. CYTOKINES IN BREAST CANCER

Cytokines are small proteins that play a role in cell-to-cell communication by paracrine or autocrine signaling. They were named based on their function and location, including lymphokines, which are produced by lymphocytes; interleukins, which target leukocytes signaling; interferons, which interfere with virion infections. In addition to these, there are adipokines, cytokines secreted by adipose tissue (such as, leptin, TNFα, and adiponectin). However, the basis of this nomenclature is no longer valid. As described later in this chapter these molecules are multifaceted and their role is not just limited to the functions mentioned above. Considering their significance, cytokines are well regulated not only at the level of transcription and translation, but also at posttranslational level. Mode of cytokine action and biological response depends on their concentration, localization, and the target cell type. Cytokines were thought to induce immune responses to foreign threats or inflammation and play prominent roles in human biology and diseases. However, they often serve as double-edged swords. Cytokines like IFNγ, IL-2, and TNF-β, produced by Type 1 T helper (TH1) cells, activate macrophages, and IL-4, IL-5, IL-10, and IL-13 produced by type 2 Th (TH2) cells inhibit macrophage functions.[9] Imbalance in cytokine can lead to either hyper or hypoactivation of immune system leading to autoimmune disease or immune suppressive state, which favors infections and neoplastic growth. Cancer cells communicate with the host primarily via cytokines and utilize this communication system to shape tumor microenvironment and promote metastasis by facilitating tumor dissemination, EMT, motility and invasion.[10] However, the complex network of myriad of cytokines often renders it difficult to understand if the observed change in cytokine production and function is a cause or a consequence of the neoplastic growth or immune response against cancer development. So, here we discuss both the promoting and suppressing roles of different cytokines in BrCa.

2.1 Role of Interleukins in Promoting Breast Cancer

The role of interleukins in BrCa has been very well appreciated.[11–15] Plasma levels of IL6 and IL8 (now also called CXCL8) in BrCa patients are higher compared to normal healthy donors and positively correlate with BrCa stage and mortality.[11,12,16] These two proangiogeniccytokines contribute to higher mortality due to their potential role in development of chemoresistance in BrCa.[13,17] IL6 has been shown to support BrCa cell survival and mamosphere formation in vitro emphasizing its clinical association with early and advanced disease; suggesting its role in pathogenesis of BrCa.[14,15,18] A variant in the IL6 gene (C>T at IL6 rs2069861) has been associated with BrCa in older women, while rs1800795 GG polymorphism is associated with increased risk of BrCa when in conjunction with central adiposity.[19,20] Estrogen and epidermal growth factor (EGF) influence level of IL8 in hormone positive BrCa tissue and cell lines.[21] However, there are elevated circulatory levels of IL8 reported in HER2+[22] and TNBC tumors.[23] Further studies probing the role of IL8 in different BrCa subtypes would be of interest.

Besides IL6 and IL8, IL1β also contributes to BrCa progression by promoting migratory potential and higher levels of IL1β in BrCa tissue correlate with disease relapse.[23–28] Furthermore, IL1β promotes BrCa by increasing vascular endothelial growth factor (VEGF) via HIF-1α.[24,25] It is proclaimed that IL4 also accelerates BrCa progression and its levels are higher in malignant tumors compared to benign and therefore, IL4 is associated with BrCa specific mortality.[29] Whereas IL7 that promotes growth and proliferation is associated with poor prognosis of BrCa.[30,31] A splice variant in IL7 (IL-7δ5) induces EMT and metastasis in BrCa.[32]

Tumor cells produce IL10 to progress by supporting suppressor T cell expansion and higher IL10 is associated with poor BrCa prognosis.[33,34] Polymorphisms in the IL10 promoter (-1082AA genotype) are associated with lymph node metastasis and larger tumors.[35] In addition to these interleukins, IL11 and IL17 also contribute in BrCa progression by supporting cell growth, proliferation and migration.[36–38] Other than promoting cell migration and proliferation, required for cancer cell to progress and metastasize, IL17A promotes tumor by enriching suppressor T cell in tumor microenvironment [39] (Fig. 1). Higher expression of IL13 was observed in BrCa tissue compared to adjacent tissue in patients with nodal involvement.[40] Role of IL13 specifically in BrCa is not well defined but higher IL13 at tumor site indicates dominant TH2 environment, which is known to support tumor growth. Additionally, inhibition of lung metastasis after IL13Rα2 depletion suggests its involvement in lung metastasis.[41] However, not all cytokines are tumor promoting, IL15 inhibits protumorigenic macrophages by activating natural killer (NK) cells and enhances efficacy of anti-Her2 therapy.[42] In addition, it also has been shown to promote tumor

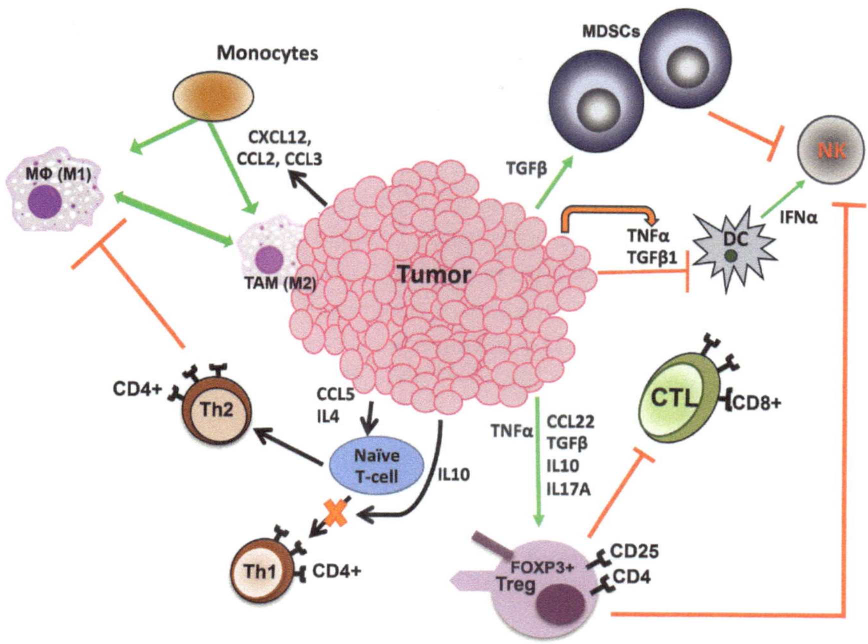

Fig. 1 Tumor-induced impairment of host immunity. Tumor-secreted cytokine and chemokine play a key role in TH1 versus TH2 differentiation and suppress innate and adaptive immune system by expanding suppressor cell phenotype. *MDSCs*, Myeloid derived suppressor cells; *NK*, natural killer; *TAM*, tumor-associated macrophages.

apoptosis and suppress tumor metastases.[43–45] Thus altered expression of interleukins have been correlated with BrCa risk, survival, and overall prognosis.

2.2 Transforming Growth Factorβ and Breast Cancer

Transforming growth factor beta (TGFβ) family of cytokines consists of three isoforms: β1, β2, and β3, which function via TGFβ receptor (TGFBR). In addition to its role in normal breast development, TGFβ also contribues in BrCa progression.[46,47] Autocrine and paracrine TGFβ signaling have been shown to play role in growth of BrCa.[48,49] Antitumor as well as tumor promoting role of TGFβ depends on the stage of disease it is activated. It prevents/delays BrCa progression, when activated in precancerous lesions by maintaining genomic stability, promoting apoptosis and cellular differentiation. However, it supports BrCa progression by promoting metastasis, angiogenesis and immune evasion if activated at late stage of the disease. TGFβ is higher in hormone independent BrCa while ER positive tumors show increase in TGFβ1 secretion followed by increase in TGFβ2 with antiestrogen therapy.[50] In addition to this, TGF-β also suppress NK cell function by activating suppresor cells i.e. regulatory T cell (Treg) and myeloid derived suppressor cells (MDSCs).[51]

Studies aimed to dissect dual effects of TGFβ on BrCa show that estrogen mediated transcriptional activities in BrCa is repressed by TGFβ. However, this effect is lost with over expression of Smad 3 or Smad 4 inhibition.[52] Therefore, it is often important to consider BrCa molecular subtyping when elucidating the roles of cytokines in progression. The multiple functions of TGFβ are distinguished by the three isoforms, all of which are elevated in advanced stage BrCa (TGFβ1,[53,54] TGFβ2[55] and TGFβ3[53]). TGFβ1 has been associated with better prognosis in triple negative breast cancer (TNBC).[56] This could be related to its ability to inhibit BrCa cell proliferation by increasing p21/waf1/cip (cell cycle inhibitor) nuclear accumulation.[57] TGFβ2 mRNA has been correlated with early relapse of BrCa[55] and TGFβ3 with worse prognosis.[53]

In vivo studies show TGFβ increases VEGF and CXCR4 expression[58], both of which are known to have proangiogenic effect. More specifically, TGFβ1 elevates VEGF expression in invasive BrCa cell lines, such effect is not seen in noninvasive cells.[59] Antagonizing TGFβ inhibits distant metastasis [60], while stable transfection of TGFβ results in tissue invasion and higher lung metastasis[49]. Reports show that it also enhances bone metastasis by stimulating IL11, connective tissue growth factor and parathyroid hormone-related protein (PTHrP).[60,61] Coculture of cancer-associated fibroblast (CAF) with BrCa cell lines transformed them to more aggressive phenotypes: increased migration and invasion with enhanced EMT via paracrine TGF-β signaling. These significant contributions of TGFβ to BrCa make it an attractive therapeutic target and potential biomarker.

2.3 Contribution of Interferon in Breast Cancer

Interferons are classified into three classes Type I (IFN-α, β, ε, κ, ω), Type II (IFN-γ), and Type III (IFN-λ) based on the differences in their properties, sequence, corresponding receptor, and cells producing them. Immune destruction in advanced stages of BrCa is partially contributed by defective interferon signaling (Fig. 1). Specifically IFN-α signaling in T and B cells and IFN-γ signaling in B cells is impaired.[62] IFN-α signaling defects in lymphocytes of early and advanced staged BrCa is associated with a decrease in five major interferon stimulated genes (STAT1, IFI44, IFIT1, IFIT2, and MX1). Also, advanced BrCa patients' blood has high-tumor-associated plasmacytoid dendritic cells (taPDC). Normally PDCs produce IFN-α in response to toll like receptors (TLRs), however, these taPDCs, which infiltrate advanced BrCa have a compromised IFN-α producing capacity resulting in a Treg favoring tumor environment.[63] Therefore, IFN therapies were designed for BrCa treatment, however not all gave satisfactory results. Combination therapies of various IFNs have been tested. IFN-β, but not -α, with tamoxifen gave promising outcomes in otherwise tamoxifen resistant patients. IFN-α2b in combination with high doses of Ionidamine and Epirubicin was also beneficial. Treatment of large BrCa patient cohort with poly (A:U) that stimulates production of cytokines, including IFN was associated with low toxicity and longer disease free survival. On the other hand, survival and resistance of BrCa cells against IFNB1 treatment is supported by IFNB1-induced autophagy.[64]

Studies show that IFN-γ produced by $CD8^+$ T cells promotes BrCa stem cells by causing Her2/neu antigen loss rendering the cancer cells resistance against Her2/neu targeted immunotherapy and relapse.[65] However, a correlation of IFNγ expression with favorable outcome in BrCa has also been reported.[66] This could be associated with inhibition of TH2 cell proliferation by IFNγ. Thus, interferon signaling is majorly affected in host immune system of BrCa patients. Careful restoration of appropriate IFN signaling, keeping in mind hormone receptor status of BrCa cells will affect the outcome of interferon therapies and immune clearance[67] (Table 1).

2.4 Tumor Necrosis Factor in Breast Cancer

This superfamily of cytokines was named after its capacity to kill tumor cells. Although it consists of other cytokines, role of TNFα is the most extensively addressed in cancer. Tissues of breast carcinoma have higher TNFα[23,68] and it has been shown to increase the migratory potential of BrCa cells. Coexpression of IL6 and TNFα have been associated with lymph node involvement and shorter survival,[70] while jointly RAS and TNFα increased angiogenesis in BrCa.[69] Lymph node invasion in obese BrCa patients[68,71] and decreased survival in HER2+[71] have been associated with higher TNFα. This is possibly because it synergizes with TGFβ1 to block IFNγ production by dendritic cells[72] (Fig. 1), which is required to prime NK cells. Studies also show that it influences the conversion of naïve T cells to Tregs.[75] Certain SNPs in TNFα have been associated with BrCa. Like the TT genotype of TNFSF10 (rs1131532) was correlated with low overall survival in patients with invasive BrCa.[76] The TNFα 308 polymorphism has been shown to be associated with high recurrences, metastases, and worst

Table 1 Cytokines Associated With BrCa Progression.

Cytokinez	Expression in Breast Cancer	Role in Breast Cancer
IL1β	Elevated[23]	Increases migratory potential;[24,25] associated with disease relapse[28]
IL6	Elevated[11]	Promotes migration,[26] angiogenesis,[13] survival and mamosphere formation;[17,18] correlates with poor prognosis[14–16]
IL7	—	Promotes growth and proliferation; associated with worse prognosis[30,31]
IL8 (CXCL8)	Elevated[12]	Promotes angiogensis[17]
IL10	—	Marker for poor prognosis[33,34]
IL11	—	Promotes migration, growth and prolifartion;[38] associated with bone metastasis[61]
IL13	Elevated[40]	Enhances lung metastasis[41]
IL15	—	Promotes apoptosis and reduces metastases,[43–45] stimulates NK activity;[42] enhances Trastuzumab[42]
IL17	—	Promotes proliferation, growth and migration;[36–38] promotes Tregs[39]
TNFα	Elevated[12,23,28,68]	Promotes angiogenesis;[69] associated with lymph node involvement;[70,71] impairs IFNα production by paDC[72]
TGFβ[a]	Elevated[73]	Promotes angiogenesis by increasing VEGF and CXCR4 expression;[58] promotes metastasis;[49,60] marker of early age of onset[73]
TGFβ1	Elevated[54]	Inhibits BrCa proliferation,[57] increases VEGF;[59] associated with better prognosis in TNBC[56]
TGFβ2	Elevated[55]	Correlated with early relapse[55]
TGFβ3	Elevated[74]	Correlated with worse prognosis[53]

BrCa, Breast cancer; EMT, epithelial to mesenchymal transition; TGFβ transforming growth factor beta; TNBC, triple-negative breast cancer.
[a]Levels of TGF β superfamily should be interpreted with respect to all three (TGFβ1, 2 and 3).

overall survival in BrCa patient.[77] Middle-aged women with nonsynonymous coding SNP TNFRSF1A (Rs767455 T>C) are at higher risk of developing BrCa.[19] Thus, TNFα along with TGFβ may work upstream of damaging IFNγ signaling in BrCa patients.

3. CHEMOKINES AND BREAST CANCER PATHOGENESIS

Chemokines make up the largest group of cytokines that primarily lead to chemotaxis of immune cells. About 50 different chemokines execute their function via 25 different 7 trans membrane G-protein coupled receptors. Some receptors are constitutively expressed maintaining homeostasis among tissues and cells, while some are induced. There are four subclasses based on the motif of their first two cysteines, including CXC, CC, C, or CX3C.[78] The CXC group that stimulates neutrophils is further divided based on presence or absence of the Glu-Leu-Arg (ELR) motif preceding the CXC domain.[79,80] The ELR+ chemokines are potent stimulators of angiogenesis, while ELR-chemokines are antiangiogenic.[81] The CC chemokines represented by two amino terminal cysteines primarily target monocytes and leukocytes.[79,82] The C chemokine subgroup, which consist of a single NH2-terminal cysteine is solely represented by lymphotactin/XCL1.[83,84] Lastly, the CX3C group (fractalkine) which has three amino acids between the first two cysteines, is a chemo-attractant for T cells and monocytes.

In addition to governing the immune cell trafficking between or within tissues, these chemotactic molecules play significant role in several biological processes and often regulate multiple signaling pathways. They can modulate host–tumor interactions and facilitate immune escape of cancer cells by recruiting Tregs and tumor-associated macrophages (TAMs) generating an immunosuppressive tumor microenvironment. Besides, they also support cancer cells in their metastatic journey toward secondary tumor sites.[85] This is evident from the fact that multiple chemokines are upregulated in various cancers and are prominent tumor promoters[86–89] (Fig. 2). Some chemokines that are pathologically relevant to BrCa are shown in Tables 2 and 3.

3.1 CXC Chemokine/Receptor in Breast Cancer

3.1.1 CXCR4/CXCR7: CXCL12

Members of the CXC family of chemokines are highly studied in cancer, including BrCa.[133,134] Contribution of CXCL12, a natural ligand for CXCR4

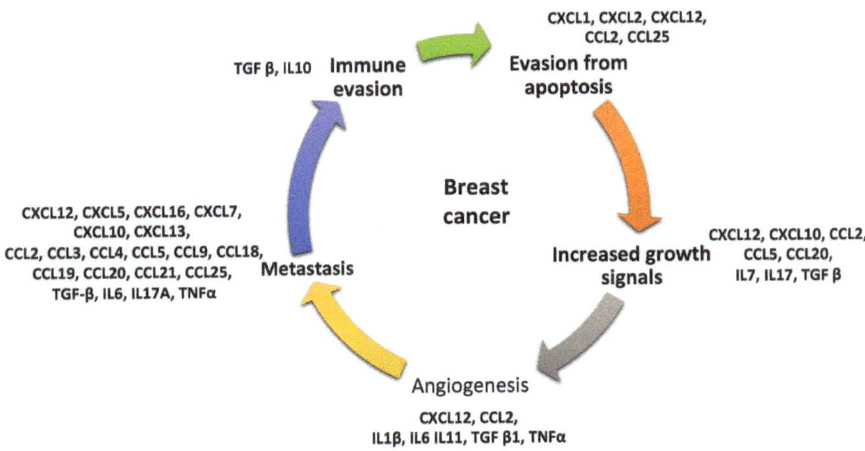

Fig. 2 Chemokines and cytokines in BrCa progression. Contribution of chemokines and cytokines in key steps of BrCa progression and metastasis.

and CXCR7, in BrCa pathogenesis is evident from the fact that CXCL12 is highly expressed in cancer associated firbroblast (CAFs) compared to fibroblast obtained from non cancerous region. Additionally, frequency of MCF7 xenograft tumor is higher when they are mixed with CAFs.[104] CXCL12

Table 2 Expression and Significance of CXC Chemokines in BrCa.

CXC Chemokine	Expression in Breast Cancer	Role in Breast Cancer
CXCL1	Elevated[90]	Increases chemoresistance;[90] stromal CXCL1 is a poor prognostic marker[91]
CXCL2	Elevated[90]	Increases chemoresistance[90]
CXCL3	Elevated[92]	Associated with metastasis[92]
CXCL5	—	Increases migration and invasion[93]
CXCL9	Elevated[23]	Associated with development of BrCa[23]
CXCL10	Elevated[23]	Promotes growth, motility [23],[94] and proliferation[95]
CXCL12	—	Increases invasion[96], mamosphere formation,[97] metastasis,[98–102] chemoresistance;[103] CXCL12 production by stromal fibroblasts promotes angiogenesis[104]
CXCL13	Elevated[92]	Associated with metastasis[92]
CXCL16	Elevated	Transmembrane CXCL16 suppresses migration and induces apoptosis;[105] soluble CXCL16 increases migration and invasion[106]

Table 3 Association of CC Chemokines in BrCa Progression.

CC-Chemokine	Expression in Breast Cancer	Role in Breast Cancer
CCL2	Elevated[23,28,107]	Increases survival and motility,[108] stem cell renewal;[109,110] recruits tumor promoting macrophages;[111,112] promotes angiogenesis;[113] negative regulator of autophagy and necrosis[107]
CCL5	Elevated[28,114]	Promotes proliferation[115–117] and invasion;[118] associated with advanced stage tumors[119,120]
CCL8	Elevated[121]	Promotes metastasis[121]
CCL9	—	Promotes invasion[118]
CCL18	Elevated[11,23]	Production by TAM promotes metastasis[11]
CCL19	Elevated[122,123]	Mediates metastasis;[122] associated with aggressive disease[123]
CCL20	—	Induces proliferation,[124–126] migration and invasion[124–126]
CCL21	Elevated[127]	Promotes invasion and metastasis[122,127]
CCL22	Elevated[23]	Recruits Tregs;[128–130] associated with susceptibility to BrCa[23]
CCL25	Elevated[131]	Promotes chemo-resistance,[132] migration and invasion[131]

produced by BrCa cells promotes tumor development in vivo by recruiting macrophages, which are responsible for intensifying CXCL12 mediated signaling and making tumor more invasive.[96] However, CXCL12 mediated response in immune deficient and immune competent BrCa model models are not the same. Higher CXCL12 at primary tumor site in syngenic BrCa model prevents metastasis and tumor growth by promoting high CTL response at tumor vicinity.[135] Where as low CXCL12 at primary site facilitates metastasis by creating CXCL12 gradient between primary and metastatic site.[136] As mentioned above, CXCL12 is ligand shared by CXCR4 and CXCR7 and its biological response in BrCa depends on which of the two receptors it binds and activates (Figs. 1 and 2).[137]

Chemokine receptor CXCR4 plays a role in progression of many different cancers[138] and its expression correlates with metastases, chemoresistance and overall worse prognosis in BrCa.[139–141] Activation of CXCR4 with CXCL12 supports BrCa progression in vitro and in vivo. It promotes mamosphere formation.[97] Expression and activation of CXCR4 promotes EMT, activates ERK1/2 and MMPs, which are required for BrCa cell to survive, migrate and invade.[100] Therefore, blocking or downregulating CXCR4 decreases metastases[98–102] and delays growth of BrCa cells in mice.[3,13] Inhibition of CXCR4-CXCL12 axis in CXCR4 expressing BrCa cells improves *in vivo* efficacy of Cisplatin by lowering p53 and BCL2.[103]

CXCR7 is another receptor that is activated by CXCL12 and is highly expressed in breast tumors.[142] Estrogen regulates this chemokine signaling network by repressing CXCR7 expression via demobilizing NFKβ at the promoter of the gene, while it enhances CXCL12 and CXCR4.[143] Overexpression of CXCR7 enhances BrCa progression by enhancing cell proliferation, growth as well as migration.[142,144–147] Targeting CXCR7 from tumor vascular endothelial, which express high CXCR7,[146] resulted in significantly greater recurrence, increased tumor cells and spontaneous metastasis in mice.[148] CXCR7 also regulates CXCR4 function by promoting intracellular accumulation and degradation of CXCL12 in BrCa cells.[149] Further studies are required to understand the precise influence of CXCR7 on CXCR4-CXCL12 signaling and vice versa.

3.1.2 CXCR6: CXCL16

The CXCR6 chemokine receptor has recently been implicated in cancer. Its ligand CXCL16 is membrane bound, which is released once cleaved by protease. Membrane and soluble form of CXCL16 have contrasting functions with respect to cancer progression. Chen *et al.* found that transmembranous CXCL16 suppresses BrCa migration and induces apoptosis,[105] while unpublished data from our lab shows elevated CXCR6 in BrCa tissues and cells (MCF-7 and MDA-MB-231) as compared to nonneoplastic tissue and immortalized breast epithelial cells (MCF-10a). Expression of disintegrin and metalloprotease 10 (ADAM10) is significantly elevated in MCF-7 and MDA-MB-231, and correlates with the increased levels of surface CXCL16. Treatment with CXCL16 increases BrCa cell metastatic potential via Src, FAK, and ERK1/2 kinases. These kinases are involved in F-actin polymerization, as well as underlie the invasive potential of BrCa cells. CXCL16/CXCR6 mediated BrCa progression by a pERK ½ dependent mechanism has also been confirmed by another group of researchers.[106] Thus, stimulation of CXCR6 with soluble CXCL16 facilitates cytoskeletal rearrangement during BrCa cell migration and invasion.

3.2 CC Chemokine/Receptor in Breast Cancer Pathogenesis

3.2.1 CCR9: CCL25

Our group was first to establish a role of CCR9 in breast, prostate, ovarian and lung cancer.[87,89,150–152] BrCa tissue express significantly higher CCR9 compared to nonneoplastic tissue. Similarly, CCR9 is highly expressed by the aggressive BrCa cell line (MDA-MD-231) compared to the less aggressive MCF-7. Stimulation of CCR9 with its ligand CCL25 promotes metastatic process by supporting cell migration, invasion and modulating MMP, expression required for invasion. Experiments by another group show increased migration and invasion of BrCa cells via regulation of certain EMT markers after stimulation with CCL25.[131,153] This chemokine and receptor play important role in clonal selection of T cell; T cells which express CCR9 have better survival advantage over T cell that do not express CCR9. We have established that CCR9 expressed in BrCa plays an important role in disease progression and outcome by supporting cell survival and inhibiting proapoptotic molecular cascade. Using CCR9 activation, BrCa cells activate PI3K pathway and develop resilience against Cisplatin-induced apoptosis.[132] (Fig. 2 and Table 3). Thus, in addition to supporting metastatic process, CCR9-CCL25 axis underlies the resistance of BrCa cells against Cisplatin treatment.

3.2.2 CCR2: CCL2

Another CC chemokine receptor axis implicated in the progression of BrCa is CCR2-CCL2.[154] Studies show that expression of CCL2 is enhanced in BrCa tissues and plasma when compared to healthy controls.[23,28,107] Moreover, CCL2 has been reported higher in tissues from luminal B tumors compared to tissues of luminal A (less invasive) tumors.[107] Silencing of CCL2 has been shown to decrease cancer stem cell renewal[109,110] and cell proliferation, while it increases autophagy and necrosis.[107] It has also been shown to induce BrCa survival and motility via Smad3 signaling.[108] It is key in recruitment of CCR2 expressing inflammatory monocytes in the tumor vicinity (Fig. 1),[111,112] which stimulates angiogenesis (Fig. 2).[113] This effect has been shown to be intensified by obesity[129] and estradiol.[155] Therefore, CCL2 is a poor prognostic marker for BrCa[156] that not only increases the metastatic proficiency, but could also play a role in stromal-tumor cell cross talk in early stage cancer[14], given the fact that fibroblast cocultured with BrCa cells produce higher CCL2.[110] Activation of CCR2 by CCL2 in TAMs leads to secretion of CCL3 that promotes seeding of BrCa cells at metastatic sites.[111] Like CCL2, CCL3 recruits tumor-promoting macrophages (Fig. 1). Thus, CCR2-CCL2/CCL3 contributes significantly to the tumor microenvironment landscape along with increasing metastatic capacity of BrCa cells.

3.2.3 CCR5: CCL5

In addition to CCL2, BrCa tissue express high level of CCL5, a natural ligand for CCR5, compared to normal matched tissues.[28,114] Higher expression of CCR5 is associated with advanced stage tumors.[114,119] In fact, the polymorphism of CCL5 403 G>A is reported a risk factor for BrCa.[157] Signaling of CCL5 through CCR5 enhances BrCa proliferation[115–117] and promotes invasion[118] (Table 3). Consequently, Maraviorc, a CCR5 antagonist, when used in BrCa was shown to cause decreased invasion and pulmonary metastasis[158] (Fig. 2).

3.3 Other Chemokines and Their Corresponding Receptors in Breast Cancer

Several other chemokines that are reported elevated in BrCa when compared to normal patients and involved in supporting molecular mechanisms cancer cells employ to grow and achieve metastatic goals are: CXCL1[90], CXCL2[90], CXCL3[92], CXCL7[159], CXCL9[23], CXCL10[23], CXCL11[23], CXCL13[160], CCL7[23], CCL8[121], CCL19[122] and CCL21.[122,127] Furthermore, studies show that BrCa migration and invasion is enhanced by: CXCL5[93], CXCL7[159], CXCL10[94], CXCL13[161], CCL9[118], CCL20[124–126] and CCL21.[122] Proliferation of BrCa cells is reported to be increased by CCL20[124,126] and CXCL10[95] (Tables 2 and 3). As discussed, some chemokines facilitate immune escape by recruitment of TAMs,[162] whereas others like CXCL1 and CXCL2[90] have shown to develop resistance to chemotherapy. It is proclaimed that expression of CXCL1[91] is associated with poor prognosis and blocking CXCR1 and CXCR2 (receptors for IL8) enhances the efficacy of HER2 directed therapy.[163] CXCL9 is another marker reported for risk of developing BrCa.[23]

While expression was reported to be low in normal breast tissues, CCR7 expression was high in tissues of breast carcinoma.[134] Studies show that CCR7 induces chemotactic and invasive responses in BrCa cells by promoting actin polymerization and pseudopodia formation.[134] One of the ligands for CCR7, CCL19 has been correlated with aggressive disease[123] and elevated in BrCa,[123] increased migration, invasion, and proliferation of BrCa cells.[127,128,164]

Plasma CCL18 in BrCa is also reported higher in patients with advanced stage when compared to early stage cancer.[11,23] Paracrine signaling of

CCL18 from TAMs increases metastases, consequently expression of CCL18 is associated with poor prognosis.[11] The chemokine CCL22 which is reported to be elevated in BrCa, has also been associated with susceptibility to BrCa[23] and Treg enrichment.[130,165,166] Polymorphism in CCL22 (rs223818 CC genotype and allele C), which increases its expression, is reported to be associated with susceptibility to BrCa.[167]

Chemokine decoy receptors modulate leukocyte recruitment in tumors, either by removing, transporting, or concentrating the ligands.[168] In BrCa chemokine decoy receptor D6 overexpression has been shown to inhibit CCL2, CCL5, and TAM–associated invasiveness. This positively correlates with disease-free survival.[169] Due to their antiinflammatory properties and based on these findings further studies investigating the antitumor roles of chemokine decoy receptors in BrCa would be interesting. Overwhelming literature on modulated chemokine signaling during BrCa points out the need to elucidate a master regulator of these crucial molecules to gain power to tame them.

4. CONCLUDING REMARKS

Plethora of evidence presented here, including in vivo and in vitro studies, acknowledge altered levels and functions of vital cytokines and chemokines in BrCa. Systemic and local pool of cytokines and chemokines expressed by cancer cells and immune cells ultimately determines the disease progression and prognosis. Therefore, regimes should be designed carefully after scrutinizing the site of change and, the cause and effect of manipulating these multifaceted and significant biomolecules to ensure maximum effectiveness and minimal side effect of the therapy,

REFERENCES

1. DeSantis CE, Fedewa SA, Goding Sauer A, Kramer JL, Smith RA, Jemal A. Breast cancer statistics, 2015: convergence of incidence rates between black and white women. *CancerJ Clin.* 2016;66:31–42.
2. Dignam JJ, Mamounas EP. Obesity and breast cancer prognosis: an expanding body of evidence. *Ann Oncol.* 2004;15:850–851.
3. Brisken C, Hess K, Jeitziner R. Progesterone and overlooked endocrine pathways in breast cancer pathogenesis. *Endocrinology.* 2015;156(10):3442–3450.
4. Leong AS, Zhuang Z. The changing role of pathology in breast cancer diagnosis and treatment. *Pathobiology.* 2011;78(2):99–114.
5. Kalluri R, Weinberg RA. The basics of epithelial-mesenchymal transition. *J Clin Invest.* 2009;119:1420–1428.

6. Weis SM, Cheresh DA. Tumor angiogenesis: molecular pathways and therapeutic targets. *Nat Med.* 2011;17:1359–1370.
7. Vinay DS, Ryan EP, Pawelec G, et al. Immune evasion in cancer: mechanistic basis and therapeutic strategies. *Semin Cancer Biol.* 2015;35:S185–S198.
8. Dunn GP, Bruce AT, Ikeda H, Old LJ, Schreiber RD. Cancer immunoediting: from immunosurveillance to tumor escape. *Nat Immunol.* 2002;3:991–998.
9. Romagnani S. Th1/Th2 cells. *Inflammatory bowel diseases.* 1999;5:285–294.
10. Yamaguchi H, Condeelis J. Regulation of the actin cytoskeleton in cancer cell migration and invasion. *Biochim Biophys Acta.* 2007;1773:642–652.
11. Narița D, Seclaman E, Ursoniu S, Ilina R, Cireap N, Anghel A. Expression of CCL18 and interleukin-6 in the plasma of breast cancer patients as compared with benign tumor patients and healthy controls. *Rom J Morphol Embryol.* 2011;52:1261–1267.
12. Razmkhah M, Jaberipour M, Hosseini A, Safaei A, Khalatbari B, Ghaderi A. Expression profile of IL-8 and growth factors in breast cancer cells and adipose-derived stem cells (ASCs) isolated from breast carcinoma. *Cell Immunol.* 2010;265:80–85.
13. Meads MB, Gatenby RA, Dalton WS, et al. Environment-mediated drug resistance: a major contributor to minimal residual disease. *Nat Rev Cancer.* 2009;9:665–674.
14. Labovsky V, Martinez LM, Davies KM, et al. Association between ligands and receptors related to the progression of early breast cancer in tumor epithelial and stromal cells. *Clin Breast Cancer.* 2015;15:e13–e21.
15. Sağlam Öul., Ünal ZS, Subaşı C, Ulukaya E, Karaöz E. IL-6 originated from breast cancer tissue-derived mesenchymal stromal cells may contribute to carcinogenesis. *Tumour Biol.* 2015;36:5667–5677.
16. Slattery ML, Herrick JS, Torres-Mejia G, et al. Genetic variants in interleukin genes are associated with breast cancer risk and survival in a genetically admixed population: the Breast Cancer Health Disparities Study. *Carcinogenesis.* 2014;35(8):1750–1759.
17. Azenshtein E, Meshel T, Shina S, Barak N, Keydar I, Ben-Baruch A. The angiogenic factors CXCL8 and VEGF in breast cancer: regulation by an array of pro-malignancy factors. *Cancer Lett.* 2005;217:73–86.
18. Sansone P, Storci G, Tavolari S, et al. IL-6 triggers malignant features in mammospheres from human ductal breast carcinoma and normal mammary gland. *J Clin Invest.* 2007;117:3988–4002.
19. Madeleine MM, Johnson LG, Malkki M, et al. Genetic variation in proinflammatory cytokines IL6, IL6R, TNF-region, and TNFRSF1A and risk of breast cancer. *Breast Cancer Res Treat.* 2011;129:887–899.
20. Slattery ML, Curtin K, Sweeney C, et al. Modifying effects of IL-6 polymorphisms on body size-associated breast cancer risk. *Obesity.* 2008;16:339–347.
21. Bendrik C, Dabrosin C. Estradiol increases IL-8 secretion of normal human breast tissue and breast cancer in vivo. *J Immunol.* 2009;182:371–378.
22. Vazquez-Martin A, Colomer R, Menendez JA. Her-2/neu-induced "Cytokine Signature" in Breast Cancer. *Adv Exp Med Biol.* 2008;617:311–319.
23. Narita D, Seclaman E, Anghel A, et al. Altered levels of plasma chemokines in breast cancer and their association with clinical and pathological characteristics. *Neoplasma.* 2016;63:141–149.
24. Filippi I, Carraro F, Naldini A. Interleukin-1 β affects MDAMB231 breast cancer cell migration under hypoxia: role of HIF-1 α and NF κ B transcription factors. *Mediators Inflamm.* 2015;2015:1–10.
25. Naldini A, Filippi I, Miglietta D, Moschetta M, Giavazzi R, Carraro F. Interleukin-1β regulates the migratory potential of MDAMB231 breast cancer cells through the hypoxia-inducible factor-1α. *Eur J Cancer.* 2010;46:3400–3408.
26. Arihiro K, Oda H, Kaneko M, Inai K. Cytokines facilitate chemotactic motility of breast carcinoma cells. *Breast Cancer.* 2000;7:221–230.

27. Verhasselt B, Van Damme J, van Larebeke N, et al. Interleukin-1 is a motility factor for human breast carcinoma cells in vitro: additive effect with interleukin-6. *Eur J Cell Biol.* 1992;59:449–457.
28. Soria G, Ofri-Shahak M, Haas I, et al. Inflammatory mediators in breast cancer: coordinated expression of TNFα andamp; IL-1β with CCL2 andamp; CCL5 and effects on epithelial-to-mesenchymal transition. *BMC Cancer.* 2011;11:130.
29. Rohani Borj M, Andalib AR, Mohammadi A, et al. Evaluation of IL-4, IL-17, and IFN-γ levels in patientswith breast cancer. *Int J Basic Sci Med.* 2017;2(1):20–24.
30. Al-Rawi MA, Rmali K, Mansel RE, Jiang WG. Interleukin 7 induces the growth of breast cancer cells through a wortmannin-sensitive pathway. *Br J Surg.* 2004;91 (1):61–68.
31. Al-Rawi MA, Rmali K, Watkins G, Mansel RE, Jiang WG. Aberrant expression of interleukin-7 (IL-7) and its signalling complex in human breast cancer. *Eur J Cancer.* 2004;40(4):494–502.
32. Yang J, Zeng Z, Peng Y, Chen J, Pan L, Pan D. IL-7 splicing variant IL-7δ5 induces EMT and metastasis of human breast cancer cell lines MCF-7 and BT-20 through activation of PI3K/Akt pathway. *Histochem Cell Biol.* 2014;142(4):401–410.
33. Bhattacharjee HK, Bansal VK, Nepal B, Srivastava S, Dinda AK, Misra MC. Is interleukin 10 (IL10) expression in breast cancer a marker of poor prognosis? *Indian J Surg Oncol.* 2016;7(3):320–325.
34. Zhao S, Wu D, Wu P, Wang Z, Huang J. Serum IL-10 predicts worse outcome in cancer patients: a meta-analysis. *PLoS One.* 2015;10(10):e0139598.
35. Kong F, Liu J, Liu Y, Song B, Wang H, Liu W. Association of interleukin-10 gene polymorphisms with breast cancer in a Chinese population. *J Exp Clin Cancer Res.* 2010;29(1):72.
36. Nam JS, Terabe M, Kang MJ, et al. Transforming growth factor beta subverts the immune system into directly promoting tumor growth through interleukin-17. *Cancer Res.* 2008;68(10):3915–3923.
37. Kim G, Khanal P, Lim SC, et al. Interleukin-17 induces AP-1 activity and cellular transformation via upregulation of tumor progression locus 2 activity. *Carcinogenesis.* 2013;34(2):341–350.
38. Cui M, Zhang Q, Qiu X, Jin Feng. The role of IL-11 and IL-17Ra in angiogenesis of breast cancer. *Int J Clin Exp Pathol.* 2016;9(11):11682–11687.
39. Benevides L, Cardoso CR, Tiezzi DG, Marana HR, Andrade JM, Silva JS. Enrichment of regulatory T cells in invasive breast tumor correlates with the upregulation of IL-17A expression and invasiveness of the tumor. *Eur J Immunol.* 2013;43(6):1518–1528.
40. Srabovici N, Mujagic Z, Mujanovic-Mustedanagic J, Muminovic Z, Softic A, Begic L. Interleukin 13 expression in the primary breast cancer tumour tissue. *Biochem Med.* 2011;21(2):131–138.
41. Papageorgis P, Ozturk S, Lambert AW, et al. Targeting IL13Ralpha2 activates STAT6-TP63 pathway to suppress breast cancer lung metastasis. *Breast Cancer Res.* 2015;17(1):98.
42. Wege AK, Weber F, Kroemer A, Ortmann O, Nimmerjahn F, Brockhoff G. IL-15 enhances the anti-tumor activity of trastuzumab against breast cancer cells but causes fatal side effects in humanized tumor mice (HTM). *Oncotarget.* 2017;8(2): 2731–2744.
43. Gillgrass AE, Chew MV, Krneta T, Ashkar AA. Overexpression of IL-15 promotes tumor destruction via NK1.1+ cells in a spontaneous breast cancer model. *BMC Cancer.* 2015;15:293.
44. Croci S, Nanni P, Palladini A, et al. Interleukin-15 is required for immunosurveillance and immunoprevention of HER2/neu-driven mammary carcinogenesis. *Breast Cancer Res.* 2015;17(1):70.

45. Gillgrass A, Gill N, Babian A, Ashkar AA. The absence or overexpression of IL-15 drastically alters breast cancer metastasis via effects on NK Cells, CD4 T cells, and macrophages. *J Immunol.* 2014;193(12):6184–6191.
46. Xu C, Wang Z, Cui R, et al. Co-expression of parathyroid hormone related protein and TGF-beta in breast cancer predicts poor survival outcome. *BMC Cancer.* 2015;15:925.
47. Moses H, Barcellos-Hoff MH. TGF—biology in mammary development and breast cancer. *Cold Spring Harbor Perspect Biol.* 2011;3:a003277.
48. Ammanamanchi S, Tillekeratne MPM, Ko TC, Brattain MG. Endogenous control of cell cycle progression by autocrine transforming growth factor beta in breast cancer cells. *Cancer Res.* 2004;64:2509–2515.
49. Tobin SW, Douville K, Benbow U, Brinckerhoff CE, Memoli VA, Arrick BA. Consequences of altered TGF-β expression and responsiveness in breast cancer: evidence for autocrine and paracrine effects. *Oncogene.* 2002;21:108–118.
50. Knabbe C, Kopp A, Hilgers W, et al. Regulation and role of TGF beta production in breast cancer. *Ann N Y Acad Sci.* 1996;784:263–276.
51. Bierie B, Moses HL. Transforming growth factor beta (TGF-β) and inflammation in cancer. *Cytokine Growth Factor Rev.* 2010;21:49–59.
52. Ren Y, Wu L, Frost AR, Grizzle W, Cao X, Wan M. Dual effects of TGF-β on ERα-mediated estrogenic transcriptional activity in breast cancer. *Mol Cancer.* 2009;8:111.
53. Auvinen P, Lipponen P, Johansson R, Syrjänen K. Prognostic significance of TGF-beta 1 and TGF-beta 2 expressions in female breast cancer. *Anticancer Res.* 1995;15: 2627–2631.
54. Ivanović V, Todorović-Raković N, Demajo M, et al. Elevated plasma levels of transforming growth factor-beta 1 (TGF-beta 1) in patients with advanced breast cancer: association with disease progression. *Eur J Cancer.* 2003;39:454–461.
55. Dave H, Trivedi S, Shah M, Shukla S. Transforming growth factor beta 2: a predictive marker for breast cancer. *Indian J Exp Biol.* 2011;49:879–887.
56. Panis C, Herrera AC, Victorino VJ, Aranome AMF, Cecchini R. Screening of circulating TGF-β levels and its clinicopathological significance in human breast cancer. *Anticancer Res.* 2013;33:737–742.
57. Mazars P, Barboule N, Baldin V, Vidal S, Ducommun B, Valette A. Effects of TGF-beta 1 (transforming growth factor-beta 1) on the cell cycle regulation of human breast adenocarcinoma (MCF-7) cells. *FEBS Lett.* 1995;362:295–300.
58. Dunn LK, Mohammad KS, Fournier PG, et al. Hypoxia and TGF-beta drive breast cancer bone metastases through parallel signaling pathways in tumor cells and the bone microenvironment. *PLoS One.* 2009;4(9):e6896.
59. Ganapathy V, Ge R, Grazioli A, et al. Targeting the transforming growth factor-β pathway inhibits human basal-like breast cancer metastasis. *Mol Cancer.* 2010;9:122.
60. Kang Y, Siegel PM, Shu W, et al. A multigenic program mediating breast cancer metastasis to bone. *Cancer Cell.* 2003;3:537–549.
61. Lindemann RK, Ballschmieter P, Nordheim A, Dittmer J. Transforming growth factor regulates parathyroid hormone-related protein expression in MDA-MB-231 breast cancer cells through a novel Smad/Ets synergism. *J Biol Chem.* 2001;276:46661–46670.
62. Critchley-Thorne RJ, Simons DL, Yan N, et al. Impaired interferon signaling is a common immune defect in human cancer. *Proc Natl Acad Sci USA.* 2009;106 (22):9010–9015.
63. Sisirak V, Faget J, Gobert M, et al. Impaired IFN-alpha production by plasmacytoid dendritic cells favors regulatory T-cell expansion that may contribute to breast cancer progression. *Cancer Res.* 2012;72(20):5188–5197.
64. Ambjørn M, Ejlerskov P, Liu Y, Lees M, Jäättelä M, Issazadeh-Navikas S. IFNB1/ interferon-β-induced autophagy in MCF-7 breast cancer cells counteracts its proapoptotic function. *Autophagy.* 2013;9(3):287–302.

65. Kmieciak M, Knutson K, Manjili M. The IFN-γ/IFN-γ Rα axis is involved in immunoediting of breast cancer that leads to the de novo generation of a novel breast cancer stem cell (BCSC)-like tumor variant and subsequent tumor recurrence (100.12). *J Immunol.* 2010;184(suppl 1):100.112.
66. Zhu X, Du L, Feng J, Ling Y, Xu S. Clinicopathological and prognostic significance of serum cytokine levels in breast cancer. *Clin Lab.* 2014;60:1145–1151.
67. Mostafa AA, Codner D, Hirasawa K, et al. Activation of ERα signaling differentially modulates IFN-γ induced HLA-class II expression in breast cancer cells. *PLoS One.* 2014;9(1):e87377.
68. Alokail MS, Al-Daghri NM, Mohammed AK, Vanhoutte P, Alenad A. Increased TNF α, IL-6 and ErbB2 mRNA expression in peripheral blood leukocytes from breast cancer patients. *Med Oncol.* 2014;31:38.
69. Leibovich-Rivkin T, Liubomirski Y, Meshel T, et al. The inflammatory cytokine TNFα cooperates with Ras in elevating metastasis and turns WT-Ras to a tumor-promoting entity in MCF-7 cells. *BMC Cancer.* 2014;14:158.
70. Tripsianis G, Papadopoulou E, Anagnostopoulos K, et al. Coexpression of IL-6 and TNF-α: prognostic significance on breast cancer outcome. *Neoplasma.* 2014;61: 205–212.
71. Tripsianis G, Papadopoulou E, Romanidis K, et al. Overall survival and clinicopathological characteristics of patients with breast cancer in relation to the expression pattern of HER-2, IL-6, TNF-α and TGF-β1. *Asian Pac j Cancer Prevent.* 2013;14: 6813–6820.
72. Sisirak V, Vey N, Goutagny N, et al. Breast cancer-derived transforming growth factor-β and tumor necrosis factor-α compromise interferon-α production by tumor-associated plasmacytoid dendritic cells. *Int J Cancer.* 2013;133:771–778.
73. Figueroa JD, Flanders KC, Garcia-Closas M, et al. Expression of TGF-β signaling factors in invasive breast cancers: relationships with age at diagnosis and tumor characteristics. *Breast Cancer Res Treat.* 2010;121:727–735.
74. Ghellal A, Li C, Hayes M, Byrne G, Bundred N, Kumar S. Prognostic significance of TGF beta 1 and TGF beta 3 in human breast carcinoma. *Anticancer Res.* 2000;20 (6b):4413–4418.
75. Tran DQ. TGF: the sword, the wand, and the shield of FOXP3+ regulatory T cells. *J Mol Cell Biol.* 2012;4:29–37.
76. Jung JH, Chae YS, Moon JH, et al. TNF superfamily gene polymorphism as prognostic factor in early breast cancer. *J Cancer Res Clin Oncol.* 2010;136:685–694.
77. Korobeinikova E, Myrzaliyeva D, Ugenskiene R, et al. The prognostic value of IL10 and TNF alpha functional polymorphisms in premenopausal early-stage breast cancer patients. *BMC Genet.* 2015;16:70.
78. Rossi D, Zlotnik A. The biology of chemokines and their receptors. *Annu Rev Immunol.* 2000;18:217–242.
79. Graves DT, Jiang Y. Chemokines, a family of chemotactic cytokines. *Crit Rev Oral Biol Med.* 1995;6:109–118.
80. Vandercappellen J, Van Damme J, Struyf S. The role of CXC chemokines and their receptors in cancer. *Cancer Lett.* 2008;267:226–244.
81. Strieter RM, Polverini PJ, Kunkel SL, et al. The functional role of the ELR motif in CXC chemokine-mediated angiogenesis. *J Biol Chem.* 1995;270:27348–27357.
82. Thorburn Nee Krasna E, Kolesar L, Brabcova E, et al. CXC and CC chemokines induced in human renal epithelial cells by inflammatory cytokines. *APMIS.* 2009;117:477–487.
83. Dong C, Chua A, Ganguly B, Krensky AM, Clayberger C. Glycosylated recombinant human XCL1/lymphotactin exhibits enhanced biologic activity. *J Immunol Methods.* 2005;302:136–144.

84. Volkman BF, Liu TY, Peterson FC. Chapter 3. Lymphotactin structural dynamics. *Methods Enzymol.* 2009;461:51–70.
85. Singh R, Lillard Jr JW, Singh S. Chemokines: key players in cancer progression and metastasis. *Front Biosci.* 2011;3:1569–1582.
86. El-Haibi CP, Sharma P, Singh R, et al. Differential G protein subunit expression by prostate cancer cells and their interaction with CXCR5. *Mol Cancer.* 2013;12:64.
87. Gupta P, Sharma PK, Mir H, et al. CCR9/CCL25 expression in non-small cell lung cancer correlates with aggressive disease and mediates key steps of metastasis. *Oncotarget.* 2014;5(20):10170–10179.
88. Singh R, Kapur N, Mir H, Singh N, Lillard Jr JW, Singh S. CXCR6-CXCL16 axis promotes prostate cancer by mediating cytoskeleton rearrangement via Ezrin activation and alphavbeta3 integrin clustering. *Oncotarget.* 2016;7(6):7343–7353.
89. Singh R, Stockard CR, Grizzle WE, Lillard Jr JW, Singh S. Expression and histopathological correlation of CCR9 and CCL25 in ovarian cancer. *Int J Oncol.* 2011;39 (2):373–381.
90. Feliciano P. CXCL1 and CXCL2 link metastasis and chemoresistance. *Nat Genet.* 2012;44(8):840.
91. Zou A, Lambert D, Yeh H, et al. Elevated CXCL1 expression in breast cancer stroma predicts poor prognosis and is inversely associated with expression of TGF-β signaling proteins. *BMC Cancer.* 2014;14(1):781.
92. See AL, Chong PK, Lu SY, Lim YP. CXCL3 is a potential target for breast cancer metastasis. *Curr Cancer Drug Targets.* 2014;14(3):294–309.
93. Hsu YL, Hou MF, Kuo PL, Huang YF, Tsai EM. Breast tumor-associated osteoblast-derived CXCL5 increases cancer progression by ERK/MSK1/Elk-1/Snail signaling pathway. *Oncogene.* 2013;32(37):4436–4447.
94. Jin WJ, Kim B, Kim D, et al. NF-kappaB signaling regulates cell-autonomous regulation of CXCL10 in breast cancer 4T1 cells. *Exp Mol Med.* 2017;49(2):e295.
95. Datta D, Flaxenburg JA, Laxmanan S, et al. Ras-induced modulation of CXCL10 and its receptor splice variant CXCR3-B in MDA-MB-435 and MCF-7 cells: relevance for the development of human breast cancer. *Cancer Res.* 2006;66(19):9509–9518.
96. Boimel PJ, Smirnova T, Zhou ZN, et al. Contribution of CXCL12 secretion to invasion of breast cancer cells. *Breast Cancer Research.* 2012;14:R23.
97. Ablett MP, O'Brien CS, Sims AH, Farnie G, Clarke RB. A differential role for CXCR4 in the regulation of normal versus malignant breast stem cell activity. *Oncotarget.* 2014;5:599–612.
98. Liu X, Xiao Q, Bai X, et al. Activation of STAT3 is involved in malignancy mediated by CXCL12-CXCR4 signaling in human breast cancer. *Oncol Rep.* 2014;32: 2760–2768.
99. Kang H, Watkins G, Parr C, Douglas-Jones A, Mansel RE, Jiang WG. Stromal cell derived factor-1: its influence on invasiveness and migration of breast cancer cells in vitro, and its association with prognosis and survival in human breast cancer. *Breast Cancer Res.* 2005;7:R402–R410.
100. Sobolik T, Su Y-j, Wells S, Ayers GD, Cook RS, Richmond A. CXCR4 drives the metastatic phenotype in breast cancer through induction of CXCR2 and activation of MEK and PI3K pathways. *Mol Biol Cell.* 2014;25:566–582.
101. Liang Z, Wu H, Reddy S, et al. Blockade of invasion and metastasis of breast cancer cells via targeting CXCR4 with an artificial microRNA. *Biochem Biophys Res Commun.* 2007;363(3):542–546.
102. Liang Z, Wu T, Lou H, et al. Inhibition of breast cancer metastasis by selective synthetic polypeptide against CXCR4. *Cancer Res.* 2004;64(12):4302–4308.
103. Liang S, Peng X, Li X, et al. Silencing of CXCR4 sensitizes triple-negative breast cancer cells to cisplatin. *Oncotarget.* 2015;6:1020–1030.

104. Orimo A, Gupta PB, Sgroi DC, et al. Stromal fibroblasts present in invasive human breast carcinomas promote tumor growth and angiogenesis through elevated SDF-1/CXCL12 secretion. *Cell.* 2005;121:335–348.
105. Fang Y, Henderson Jr FC, Yi Q, Lei Q, Li Y, Chen N. Chemokine CXCL16 expression suppresses migration and invasiveness and induces apoptosis in breast cancer cells. *Mediators Inflamm.* 2014;478641.
106. Xiao G, Wang X, Wang J, et al. CXCL16/CXCR6 chemokine signaling mediates breast cancer progression by pERK1/2-dependent mechanisms. *Oncotarget.* 2015;6(16):14165–14178.
107. Fang WB, Yao M, Jokar I, et al. The CCL2 chemokine is a negative regulator of autophagy and necrosis in luminal B breast cancer cells. *Breast Cancer Res Treat.* 2015;150:309–320.
108. Fang WB, Jokar I, Zou A, Lambert D, Dendukuri P, Cheng N. CCL2/CCR2 chemokine signaling coordinates survival and motility of breast cancer cells through Smad3 protein- and p42/44 mitogen-activated protein kinase (MAPK)-dependent mechanisms. *J Biol Chem.* 2012;287:36593–36608.
109. Fang WB, Yao M, Brummer G, et al. Targeted gene silencing of CCL2 inhibits triple negative breast cancer progression by blocking cancer stem cell renewal and M2 macrophage recruitment. *Oncotarget.* 2016;7(31):49349–49367.
110. Tsuyada A, Chow A, Wu J, et al. CCL2 mediates cross-talk between cancer cells and stromal fibroblasts that regulates breast cancer stem cells. *Cancer Res.* 2012;72:2768–2779.
111. Kitamura T, Qian B-Z, Soong D, et al. CCL2-induced chemokine cascade promotes breast cancer metastasis by enhancing retention of metastasis-associated macrophages. *J Exp Med.* 2015;212:1043–1059.
112. Qian B-Z, Li J, Zhang H, et al. CCL2 recruits inflammatory monocytes to facilitate breast-tumour metastasis. *Nature.* 2011;475:222–225.
113. Bonapace L, Coissieux M-M, Wyckoff J, et al. Cessation of CCL2 inhibition accelerates breast cancer metastasis by promoting angiogenesis. *Nature.* 2014;515:130–133.
114. Hartmann MC, Dwyer RM, Costello M, et al. Relationship between CCL5 and transforming growth factor-β1 (TGFβ1) in breast cancer. *Eur J Cancer.* 2011;47:1669–1675.
115. Kuang J-X, Wang W-X, Sun S-R, Wang W-R, Yao X-L. Effects of down-regulation of CC chemokine ligand 5 (CCL5) on proliferation of human breast cancer cells in vitro. *Zhonghua Zhong Liu za Zhi [Chin J Oncol].* 2011;33:174–177.
116. Gao D, Rahbar R, Fish EN. CCL5 activation of CCR5 regulates cell metabolism to enhance proliferation of breast cancer cells. *Open Biol.* 2016;6:160122.
117. Yasuhara R, Irié T, Suzuki K, et al. The β-catenin signaling pathway induces aggressive potential in breast cancer by up-regulating the chemokine CCL5. *Exp Cell Res.* 2015;338:22–31.
118. Swamydas M, Ricci K, Rego SL, Dreau D. Mesenchymal stem cell-derived CCL-9 and CCL-5 promote mammary tumor cell invasion and the activation of matrix metalloproteinases. *Cell Adhes Migr.* 2013;7(3):315–324.
119. Zhang Y, Yao F, Yao X, et al. Role of CCL5 in invasion, proliferation and proportion of CD44+/CD24- phenotype of MCF-7 cells and correlation of CCL5 and CCR5 expression with breast cancer progression. *Oncol Rep.* 2009;21:1113–1121.
120. Yaal-Hahoshen N, Shina S, Leider-Trejo L, et al. The chemokine CCL5 as a potential prognostic factor predicting disease progression in stage II breast cancer patients. *Clinical cancer research an official journal of the American Association for Cancer Research.* 2006;12:4474–4480.
121. Farmaki E, Chatzistamou I, Kaza V, Kiaris H. A CCL8 gradient drives breast cancer cell dissemination. *Oncogene.* 2016;35(49):6309–6318.

122. Wilson JL, Burchell J, Grimshaw MJ. Endothelins induce CCR7 expression by breast tumor cells via endothelin receptor A and hypoxia-inducible factor-1. *Cancer Res.* 2006;66(24):11802–11807.
123. Cassier PA, Treilleux I, Bachelot T, et al. Prognostic value of the expression of C-Chemokine Receptor 6 and 7 and their ligands in non-metastatic breast cancer. *BMC Cancer.* 2011;11:213.
124. Boyle ST, Faulkner JW, McColl SR, Kochetkova M. The chemokine receptor CCR6 facilitates the onset of mammary neoplasia in the MMTV-PyMT mouse model via recruitment of tumor-promoting macrophages. *Mol Cancer.* 2015;14(1):115.
125. Kim KY, Baek A, Park YS, et al. Adipocyte culture medium stimulates invasiveness of MDA-MB-231 cell via CCL20 production. *Oncol Rep.* 2009;22(6):1497–1504.
126. Marsigliante S, Vetrugno C, Muscella A. CCL20 induces migration and proliferation on breast epithelial cells. *J Cell Physiol.* 2013;228:1873–1883.
127. Cunningham HD, Shannon LA, Calloway PA, et al. Expression of the C-C chemokine receptor 7 mediates metastasis of breast cancer to the lymph nodes in mice. *Transl Oncol.* 2010;3(6):354–361.
128. Su ML, Chang TM, Chiang CH, et al. Inhibition of chemokine (C-C motif) receptor 7 sialylation suppresses CCL19-stimulated proliferation, invasion and anti-anoikis. *PLoS One.* 2014;9(6):e98823.
129. Arendt LM, McCready J, Keller PJ, et al. Obesity promotes breast cancer by CCL2-mediated macrophage recruitment and angiogenesis. *Cancer Res.* 2013;73: 6080–6093.
130. Ménétrier-Caux C, Faget J, Biota C, Gobert M, Blay J-Y, Caux C. Innate immune recognition of breast tumor cells mediates CCL22 secretion favoring Treg recruitment within tumor environment. *Oncoimmunology.* 2012;1(5):759–761.
131. Johnson-Holiday C, Singh R, Johnson E, et al. CCL25 mediates migration, invasion and matrix metalloproteinase expression by breast cancer cells in a CCR9-dependent fashion. *Int J Oncol.* 2011;38(5):1279–1285.
132. Johnson-Holiday C, Singh R, Johnson EL, et al. CCR9-CCL25 interactions promote cisplatin resistance in breast cancer cell through Akt activation in a PI3K-dependent and FAK-independent fashion. *World J Surg Oncol.* 2011;9:46.
133. Belperio JA, Keane MP, Arenberg DA, et al. CXC chemokines in angiogenesis. *J Leukoc Biol.* 2000;68:1–8.
134. Müller A, Homey B, Soto H, et al. Involvement of chemokine receptors in breast cancer metastasis. *Nature.* 2001;410:50–56.
135. Williams SA, Harata-Lee Y, Comerford I, Anderson RL, Smyth MJ, McColl SR. Multiple functions of CXCL12 in a syngeneic model of breast cancer. *Mol Cancer.* 2010;9:250.
136. Fridrichova I, Smolkova B, Kajabova V, et al. CXCL12 and ADAM23 hypermethylation are associated with advanced breast cancers. *Transl Res.* 2015;165:717–730.
137. Hernandez L, Magalhaes MA, Coniglio SJ, Condeelis JS, Segall JE. Opposing roles of CXCR4 and CXCR7 in breast cancer metastasis. *Breast Cancer Res.* 2011;13:R128.
138. Balkwill F. The significance of cancer cell expression of the chemokine receptor CXCR4. *Semin Cancer Biol.* 2004;14:171–179.
139. Hung C-S, Su H-Y, Liang H-H, et al. High-level expression of CXCR4 in breast cancer is associated with early distant and bone metastases. *Tumor Biol.* 2014;35:1581–1588.
140. De Luca A, D'Alessio A, Gallo M, Maiello M, Bode A, Normanno N. Src and CXCR4 are involved in the invasiveness of breast cancer cells with acquired resistance to lapatinib. *Cell Cycle.* 2014;13:148–156.
141. Sun Y, Mao X, Fan C, et al. CXCL12-CXCR4 axis promotes the natural selection of breast cancer cell metastasis. *Tumor Biol.* 2014;35:7765–7773.

142. Salazar N, Muñoz D, Kallifatidis G, Singh RK, Jordà M, Lokeshwar BL. The chemokine receptor CXCR7 interacts with EGFR to promote breast cancer cell proliferation. *Mol Cancer.* 2014;13:198.
143. Kerdivel G, Boudot A, Pakdel F. Estrogen represses CXCR7 gene expression by inhibiting the recruitment of NFκB transcription factor at the CXCR7 promoter in breast cancer cells. *Biochem Biophys Res Commun.* 2013;431:729–733.
144. Tang X, Li X, Li Z, et al. Downregulation of CXCR7 inhibits proliferative capacity and stem cell-like properties in breast cancer stem cells. *Tumor Biol.* 2016;37:13425–13433.
145. Luker KE, Lewin SA, Mihalko LA, et al. Scavenging of CXCL12 by CXCR7 promotes tumor growth and metastasis of CXCR4-positive breast cancer cells. *Oncogene.* 2012;31:4750–4758.
146. Miao Z, Luker KE, Summers BC, et al. CXCR7 (RDC1) promotes breast and lung tumor growth in vivo and is expressed on tumor-associated vasculature. *Proc Natl Acad Sci.* 2007;104:15735–15740.
147. Boudot A, Kerdivel G, Habauzit D, et al. Differential estrogen-regulation of CXCL12 chemokine receptors, CXCR4 and CXCR7, contributes to the growth effect of estrogens in breast cancer cells. *PLoS One.* 2011;6:e20898.
148. Stacer AC, Fenner J, Cavnar SP, et al. Endothelial CXCR7 regulates breast cancer metastasis. *Oncogene.* 2016;35:1716–1724.
149. Luker KE, Steele JM, Mihalko LA, Ray P, Luker GD. Constitutive and chemokine-dependent internalization and recycling of CXCR7 in breast cancer cells to degrade chemokine ligands. *Oncogene.* 2010;29:4599–4610.
150. Johnson EL, Singh R, Johnson-Holiday CM, et al. CCR9 interactions support ovarian cancer cell survival and resistance to cisplatin-induced apoptosis in a PI3K-dependent and FAK-independent fashion. *J Ovarian Res.* 2010;3:15.
151. Johnson EL, Singh R, Singh S, et al. CCL25-CCR9 interaction modulates ovarian cancer cell migration, metalloproteinase expression, and invasion. *World J Surg Oncol.* 2010;8:62.
152. Sharma PK, Singh R, Novakovic KR, Eaton JW, Grizzle WE, Singh S. CCR9 mediates PI3K/AKT-dependent antiapoptotic signals in prostate cancer cells and inhibition of CCR9-CCL25 interaction enhances the cytotoxic effects of etoposide. *Int J Cancer.* 2010;127(9):2020–2030.
153. Zhang Z, Sun T, Chen Y, et al. CCL25/CCR9 signal promotes migration and invasion in hepatocellular and breast cancer cell lines. *DNA Cell Biol.* 2016;35(7):348–357.
154. Deshmane SL, Kremlev S, Amini S, Sawaya BE. Monocyte chemoattractant protein-1 (MCP-1): an overview. *J Interferon Cytokine Res.* 2009;29:313–326.
155. Svensson S, Abrahamsson A, Rodriguez GV, et al. CCL2 and CCL5 are novel therapeutic targets for estrogen-dependent breast cancer. *Clin Cancer Res.* 2015;21:3794–3805.
156. Wang J, Zhuang Z-G, Xu S-F, et al. Expression of CCL2 is significantly different in five breast cancer genotypes and predicts patient outcome. *Int J Clin Exp Med.* 2015;8:15684–15691.
157. Eskandari-Nasab E, Hashemi M, Ebrahimi M, et al. Evaluation of CCL5-403 Gandgt;A and CCR5 Δ32 gene polymorphisms in patients with breast cancer. *Cancer Biomark.* 2014;14:343–351.
158. Velasco-Velazquez M, Jiao X, De La Fuente M, et al. CCR5 antagonist blocks metastasis of basal breast cancer cells. *Cancer Res.* 2012;72:3839–3850.
159. Tang Z, Yu M, Miller F, Berk RS, Tromp G, Kosir MA. Increased invasion through basement membrane by CXCL7-transfected breast cells. *Am J Surg.* 2008;196(5):690–696.
160. Panse J, Friedrichs K, Marx A, et al. Chemokine CXCL13 is overexpressed in the tumour tissue and in the peripheral blood of breast cancer patients. *Br J Cancer.* 2008;99(6):930–938.

161. Biswas S, Sengupta S, Roy Chowdhury S, et al. CXCL13-CXCR5 co-expression regulates epithelial to mesenchymal transition of breast cancer cells during lymph node metastasis. *Breast Cancer Res Treat.* 2014;143(2):265–276.
162. Elbaz M, Nasser MW, Ravi J, et al. Modulation of the tumor microenvironment and inhibition of EGF/EGFR pathway: novel anti-tumor mechanisms of Cannabidiol in breast cancer. *Mol Oncol.* 2015;9(4):906–919.
163. Singh JK, Farnie G, Bundred NJ, et al. Targeting CXCR1/2 significantly reduces breast cancer stem cell activity and increases the efficacy of inhibiting HER2 via HER2-dependent and -independent mechanisms. *Clin Cancer Res.* 2013;19:643–656.
164. Xu B, Zhou M, Qiu W, Ye J, Feng Q. CCR7 mediates human breast cancer cell invasion, migration by inducing epithelial-mesenchymal transition and suppressing apoptosis through AKT pathway. *Cancer Med.* 2017;6(5):1062–1071.
165. Freier CP, Kuhn C, Endres S, et al. FOXP3+ cells recruited by CCL22 into breast cancer correlates with less tumor nodal infiltration. *Anticancer Res.* 2016;36(6):3139–3145.
166. Li YQ, Liu FF, Zhang XM, Guo XJ, Ren MJ, Fu L. Tumor secretion of CCL22 activates intratumoral Treg infiltration and is independent prognostic predictor of breast cancer. *PLoS One.* 2013;8(10):e76379.
167. Jafarzadeh A, Fooladseresht H, Minaee K, et al. Higher circulating levels of chemokine CCL22 in patients with breast cancer: evaluation of the influences of tumor stage and chemokine gene polymorphism. *Tumour Biol.* 2015;36(2):1163–1171.
168. Bonecchi R, Savino B, Borroni EM, Mantovani A, Locati M. *Chemokine Decoy Receptors: Structure–Function and Biological Properties*. Berlin Heidelberg: Springer; 2010.
169. Wu F-Y, Ou Z-L, Feng L-Y, et al. Chemokine decoy receptor D6 plays a negative role in human breast cancer. *Mol Cancer Res.* 2008;6:1276–1288.

CHAPTER FIVE

Cancer Stem Cells and Metastasis

Sushmita Bose Nandy*, Rajkumar Lakshmanaswamy†,[1]
* Center of Emphasis in Cancer Research, Paul L. Foster School of Medicine, Texas Tech University Health Sciences Center El Paso, El Paso, TX, United States
† Center of Emphasis in Cancer Research, Paul L. Foster School of Medicine, Graduate School of Biomedical Sciences, Texas Tech University Health Sciences Center El Paso, El Paso, TX, United States
[1] Corresponding author. E-mail address: rajkumar.lakshmanaswamy@ttuhsc.edu

Contents

Progress in Molecular Biology and Translational Science, Volume 151
ISSN 1877-1173
http://dx.doi.org/10.1016/bs.pmbts.2017.07.007

Abstract

Decades of cancer research have led to substantial progress in the treatment of primary breast cancers. Despite of the advancement in this field, treating a metastatic disease has remained a mammoth task. One of the possible theories explaining metastatic disease involves the cancer stem cells (CSCs). CSCs have been shown to be an integral part of solid tumors. The process of metastasis involves the fine orchestration between CSCs and other microenvironmental factors. This chapter will provide an overview about the process of metastasis and the interactive role of CSCs with the components of the microenvironment in each phase of the metastatic cascade.

1. WHAT IS METASTASIS?

Metastasis is a Greek word meaning "displacement" (*Meta*, next; *stasis*, placement). Functionally, tumor cells begin the process of metastasis by invading the tissue around the site of the primary tumor. The tumor cells then enter the bloodstream directly or through the lymphatic system. Tumor cells, when they encounter a capillary bed, extravasate the bloodstream and land onto a distant tissue parenchyma, or the "foreign land." Almost 100 years ago, Paget described metastasis as the interaction of "seeds" (tumor cells) and "soil" (site of metastasis). The metastatic cascade can be broadly separated into three main phenotypic events: invasion, intravasation, and extravasation.

To facilitate the invasion of cancer cells, loss of cell–cell adhesion is vital. Loss of cell–cell adhesion is achieved by reduction in the levels of cell adhesion proteins. The loss of the tumor cells' capacity to adhere to each other helps them to dissociate from the primary tumor. In addition, loss of cell adhesion proteins also alters the cell–matrix adhesion properties, which further favors the process of invasion for the tumor cells. The extracellular matrix (ECM), a component of the tumor microenvironment, is a key player in the movement of tumor cells out of the primary site. The metastatic tumor cells attain the ability to release substances to degrade the matrix, which can facilitate the movement of tumor cells in the matrix.

Intravasation mainly indicates the infiltration of tumor cells into the blood vessel to finally enter the main blood circulation. The tumor cells perform the process of intravasation directly through the circulatory system or indirectly through the lymphatic system. Intravasation and extravasation through the blood circulatory system requires coordination between tumor angiogenesis and endothelial cells lining the blood vessels. The peripheral network of intratumoral blood vessels provides an easy route for the detached tumor cells to enter the blood circulation. The lymph nodes are the primary site for lymphatic drainage from all areas of the breast, and therefore the lymphatic system is important in metastasis. As the tumor's size increases, the interstitial fluid pressure increases and is released to be taken up by the lymphatic system because of its permeable nature. The flow rate in the lymphatic vessels is comparatively 100–500 times slower than blood circulation and is accompanied by lesser shearing force due to vasodilation, which makes the lymphatic system a superior route for tumor cell dissemination.

In the last phase of the metastatic cascade, these circulating tumor cells (CTCs) exit the circulatory system and land on the target organ parenchyma. Like the other steps in the whole process of metastasis, extravasation is a dynamic event that involves endothelial cells in the lining of the blood vessel. The CTCs are caught in the thinner capillaries closer to the tissues and connect to the endothelial cells, which help them migrate and successfully land on the tissue parenchymal.

The foreign tissue demonstrates congenial behavior toward the metastasized tumor cells and supports their colonization. Colonization to generate secondary tumor foci demands the clonogenic expansion of these metastasized tumor cells. The resulting deadly metastatic colonies are the root cause for cancer mortality. Specific cancers have affinity toward specific organs for metastasis.

2. THE CONCEPT OF CANCER STEM CELLS

Research on stem cells started in 1961 with the classical experiment conducted by Till and McCulloch, where they demonstrated that a single hematopoietic cell had multilineage differentiation potential while retaining its ability to self-renew. The properties of multilineage differentiation and self-renewal are often considered cardinal properties of stem cells. Self-renewal is a very important property in both malignant and nonmalignant cells. The biological process of cell division in stem cells is a key for maintaining the ability to self-renew. Upon cell division, a self-renewing stem cell can undergo two key cell division fates: (1) symmetric division, giving rise to two daughter stem cells and (2) asymmetric division, giving rise to a stem cell and a differentiated daughter cell. The gold standard for testing a cell's stemness is its ability to maintain long-term clonal growth in functional repopulation assays.

The molecular programs underlying the state of stem cell are very complicated and barely discernible. It is suggested that epigenetic states, modifiers, and transcription factors orchestrate to maintain self-renewing stem cells. In theory, both nonmalignant and malignant stem cells can self-renew; however, the capability is irregular in the latter. For many cancers, cancer stem cells (CSC) can be isolated from the bulk cell population and can have clonal long-term repopulation and self-renewing capacity—distinct features that define CSCs.[1] However, in some cancers, CSCs are indistinguishable from non-CSCs, and the bulk population of cells in those cancer cells shows traits of stemness. These tumors demonstrate more homogenous populations of cells and shallow hierarchy.

Isolation and identification of breast CSCs is a significant challenge, as no single marker can identify them. Surface markers have been used to identify CSCs, and these surface markers not only contribute to cell interactions but also provide them with unique properties. Breast CSCs were first identified and isolated using the $CD44^+/CD24^{-/low}$ Lin^- phenotype.[2] Since then, this phenotype has been used as a reliable phenotype for the isolation of breast CSCs.[3–6] Both CD44 and CD24, as expressed on cell surfaces, are glycoproteins. CD44 is actually a receptor for hyaluronan and contributes significantly toward increased breast cancer cell proliferation, angiogenesis,[7] adhesion, motion, migration, and invasion.[8] Furthermore, interaction of CD44 with osteopontin favors tumor progression.[9] CD24 expressed at low levels contributes to tumor progression and metastasis.[10]

The other recently recognized marker, aldehyde dehydrogenase (ALDH),[11] consists of a family of cytosolic enzymes involved in the oxidation of intracellular aldehydes and oxidizes retinol to retinoic acid during the differentiation of rudimentary stem cells.[12] ALDH1, the dominant form of the enzyme in mammals, mediates the conversion of retinaldehydes to retinoic acid.[13] The other markers that have been used to identify breast CSCs are CD133[4] and a $CD44^+$ $CD49f^{hi}$ $CD133/2^{hi}$ phenotype found in tumorigenic cells.[14] In vivo and in vitro studies have introduced CD49f[15] and CD61[16] as breast CSC markers as well.

Stem cell associated signaling pathways, such as Notch, Hedgehog, and Wnt have been implicated in therapy resistance and has been correlated with an increase in the number of breast CSCs during/after treatments.[17] These signaling pathways are key for embryonic development and adult tissue homeostasis.[18] Notch, Hedgehog, and Wnt signaling pathways are involved in self-renewal and maintenance of an undifferentiated state of stem cells.[19,20] Any dysregulation in these pathways, result in a CSC phenotype.

Hedgehog signaling pathway is an embryonic development associated pathway that positively influences Gli1 and Ptch1 thereby leading to breast CSC proliferation.[21] The Notch pathway is important for cell differentiation process during both embryogenesis and adulthood. Activation of notch pathway leads to proproliferative and antiapoptotic behavior in cancer cells.[22] In addition, it is also believed to be a positive modulator of breast CSCs.[23] This pathway has been reported to act in breast CSCs. Some of the genes targeted by Notch pathway include cyclinD1, c-myc, CDKN1A, and HES-related repressor protein. In addition to signaling pathways, transcriptional factors are also significant. Undifferentiated state of stem cells are mainly regulated by transcription factors like Sox2, Oct4, and Nanog.[24] Sox2 expressing cells in basal-like breast carcinomas account for almost 43%.[25] Sox4, another member of Sox family has been shown to be involved in the process of epithelial-to-mesenchymal transition (EMT). Interestingly, Sox4 also seems to increase the number of cells with a $CD44^{+}/CD24^{-}$ phenotype and promote invasion and mobility of cancer cells in vivo and in vitro (Fig. 1).[26]

3. WHERE AND HOW ARE CANCER STEM CELLS INVOLVED IN METASTASIS?

Two basic theories explain the heterogeneity in a tumor: stochastic and hierarchical models. The stochastic model suggests that all cells in a tumor have the same ability to progress a tumor. In the hierarchical model, it is believed that only a few cells in a tumor are actually capable of progressing a tumor. These are the only cells that give rise to the progenitor cells, and these progenitor cells further generate the bulk of the tumor. The latter model has been emphasized after the discovery of CSCs.

Not all cells in a tumor are able to move out of the primary site and metastasize. Only a subpopulation in the circulation is believed to initiate metastatic carcinoma.[27–31] The subpopulation is thought to be of the CSC phenotype, also known as the metastasis-initiating cells (MICs). Clinical studies have demonstrated that breast cancer patients with increased CTCs have a poor prognosis. One study analyzing CTCs in breast cancer patients suggested 42% of the CTCs exhibit a CSC, such as phenotype.[31] It is also shown that these CTCs do originate from primary tumors. Another recent study suggested that MIC-containing CTC population originate from primary human luminal breast tumor and express EpCAM, CD44, CD47,

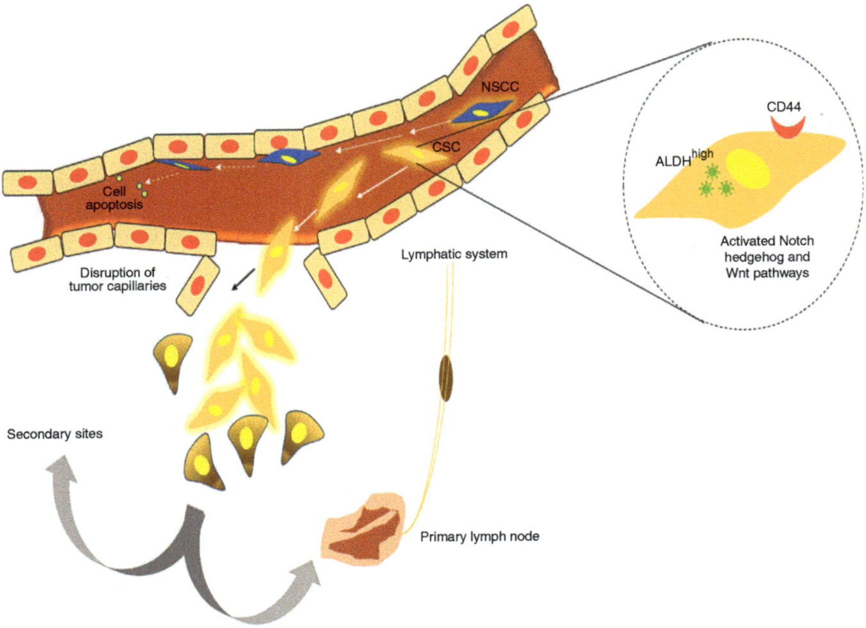

Fig. 1 Cancer stem cells (CSCs) in metastatic breast cancer. This figure illustrates a part of the process of metastasis. CSCs have an edge over their nonstem cancer cells (NSCCs) counterparts in the process of metastasis. CSCs are anoikis resistant (can survive in anchorage free conditions) as opposed to NSCCs, which tend to have a high probability to undergo cell death in the circulation. These CSCs are positive for CD44 and have high expression of aldehyde dehydrogenase (ALDH). In addition, Notch, Hedgehog, and Wnt signaling pathways are active in CSCs unlike in NSCCs. These metastatic CSCs extravasate post disruption of vessel wall and land themselves on secondary organ sites. These metastatic CSCs can also spread through the lymphatic system.

and caused lung, liver, and bone metastasis in mice. Such data indicates that analysis of functional breast CSCs and MICs within the CTCs population may be instrumental in early diagnosis and treatment of metastatic breast cancer.

CSCs possibly have a strong relationship with tumor-induced lymphangiogenesis and lymphatic metastasis. Both processes involve tumor neoangiogenesis (formation of newer blood vessels) and play a key role in tumor progression and metastasis.[32] Conditions within the tumor, such as low oxygen tension or hypoxia, promote the release of vascular endothelial growth factor (VEGF), which then promotes the development of tumor vasculature. These newly formed blood vessels in the tumor not only help the nutrients reach the innermost core of the tumor but also facilitate the dissemination of cells into circulation.[33] Interestingly, factors such as hypoxia target a specific subpopulation of cells in the tumor that have the ability to invade and metastasize.[34] Lymphangiogenesis is also considered a potential facilitator of cancer metastasis. Studies have confirmed the link between lymphatic vessel density and survival rates of cancer patients in general. It has been reported that cell populations with higher lymphatic metastatic ability are composed of cells with a higher proportion of CSC–type of phenotype and therefore have a higher clonogenic potential and greater survival abilities and can establish and propagate in a foreign microenvironment.[35]

4. THE PROCESS OF EPITHELIAL-TO-MESENCHYMAL TRANSITION AND ITS ROLE IN METASTASIS

The concept of EMT was first reported in the early 1980s by Elizabeth Hay, who noted and described the phenotypic change from epithelial-to-mesenchymal in a primitive streak of chick embryos.[36] During the process of EMT, epithelial cells lose their junctions and polarity (apicobasal), reorganize their cytoskeleton, and undergo changes in the molecular signaling programs. These alterations, by defining the cells' shape and ability of increased motility, enable the development of an invasive and metastatic phenotype.

Furthermore, gene expression profiling suggests that the $EpCAM^{+}$ $CD24^{-}CD44^{+}$ CSCs express genes associated with EMT and are relatively quiescent, whereas the $ALDH^{+}$ stem cells have an epithelial phenotype and are associated with a self-renewal state. These CSCs tend to stay in a transition state between epithelial and mesenchymal phenotypes. This kind of plasticity has been seen only in CSCs. More importantly, evidence suggests that this phenomenon of plasticity plays a key role in the ability of CSCs to metastasize. This phenomenon of epithelial/mesenchymal stem cell states can be influenced by the tumor microenvironment.

EMT has been classified into three subtypes: EMT type I, II, and III.[37–40] EMT type I includes the transition of primitive epithelial cells to mesenchymal cells during gastrulation. It is believed these mesenchymal cells do have the ability to undergo mesenchymal-to-epithelial transition (MET) and reacquire the epithelial phenotype and can form the secondary epithelia during embryonic development. EMT type II is mainly involved in wound healing and tissue regeneration. In this subtype of EMT, secondary epithelial cells generate tissue-associated fibroblasts to regenerate the injured tissue. The third type of

EMT takes place during cancer progression. This subtype of EMT favors the transition of epithelial carcinoma cells to highly motile and invasive mesenchymal cells, leading to the promotion of cancer metastasis.

EMT and MET are integral processes that are active predominantly during embryonic development and also in some physiological processes in the adult life. The same mechanisms of EMT and MET are reactivated during tumorigenesis and are used by the opportunistic CSCs to attain successful metastasis.

5. MOLECULAR BASIS OF EPITHELIAL-TO-MESENCHYMAL TRANSITION

EMT is preceded by the deconstruction of cell junctions. Cell–cell junctions are key for epithelial cell integrity. Junctions include subapical tight junctions, adherens junctions, desmosomes, and gap junctions at lateral surfaces. Deconstruction of the cell junctions involves the complete degradation or relocalization of the junction proteins. Claudin, occludin, and zona occludens 1 are the major tight junction proteins, which are severely low in expression during EMT.[41] Protein of the adherens junction, epithelial cadherin (E-cadherin), is cleaved at the plasma membrane and is degraded, leading to several prometastatic signaling cascades, such as the activation of β-catenin and p120 catenin mediated transcription.[42,43] Desmosome and gap junction degradation also occurs with reduction in the level of connexin.[44] Overall, during the process of EMT, all junction proteins are transcriptionally suppressed for a putative metastatic cancer cell to lose contact with its neighbors and have an enhanced opportunity to leave the herd and enter circulation.

Many molecular signaling pathways orchestrate to activate pro-EMT transcription factors. TGF-β, Wnt proteins, Notch, and growth factors have been noted to activate some of the basic transcription factors associated with the process of EMT both during development and cancer, such as Snail and Slug. Glycogen synthase kinase-3beta (GSK-3β) regulates the expression of SNAIL1 by phosphorylation at residues on two serine-rich motifs. Phosphorylation at Ser97 and Ser101 in the first motif facilitates the nuclear incorporation of SNAIL, and phosphorylation at Ser112, Ser116, and Ser120 in the second motif leads to its ubiquitination.[45] To influence SNAIL, many other pathways utilize its GSK-3β-mediated regulation. WNT and PI3K-AKT pathways inhibit the phosphorylation of SNAIL1 through GSK-3β to stabilize its activity.[46,47] Furthermore, Notch and NF-κB signaling interrupt the interaction between GSK-3β and SNAIL1 in an attempt to stabilize it and promote EMT.[48,49] These transcription factors can bind to the E-box DNA sequences found in the regulatory motifs of many epithelial genes and can repress their expression through epigenetic mechanisms. Interestingly, the tumor suppressor, p53 favors the repression of Slug by recruiting the latter to ubiquitin ligase.[50] Snail and Slug cooperate with other transcription factors to activate metalloproteinase (MMP) expression.[51]

The basic helix loop helix (bHLH) family of transcription factors involving TWIST1, TWIST2, and ID proteins have critical roles to play in EMT.[52,53] TWIST1 favors the activation of mesenchymal genes and suppression of epithelial genes and has been shown to repress E-cadherin, and to promote mesenchymal protein N-cadherin in cancer cells independent of SNAIL1.[54–56] Suppression of E-cadherin and activation of N-cadherin by TWIST1 involves recruitment of methyltransferases to promoter regions and creation of H4K20 histone mark.[56] Several pathways achieve TWIST activation during EMT. Under hypoxic microenvironment in the tumor, hypoxia inducible factor 1 alpha (HIF1α) can induce TWIST1 and thus promote EMT and invasion.[55]

Zinc finger E-box binding homeobox (ZEB) family of transcription factors, ZEB1 and ZEB2, can also bind to the E-box regulatory binding sequence and regulate gene expression.[52,54] ZEB1 can act as a repressor as well as an activator of gene expression. It can repress E-cadherin by recruiting C-terminal binding protein (CTBP), a corepressor.[57] It can also repress E-cadherin independent of CTBP by recruiting chromatin-binding proteins.[57] ZEB1 is activated by several signaling mechanisms.[58,59] Binding of p300 to ZEB1 transforms ZEB1 from a repressor into an activator.[59] SNAIL1 and ZEB1 can bind to E-box sequences and regulate the expression of epithelial junctions and polarity proteins (Fig. 2).[52,54]

6. THE PROCESS OF MESENCHYMAL-TO-EPITHELIAL TRANSITION AND ITS ROLE IN METASTASIS

MET is an event opposite to EMT. MET is as critical at the later stages of metastasis as EMT is at the earlier stages. The process of EMT begins at the site of primary tumor; however, the event of MET begins at a later stage of metastasis at a distant site. Our understanding of the MET process is still evolving and is currently under-investigated. MET is believed to

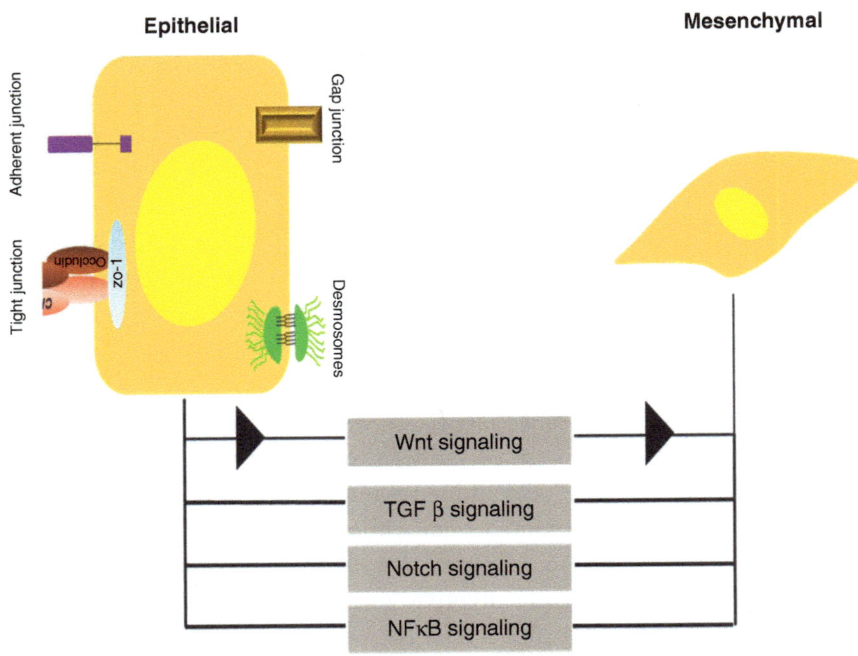

Fig. 2 Molecular basis of epithelial-to-mesenchymal transition. This figure illustrates the process of epithelial-to-mesenchymal transition. During this phase the cell–cell junctions are disrupted and the cells change from an epithelial to a mesenchymal phenotype. The signaling pathways, which are believed to influence this process are Wnt, TGF-β, Notch, and NFκB.

substantially contribute to the establishment of metastatic secondary tumors from disseminated tumor cells. Unfortunately, not much evidence exists about MET. The most important hallmark of MET is the reexpression of E-cadherin.[60] Some evidence suggests that EMT is not conducive to the generation of metastases but that reversal of EMT to MET is required. Dykxhoorn *et al.*, performed a study using four different mouse mammary tumor cell lines, 67NR, 168FARN, 4TO7, and 4T1.[61] These cell lines have similar abilities to form primary tumors in vivo but not metastasis. 4T1 cell line is the only one among these cell lines with a full metastatic potential. 67NR, 168FARN, and 4TO7 demonstrate inhibition at different metastatic checkpoints. Interestingly, the expression of mesenchymal markers N-cadherin and vimentin was the highest in the nonmetastatic cell line, 67NR, and it decreased with the increase of the metastatic potential of the other cell lines. And epithelial markers E-cadherin, CK-18, and epidermal growth factor receptor (EGFR) demonstrated a contrasting trend, with highest expression in the most metastatic cell line, 4T1. This suggests that epithelial characteristics could enhance colonization of disseminated tumor cells. It also points to the fact that the process of MET is key for colonization or for completion of a metastatic event.

In another study by Chao *et al.*, a highly metastatic breast cancer cell line, MDA-MB-231, which is negative for the epithelial marker E-Cadherin (transcriptionally repressed by hypermethylation of E-cadherin promoter), is used to create metastatic colonies in vivo.[62] Intriguingly, E-cadherin levels were highly detected in these metastatic colonies. This study demonstrated generation of E-cadherin positive cells from E-cadherin negative cells and presents more direct evidence for the requirement of MET in the reexpression of E-cadherin. Existing literature still only touches the surface of this process and more evidence is required to understand the phenomenon of MET.

7. MOLECULAR BASIS OF MESENCHYMAL-TO-EPITHELIAL TRANSITION

Cytokines and their receptors have been shown to significantly impact the molecular regulation of MET in tumor cells. FGFR2 gene expression, as a result of alternative splicing, leads to translation of two isoforms, FGFR2b and FGFR2c. FGFR2b is predominantly expressed in epithelial cells and binds to FGF10 and FGF7, whereas FGFR2c expression is favored in the mesenchymal cells and predominantly binds to FGF2. With a shift from EMT to MET, there is a class switch between the isoforms of FGFR2.[63] The mechanism of FGF/FGFR interaction was demonstrated in bladder and prostate cancer. So, what regulates the switch between FGFR2b and 2c? Fox protein family members, especially Fox-2, have been shown to regulate the FGFR2 alternate splicing through interaction with a highly conserved specific sequence element (GCAUG) present on FGFR2 pre-mRNA.[64,65] The molecular and phenotypic changes associated with MET were consistent with the induction of Fox-2.[65] Rbm35a and Rbm35b, two RNA-binding proteins in epithelial cells, were found to also regulate the alternate splicing process in the FGFR2 gene. Evidence demonstrates that induction of either of the two proteins in cells that express FGFR2c isoform would lead to a switch to FGFR2b isoform. These findings indicate that FGFR2 plays a critical role in the MET process.

Receptor tyrosine kinase family members, such as EGFR (or ErbB1) and human epidermal growth factor receptor-2 (HER2 or ErbB2), have been long associated with cancer progression. Various experimental approaches have shown that suppression of endogenous EGFR signaling leads to an increase in the expression of E-cadherin.[66,67] This finding has been linked with another finding that reveals the presence of E-cadherin in the liver metastasis with negligible expression of EGFR, which indicates that EGFR signaling negatively correlates with E-cadherin and that it also plays a key role in MET.[66] Suppression of Src homology phosphotyrosyl phosphatase 2 (SHP2), a target of receptor tyrosine kinases, has been shown to induce epithelial morphology, upregulate E-cadherin levels, and reverse to a normal breast epithelial phenotype. Inhibition of SHP2 seems to deactivate extracellular signal-regulated kinase (ERK) and phosphatidylinositol 3 kinase (PI3K)-Akt signaling induced by EGFR signaling.[68] This study indicates that shutting down ERK and PI3K-Akt signaling is key for MET to occur.

Bone morphogenetic protein 2 (BMP2), a member of TGFβ superfamily, has been shown to be overexpressed in breast and prostate carcinomas, and suppression of BMP2 has also shown to lead to activation of MET.[69]

Another signaling pathway that has been associated with MET is the Wnt/β-catenin signaling pathway. β-catenin is the effector molecule for Wnt signaling. β-catenin has been shown to be associated with cell–cell adhesion, cell migration, and E-cadherin and α-catenin interaction. β-catenin has been observed to be located predominantly in the nucleus of the dedifferentiated mesenchymal cells during invasion; however, it is localized to the membrane and cytoplasm in the polarized epithelial tumor cells in the metastases.[70] Wnt inhibitory factor 1 (WIF1) is an inhibitor of Wnt signaling, and its expression is suppressed in breast and other carcinomas.[71,72] Previous studies have demonstrated that WIF1 induces the reversal of EMT, thereby inhibiting Wnt signaling.[71] One of the Wnt signaling ligands, Frizzled 4 (FZD 4), favors EMT, and silencing FZD4 in prostate cancer cells seemed to induce MET, as indicated by the increase in expression of E-cadherin and β1-integrin (Fig. 3).[73]

8. CHEMICAL FACTORS INFLUENCING CANCER CELLS IN FAVOR OF METASTASIS

8.1 Chemokines in Metastasis

Chemokines are members of the chemotactic cytokines. They are involved in processes, such as inflammatory responses, vira–cell interactions, angiogenesis, and cancer metastasis. In the scope of the current review, chemokines

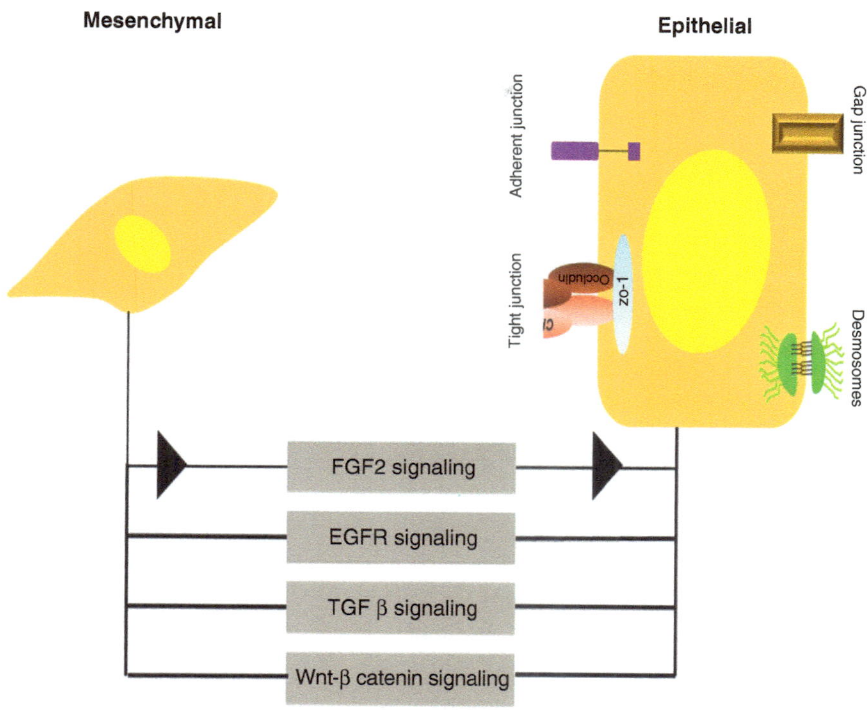

Fig. 3 Molecular basis of mesenchymal-to-epithelial transition. This illustration demonstrates the transition from mesenchymal-to-epithelial phenotypes during the process of metastasis. This transition allows the metastatic cells to regain epithelial phenotype. The epithelial cells now establish all the cell–cell junctions to form an epithelial colony. Some of the major signaling pathways believed to be involved in this phase of metastasis are FGF2, EGFR, TGF-β, and Wnt–β-catenin.

influence directional migration of disseminated cancer cells with their interaction with G-protein-coupled receptors, cytoskeletal rearrangements, and firm adhesion to endothelial cells.[74] Chemokines have 5 major subclasses: CXC (17 members), CC (28 members), CX3CL1, XCL1, and XCL2. This classification is based on their amino acid composition (cysteine residues) at the NH2-terminal part of the protein. The presence of a conserved tetracysteine motif is the major basis of classification. Chemokines interact with seven transmembrane-spanning domains containing G-protein-coupled receptors.[75] Upon binding to these receptors, chemokines activate a cascade of downstream signaling pathways, which include calcium mobilization, activation of ERKs, MAPK, PI3K, RAS, Rho-GTPases, PAK, and NFκB. The interactions between the chemokines and their receptors are highly redundant. Many of these chemokines bind to several receptors and vice versa.

Some chemokines are secreted at a higher level in breast cancer compared to nonmalignant breast epithelial cells. Global gene expression profiling using SAGE analysis has revealed that many chemokines lose their expression in carcinomas.[76] Some of the chemokines widely studied in breast cancer are CXCL8, CCL2, CCL4, and CCL5. CXCL8 is overexpressed in breast cancer[77,78] and is secreted at a higher level in the sera of breast cancer patients.[79,80] CXCL8 has also been inversely correlated with ERα levels in breast cancer, indicating the involvement of CXCL8 in metastasis and poor prognosis.[78,81,82] Interestingly, among the breast cancer cell lines, with the exception of ZR-75, most of the cell lines expressing high levels of CXCL8 are ERα-negative. Other genes located at the same locus as CXCL8 on chromosome 4q21—such as CXCL1, CXCL3, CXCL5, and CXCL6—seem to be overexpressed in CXCL8 overexpressing breast tumors. Furthermore, CXCL8 was mainly found in the epithelial cells of the tumor and was also found to increase with higher grades of breast cancer. Strong increases in its levels during metastases have also been observed.

Another chemokine axis emphasized in breast cancer metastasis is CXCL12/CXCR4.[83] Metastatic breast cancer cells are rich in CXCR4 and CCR7 chemokine receptors. The most favorable anatomical sites for secondary metastatic tumor formation in breast cancer are rich in the ligands for CXCR4. CXCL12 interacts with CXCR4, mediating cytoskeletal rearrangement (actin polymerization), and pseudopodia formation in breast cancer cells leads to the induction of chemotactic and invasive responses. Neutralization of CXCR4 inhibits breast cancer metastasis to a significant extent in animal models. Interaction of CXCR4 with stromal derived factor-1 (SDF-1), an alpha-chemokine, also has been reported to be involved in cell trafficking.[84] The redundancy in the chemokine ligand and receptor doesn't apply to SDF-1 and CXCR4, as SDF-1 exclusively binds to CXCR4, indicating a unique biological role.[85–90] Some of the organs that highly express SDF-1 are the lymph nodes, lungs, liver, and bones. It is suggested that many of the CXCR4-positive tumors—such as breast, ovarian, and prostate—have a high probability of metastasizing to one of these SDF-1 expressing anatomical sites in an SDF-1 dependent manner.[83,91–96] The expression of CXCR4 is regulated by tissue-specific paired-box transcription factors (PAX).[95] Expression of CXCR4 is also influenced by stress/hypoxia and tissue damage, such as hypoxia-inducible factor (HIF-1),[97]

NF-κB,[98] glucocorticoids,[99] TGF-β,[100] lysophosphatidylcholine,[101] VEGF,[102] IFN-α,[103] and interleukins.[104,105] The responsiveness of $CXCR^+$ stem cells toward SDF-1 gradient is affected critically by several inflammation-associated factors, such as C3a (product of C3a degradation of carboxypeptidases), hyaluronic acid, and sphingosine-1 phosphate. It is believed that tissue inflammation plays a pivotal role in tumor dissemination and progression leading to metastasis. Thus, these factors are likely to influence the migratory behavior of $CXCR^+$ stem cells. SDF-1 is a very potent chemo attractant for both nonmalignant and malignant stem cells.[85–90,92–96,106–108] $CXCR^+$ CSCs in circulation are directed by SDF-1 gradient toward metastatic niches. SDF-1 not only favors migration but also increases the average velocity of displacement of $CXCR^+$ CSCs.[95,96] In vitro studies have shown that exposure of cancer cells to SDF-1 extensively favors the formation of the leading edge required for effective migration. In addition, SDF-1-mediated cytoskeletal rearrangement in terms of increased actin fiber bundling has also been found in tumor cells.

C-C motif ligand 2 (CCL2), a chemokine which attracts monocytes toward the site of inflammation[109] has been implicated in breast cancer progression and metastasis.[110] It has been reported that CCL2 derived from tumor stroma can promote the expansion of breast CSCs.[111] The expression levels of CCL2 in the tumor microenvironment recruit tumor-associated macrophages (TAMs) and serve as an indication of early relapse.[112,113] In the context of breast CSCs, it has been suggested that CCL2 facilitates their expansion[111] not only by conversion of nonstem cancer cells to CSCs but also by leveraging the self-renewal ability of breast CSCs.[114] Mechanistically, CCL2 secretion by tumor-associated fibroblasts leads to the induction of Notch-1 in breast cancer cells, thereby affecting the self-renewal ability of the CSCs. High levels of Notch-1 and CCL1 have been correlated with poorly differentiated breast tumors prognosis.[114]

8.2 Cytokines in Metastasis

The role of cytokines in cancer progression has been widely reported. TGF-β and IL-6 are the most studied cytokines in breast cancer; both promote breast tumor metastasis.[115] The TGF-β superfamily is considered to be a master regulator of EMT. TGF-β signaling takes place with ligand (TGF-β1 or TGF-β3) binding to two receptors, Tβ1R and TβRII. Ligand binding leads to TβRII homodimer formation, leading to autophosphorylation and further activation of Tβ1R. Activated Tβ1R subsequently phosphorylates Smad2 and 3 and, along with 4, gets translocated to the nucleus to regulate the transcription of TGF-β responsive genes. TGF-β can also regulate gene expression through Smad-independent pathways, such as MAPK, ER, and JNK.[116,117]

TGF-β plays a very complicated role in several cancers. During the initial stages, it acts as a tumor suppressor; in the later stages, it acts as a tumor promoter.[118–120] TGF-β is a critical regulator of EMT, mediating cancer metastasis.[121–123] Although the role of TGF-β in CSCs is not very clear, we can see that it promotes CSCs by interacting with Wnt, HER2, and FAK.[124,125] It has been noted that CSCs have activated TGF-β signaling, which favors the expression of vimentin and CTGF.[126]

The role of IL-6 in CSC regulation has been reported. CSCs exist in a transition state, maintaining a bidirectional equilibrium between nonstem cancer cells and CSCs. Conversion of nonstem cancer cells to CSCs, as in Trastuzumab resistance leading to CSC expansion, takes place primarily because of IL-6 secretion by CSCs. In addition, there is elevation in the expression of vimentin and TWIST and the loss of E-cadherin, EpCAM, and Claudin3.[127] IL-6 is also able to induce the expression of other CSC-favoring cytokines.[127] IL-6 is produced by mesenchymal stem cells and other immune cells in the tumor microenvironment. Bone marrow–derived mesenchymal stem cells releasing IL-6 are recruited to the breast tumor site and can enhance the metastatic ability of tumor cells.[128] It is believed that IL-6–CXCL7 cytokine loop is activated in the cross-talk between mesenchymal stem cells and CSCs. Increased activity of IL-6-JAK2-Stat3 has also been reported predominantly in breast CSCs.[129]

IL-8 is another inflammatory cytokine that has been positively correlated with metastatic breast cancer. Increased expression of IL-8 is positively correlated with ER—more than ER^+ breast cancers. ER^- breast cancers have higher rates of leading to a metastatic disease. IL-8 interacts with CXCR1 to regulate cell proliferation, migration, and invasion. CXCR1 is highly expressed in breast tumor cells and is positively correlated with metastatic disease.[130] Breast CSCs have consistently shown the expression of CXCR1, indicating a role for IL-8-CXCR1 signaling. In vitro studies have shown that IL-8 promotes sphere formation and expansion of breast CSCs. It also induces chemotactic behavior of breast CSCs, which can contribute to metastasis.[131,132] It has been shown that IL-8 can be induced by chemotherapeutic drugs, such as Docetaxel, and monoclonal antibody, such as trastuzumab, indicating that IL-8 may be involved in the expansion of drug resistant CSCs posttherapy.

9. PHYSICAL FORCES INFLUENCING CANCER CELL METASTASIS

Metastatic cells encounter many physical forces throughout the whole process. The ECM imposes physical forces on primary tumor cells, and the CTCs also encounter several hemodynamic forces. Finally, when a metastatic cell lands on a foreign tissue, it must overcome the forces presented by the tissue's ECM. For a tumor cell, successfully metastasizing and colonizing at a distant site requires a lot of negotiation with the physical forces at every step. This is one of the reasons why metastasis is not a very efficient process and why only a few potent circulation tumor cells end up forming metastatic colonies. Understanding the physics behind metastasis is essential for designing diagnostics or therapeutics to inhibit the process.

During the process of intravasation and extravasation, a metastatic cell undergoes dramatic changes in shape due to cytoskeletal remodeling. Cytoplasm has exhibited elastic properties at high deformation rates and has behaved more viscously at low deformation rates.[133] Elasticity of the cytoplasm means the ability of the cytoplasm to rebound after the application of a physical force, whereas viscosity is a measure of the ability of the cytoplasm to flow under external shear.[133] The rate-limiting step for a tumor to intravasate is the deformation of the interphase nucleus and, interestingly, the nucleus is 10 times stiffer than the cytoplasm.[134,135] Elasticity of the nucleus is mainly due to the nuclear lamina[135], chromatin organizations,[136] and linkers of the nucleus and cytoskeleton (LINC) complexes.[137–139] Suppression of the LINC complex components leads to deformation of the nucleus and the softening of the cytoplasm and nucleus.[136] Mutations in lamin A/C have been found in breast cancer[140], which might lead to alterations in LINC-mediated connections between the nucleus and cytoplasm and affect the invasiveness of tumor cells.

Cancer cells are softer than nonmalignant cells, as revealed by biophysical analysis.[141,142] The softness of cancer cells is also correlated with higher metastatic probability. Invasion of metastatic tumor cells would require optimal mechanical properties. Tumor cells must be neither too soft nor too stiff to traverse through the highly cross-linked collagen fibers in the matrix. Single-cell biophysical analysis has consistently demonstrated that certain types of highly metastatic tumor cells display a wide range of mechanical properties. And these types of tumor cells may be capable of efficiently handling the dynamic nature of the physical forces at every step in the metastasis process. It seems that certain cells that display a wide range of mechanical behavior can conserve this property over several generations.

Upon entering the circulation, a CTC encounters immense physical and mechanical pressures, and several parameters, such as diameter of the blood vessel, pattern of the blood flow, and presence of shear flow determine the trajectory of the CTC. While egressing the circulation, a CTC should bind to the blood vessel wall and be arrested through different mechanisms, such as physical occlusion or cell adhesion. For physical occlusion to occur, the diameter of the blood vessel should be less than the diameter of the circulating cells ($d_{vessel} < d_{cell}$). CTCs of epithelial origin, usually >10 μm in size, can get arrested in smaller capillaries of <10 μm in diameter.[143] Extravasation through the mechanism of adhesion comes into play for CTCs exiting a bigger blood vessel. In this case, the chances or probability of a CTC getting arrested depends on the: (a) collision frequency between membrane-bound receptors and endothelial ligands and (b) time of residence at that point (depends on the shear force). CTCs travelling along the blood vessel wall have a greater chance of adhering to endothelial cells along the wall and thus have a higher probability to extravasate. It is not only the type of velocities (translational and angular) possessed by these cells traveling along the blood vessel wall that increases their chance of extravasation but also the rotational movement of these cells that facilitates the interaction of their receptors with the ligands on the endothelial cells of the blood vessel. Furthermore, receptor–ligand bonding affinity and the number of receptor–ligand pairs participating within that residence time determines the chances of that particular tumor cell to exit circulation.

Even if a CTC overcomes all the physical and mechanical pressures and settles at a distant site, it would not necessarily be able to form colonies. It has been a subject of immense interest to understand why specific types of cancer cells are more efficient in colonizing certain anatomical sites or why some types of cancer cells prefer certain organs. For example, metastatic breast cancer cells colonize the bone marrow and brain, whereas prostate cancer cells seem to form metastatic foci in bone marrow and liver. The "seed and soil" hypothesis discussed earlier explains how different cancers prefer different organs for metastasis to a large extent. However, biophysical analysis also suggests that blood flow patterns contribute a lot toward the colonization of CTCs.

10. ROLE OF THE MICROENVIRONMENT IN THE PROCESS OF METASTASIS

Tumors grow and survive in three-dimensional space within a complex microenvironment. The microenvironment sustains growth and supports the invasion and metastasis of cancer cells, and it consists mainly of stromal cells, immune cells, adipocytes, and extracellular membrane. The interaction between the cancer cells and microenvironment are considered bidirectional. The powerful cross-talk between tumor cells and the microenvironment is thought to be a major determinant of the rate of cancer progression, invasion, and metastasis.[144] It is interesting to note how the composition of the microenvironment and its proportion of certain cells, such as stromal cells, contribute toward a diverse nature of tumors.[145,146] During the course of the disease, with oncogenic signals from the tumor cells, the microenvironment constantly changes. This dynamic nature of tumor microenvironment tremendously influences the whole process of metastasis.

10.1 Tumor-Associated Macrophages

TAMs are considered one of the key regulators of tumorigenesis. Usually macrophages are known to be the effector cells of the immune system; however, in the context of tumorigenesis, they accede to the tumor cells and start supporting their growth and survival.[147] Macrophages have been observed to reside in the leading edge of tumors, and evidence suggests that they facilitate the invasion of cancer cells.[148] One signaling mechanism that has been suggested is the involvement of tumor-derived CSF-1 and macrophage-derived EGF.[148–150] These TAMs produce proteases, which favor the progression of tumor cells.[151–153] Macrophage activation is a very complex process, and they tend to change their polarization states according to the environmental condition. Macrophages range from M1 to M2 polarization states, wherein M1 type produces type I cytokines for antigen presentation and M2 type produces type II proinflammatory cytokines having protumorigenic functions.[154] Macrophages at the tumor site are thought to express the M2 phenotype.

Recently, it was discovered that TAMs help create a CSC niche through juxtacrine signaling or contact-dependent signaling.[155] Breast CSCs express high levels of CD90 (Thy-1) and EphA4 during EMT. High levels of CD90 and EphA4 enable the CSCs to interact with the TAMs, thereby activating the Src and NFκB signaling in the CSCs. NFκB activation in the CSCs leads to production of several cytokines that support the CSC niche. TAMs can activate Notch and Stat3 pathways in CSCs, leading to enhanced chances of metastasis and therapy resistance.[156] TAMs have also been shown to induce CSC-associated gene expression, such as Sox2, Oct4, Nanog, ABCG2, and Sca-1 in murine breast cancer cells.[157]

10.2 Tumor-Associated Fibroblasts

Fibroblasts are a very important component of the tumor microenvironment, as they secrete factors for the ECM. These tumor-associated fibroblasts are called cancer-associated fibroblasts (CAFs). CAFs are believed to be from the: (a) resident fibroblasts, (b) from the bone marrow–derived mesenchymal stem cells, or (c) from cancer cells with EMT phenotype.[158] CAFs have an increased level of 17-beta estradiol dehydrogenase (E2DH) activity, which promotes tumor progression.[159] CAFs have also been shown to promote tumor progression through TGF-β mediated mechanisms.[160] Evidence suggests that CAFs release 100-fold more IL-6 than nonmalignant tissue-associated fibroblasts, which can actually potentiate invasive traits.[161] CAFs promote metastasis by increasing the activity of matrix metalloproteinase (MMP14 and MMP9).[162] Release of chemokine CCL2 by the CAFs increases the number of breast CSCs, which further promotes metastasis.[111]

10.3 Endothelial Cells

Endothelial cells in the tumor microenvironment also play an important role in cancer growth and invasion. Endothelial cells derived from the bone marrow are recruited to the tumor site, where they mature and form capillaries.[163,164] Stromal derived factor-1α (SDF-1α) is abundant in bone marrow, and that is one of the major reasons why endothelial progenitor cells are present in bone marrow. However, VEGF, a proangiogenic factor, inhibits the expression of SDF-1. VEGF released by the tumor microenvironment, which decreases the levels of SDF-1α leads to the release of endothelial progenitor cells into the circulation.[165] Chemokines, such as CCL2 and CCL5 are used as a chemotactic mechanism by the tumor cells to attract the endothelial progenitor cells from the bone marrow, and subsequently the endothelial progenitor cells end up getting trapped in the tumor.[166]

Evidence suggests that endothelial cells in the tumor microenvironment regulate CSCs. Coculture of endothelial cells with breast cancer cell lines have shown to enrich breast CSCs and the metastatic potential of these cells. It is proposed that endothelial cells release Jagged1, a potent ligand for Notch-1 receptors present on the CSCs. This mediates a cross-talk between the endothelial cells and CSCs in the tumor microenvironment, supporting the survival and activity of CSCs.[167]

10.4 Tumor-Associated Adipocytes

Tumor-associated adipocytes (TAAs) are one of the key components of the breast tumor stroma[168,169] and a powerful source of adipokines for the tumor cells. TAAs are key contributors for intravasation of a tumor cell from the tumor site to the blood stream. TAAs release several prometastatic proteins and cytokines, such as MMP-11, PAI-1, collagen VI, IL-6, IL-1β, TNFa, leptin, and IGF-1.[170–176] Many of these factors have been shown to influence the survival and activity of CSCs.

10.5 Extracellular Matrix

ECM is composed of proteins, glycoproteins, proteoglycans, and polysaccharides.[177,178] The ECM contributes to both the basement membrane and interstitial matrix. The basement membrane is a specialized ECM with a distinct composition that separates the epithelial cells from the stroma. Basement membrane ECM is made up of type IV collagens, laminins, fibronectin, and linker proteins. The interstitial matrix is composed mainly of fibrillar collagens, proteoglycans, and glycoproteins. The basement membrane is more compact and less porous, whereas the interstitial matrix is charged and hydrated and determines the tensile strength of tissues.[179] The ECM contributes significantly to cell behavior regulation in terms of physical, biochemical, and biomechanical properties. Biochemical properties of the ECM also govern the type of signaling pathways activated in the cells.

ECM remodeling is one of the hallmarks of cancer progression. Fibrillar collagen and proteolysis are increased during remodeling, and denatured collagen acts as a chemoattractant for macrophages that are important cellular components of the tumor-associated stroma.[144,180] In breast cancer, increased fibrotic stromal matrix deposition with collagen (type I, III, and V), elastin, vitronectin, matricellular proteins, and oncofetal fibronectin as key components is observed. In addition, glycosaminoglycans, such as hyaluronan and chondroitin sulfate are also upregulated, whereas collagen type IV is reduced.[181] Hyaluronic acid regulates cancer cell adhesion, motility, proliferation, and differentiation through cell surface receptors, such as

CD44.[182] Levels of hyaluronic acid in breast cancer have been used to establish the prognosis, stage, and grade of a tumor.[183]

Laminins play a major role in metastatic breast cancer. LM-111 and LM-322 are majorly studied in breast cancer. LM-111 is a component of the basement membrane ECM and is degraded significantly during breast cancer progression.[181] LM-322, on the other hand, is induced by metastatic breast cancer cells in stromal cells to promote cell survival through PI3K-Akt signaling. LM-322 induction leads to enhanced migration and invasion of breast cancer cells.[182] Collagen deposition or ECM stiffness can elevate integrin signaling in cells and influence cell proliferation and survival as well as the migration of cancer cells.[184,185] ECM components have also shown to have pro- or antiapoptotic effects.[186] ECM composition is critical for normal tissue and CSC niche.

ECM contact with stem cells is important to maintain stem cell properties; functional ablation of ECM components has been shown to reduce the number of stem cells.[187–192] ECM enables the restriction of stem cells in their physical niche, the maintenance of cell polarity, and the orientation of mitotic spindles and helps to determine cell fate.[193–197] ECM also dictates the signaling pathways controlling the property of stem cell differentiation. Matricellular proteins, such as tenascin C are highly expressed in breast cancers with significantly high expression in invasion fronts.[198,199] Tenascin C and matrix metalloproteinase often colocalize to the invasive edge of the tumor. Tenascin C promotes metastasis by initiating breast tumor cells through activation of Notch and Wnt signaling.[199] Another matricellular protein, periostin, binds to bone morphogenetic protein-1 (BMP-1), leading to activation of lysyl oxidase (LOX) and thereby promoting collagen cross-linking and tissue stiffening, both central characteristics of invasive cancers.[200] ECMs are major contributors to the generation of premetastatic and metastatic niche. Both tenascin C and periostin have been associated with the promotion of metastatic niche and promotion of lung metastasis in breast cancer.[201]

10.6 Hypoxia/Reduced Oxygen Tension

Cancerous tissues differ from normal nonmalignant tissues by having low oxygen tension, and decreased levels of intratumoral oxygen favor metastatic progression of the disease. Lower oxygen levels, or hypoxia, activate the expression of hypoxia inducible factor-1 (HIF-1) in cancer and associated stromal cells. HIF-1 is a heterodimeric protein with HIF1-α and -β subunits;

where HIF-1α is oxygen regulated and HIF-1β is constitutively expressed.[202,203] Hypoxia is an independent prognostic factor in several cancers.[204] In the case of breast cancer, the median oxygen tension is 10 mmHg compared to 65 mmHg in a nonmalignant normal mammary gland. Hypoxia has been associated with poor prognosis, increased metastasis, and higher mortality rates.[205] HIF transcription factors, when in the presence of oxygen, interact with a tumor suppressor protein, VHL; this interaction leads to ubiquitination of HIF transcription factors, and a low level of them is maintained in well-oxygenated nonmalignant tissues.[206,207] In a hypoxic tumor microenvironment, this interaction is inhibited and HIF1α dimerizes with HIF1β and leads to activation of around 1000 target genes.[208,209] Hypoxia and HIFs have been shown to maintain CSC properties through regulation of stem cell genes, such as OCT4, SOX2, and Nanog.[210,211] Interestingly, hypoxia also induces stemness in differentiated progenitor and non-CSCs.[210,211] Hypoxia also induces increased levels of ROS, which helps in the survival and induction of EMT through TGF-B signaling.[212] Several other molecular mechanisms have been attributed to hypoxia-driven promotion of CSCs.[213] It has been suggested that hypoxia limits CSC differentiation, thereby expanding the pool.[213–215] The enzyme telomerase, responsible for maintaining the telomere length, is highly active in tumor cells. Hypoxia-mediated HIF1 activation has been demonstrated to regulate the expression of TERT, the telomerase gene.[216] This enables the cancer cells to increase the life-span of the tumor mass. Hypoxia also helps tumor cells evade the immune system by inhibiting NK and $CD8^+$ T cells as well as phagocytosis by macrophages.

11. STRATEGIES TO COMBAT METASTASIS THROUGH CSCs AND DESIGNING ANTIMETASTATIC THERAPY

11.1 Nanomedicine-Based Therapy to Inhibit CSCs

Anticancer drugs have demonstrated limitations in clinical translation due to poor solubility, a lesser half-life resulting in shorter circulation time, poor stability, and unfavorable distribution. Nanomedicine has shown promise in overcoming some of these limitations by improving bioavailability, decreasing systemic toxicity, and improving the efficacy of anticancer drugs in general.

There has been a surge in the availability of "nanocarriers" for efficient cancer drug delivery. Some of the nanocarriers that have been suggested are dendrimers,[217,218] polymeric micelles,[219–222] liposomes,[223] gold and other metal nanoparticals,[224–226] and nanoemulsions.[227] Some of these nanocarrier-based cancer therapeutics have been approved by Food and Drug Administration, such as Doxil (liposomal doxorubicin), Abraxane (albumin-bound paclitaxel), and Oncaspar (PEG-L-asparaginase).

Nanosystems have considered using CD44 receptors as a prime candidate for targeting breast CSCs. CD44 is highly expressed on breast CSCs. Use of anti-CD44 antibodies and paclitaxel-loaded poly [(D, L-lactide-*co*-glycolide)-*co*-PEG] micelles resulted in effective internalization of paclitaxel in breast CSCs.[228] Hyaluronic acid, which is a principal component of the ECM, is a ligand for CD44.[229–231] It was recently shown that salinomycin and hyaluronic acid-coated nanoparticles [poly (lactic *co*-glycolic acid] were efficient in CD44-mediated targeting of breast CSCs.[232] Breast CSCs have a higher uptake of hyaluronic acid than non-stem cancer cells. A nanogel drug conjugate, cholesteryl-hyaluronic acid, was demonstrated to target breast CSCs through CD44 interaction and effective internalization.[233]

11.2 Inhibition of Cancer Stem Cell–Dependent Signaling Pathways

One of the proposed strategies to root out breast CSCs effectively is the suppression of signaling pathways that fuel the existence of CSCs. Some of the pathways widely studied in this context are Wnt/β-catenin, Hedgehog, Notch, NF-κB, and Stat signaling. Wnt pathway activation in CSCs has been largely explored. Small molecule inhibitors have been developed to inhibit the pathway in breast and other cancers. ICG-001 is one of the small molecule inhibitors developed to target Wnt pathways. Several Hedgehog signaling inhibitors are in different phases of clinical trials are described in Table 1.

11.3 Immunotherapy for CSCs

One of the prerequisites for successful metastasis by cancer cells is evasion of immune surveillance.[245] In general, CSCs lack the expression of major histocompatibility complex-II (MHC-II) protein and have a lower expression of MHC-I protein.[246,247] While CSCs have also shown to lack the expression of costimulatory molecules, such as CD80 and CD86, coinhibitory molecules, such as CTLA4, PD-1/-1L, B7-H2, and B7-H3 seem to be upregulated. Proteins, such as transporter-associated antigen processing,

Table 1 Hedgehog Signaling Inhibitors are in Different Phases of Clinical Trials.

Compounds	Chemical Nature	Targets	Associated Signaling Pathway	References
Cyclopamine	Steroidal compound	Smo	Hedgehog	234,235
GDC-0449	Small molecule inhibitor	Smo	Hedgehog	236
IPI-926	Small molecule inhibitor	Smo	Hedgehog	237
BMS-833923	Small molecule inhibitor	Smo	Hedgehog	238
PF-04449913	Small molecule inhibitor	Smo	Hedgehog	239
HPI1-4	Small molecule inhibitor	Shh	Hedgehog	240
GANT58	Small molecule inhibitor	Gli	Hedgehog	241
GANT61	Small molecule inhibitor	Gli	Hedgehog	241
CWP232228	Small molecule inhibitor	β-Catenin	Wnt/β-catenin	242
Capsaicin	Natural compound	NICD	Notch	243
MR3	Natural compound (derived from resveratrol)	GSK3β	Wnt/β-catenin	244

GSK3β, Glycogen synthase kinase-3beta.

low molecular weight proteins, and β-microglobulin involved in the antigen presenting mechanism were also downmodulated in CSCs compared to their nonstem cancer cell counterparts.[248] Most of these observations have been made in melanoma and glioblastoma-derived CSCs. CSCs express several molecules that can be recognized by the immune system as tumor-associated antigens.[249] Some of the tumor associated antigens expressed by CSCs are gp100 and tyrosinase in melanoma, cancer embryonic antigen in colon cancer, mucin-1 (MUC-1) in breast cancer.[250] However, none of the tumor-associated antigens are specific for CSCs.

The role of innate immune response against CSCs has been investigated. $CD133^+$ CSCs from a brain tumor did not demonstrate expression of detectable levels of MHC I. Using immunotherapy against CSCs is a relatively new endeavor, and not much data is available on solid tumor stem cells.

11.4 Differentiation Therapy for CSCs

Reactivation of endogenous differentiation programs in CSCs is the basis of differentiation therapy. Poor cellular differentiation is one of the main traits of cancer. The neoplastic stage of cancer is due to the blockade of differentiation machinery in them. In recent times, several differentiation agents have been proposed and studied in different cancers. All-*trans*-retinoic acid and dimethyl sulfoxide were used as differentiation agents in acute promyelocytic leukemia, use of which showed terminal differentiation in promyelocytic leukemic cells. All-*trans*-retinoic acid was demonstrated to promote differentiation in undifferentiated promyelocytic leukemic cells by disrupting the promyelocytic leukemia/retinoic acid receptor α (PML/RARα) fusion protein.[251] Phorbol myristate acetate was shown to promote differentiation in leukemic cells through protein kinase C-PV.1/AP1 transcription factor activation.[252] Other agents that were used for differentiation of leukemic cells were hexamethylamine bisacetamide and Vitamin D3.[253,254]

Epigenetics, involving covalent modification of DNA (methylation of cysteine residues) and histone proteins of the nucleosomes, plays an important role in cellular differentiation and homeostasis.[255–257] In addition, the epigenetic status of a cancer determines its epithelial–mesenchymal plasticity.[258] Modulation of enzymes, such as histone methyltransferases and histone demethylases can be important in CSC differentiation. Histone deacetylase inhibitors, such as suberoylanilide hydroxamic acid has been shown to induce differentiation in breast cancer cells[259] and in endometrial carcinomas.[260] A small molecule inhibitor for EZH2 (a methyltransferase) has been proposed as a putative CSCs differentiation molecule and is under phase I/II clinical trial for lymphomas and advanced solid tumors (NCT01897571). However, use of epigenetic modifiers is still in the very early stages of research, and it is important to consider the side effects of such treatments.

An effective differentiation therapy would have to be a combination of both differentiation agents and conventional chemotherapy to efficiently target both CSCs and nonCSCs populations.

11.5 Modulation of the Tumor Microenvironment to Suppress CSCs

Tumor stroma is considered to be a major factor in providing a favorable microenvironment for tumor progression and metastasis. Another advantage of therapeutically targeting the stroma is that the stromal cells are genetically more stable than the cancer cells and are less likely to become resistant to

Table 2 FDA-Approved Anti-Metastatic Therapy

Drugs	Disease	Brand Name	FDA Approval Year
Paclitaxel Albumin-stabilized nanoparticle formulation	MBC	Abraxane	2013
Ado-trastuzumab emtansin	MBC (HER2+)	Kadcyla	2013
Anastrozole	MBC (HR+)	Arimidex	
Docetaxel	MBC	Taxotere	2004
Toremifene	MBC	Fareston	
Gemcitabine hydrochloride	MBC	Gemzar	2004
Eribulin mesylate		Halaven	2010
Ixabepilone	MBC	Ixempra	2007

therapy.[261] Endothelial cells—being important components of the tumor stroma—have been considered a key target. Receptor for VEGF, VEGFR-1, and VEGFR-2 are overexpressed in the endothelial cells of the tumor vasculature. An antibody against these receptors has been used to target the endothelial cells and thereby inhibit tumor vasculature (Table 2).

12. CONCLUSIONS

Metastasis is a very intricate phenomenon that results in a very aggressive and advanced form of cancer. Metastasis is an orchestrated show of different processes, such as EMT and MET. CSCs are key players in metastasis and are considered to be the ones to have epithelial–mesenchymal plasticity. The microenvironment at the primary and secondary tumor sites plays a decisive role in whether a cell can successfully leave its site of origin and establish itself at a distant site. To add to the complexity of metastasis, physical and chemical forces contribute significantly to the whole process. Several approaches to combat metastasis through CSCs have been proposed and tested, including modulation of the microenvironment, differentiation therapy, immunotherapy, and the inhibition of cancer stem cell-specific signaling pathways.

REFERENCES

1. Clarke MF, Fuller M. Stem cells and cancer: two faces of eve. *Cell*. 2006;124 (6):1111–1115.
2. Al-Hajj M, Wicha MS, Benito-Hernandez A, Morrison SJ, Clarke MF. Prospective identification of tumorigenic breast cancer cells. *Proc Natl Acad Sci USA*. 2003;100 (7):3983–3988.

3. Ponti D, Costa A, Zaffaroni N, et al. Isolation and in vitro propagation of tumorigenic breast cancer cells with stem/progenitor cell properties. *Cancer Res.* 2005;65(13): 5506–5511.
4. Wright MH, Calcagno AM, Salcido CD, Carlson MD, Ambudkar SV, Varticovski L. Brca1 breast tumors contain distinct CD44+/CD24- and CD133+ cells with cancer stem cell characteristics. *Breast Cancer Res.* 2008;10(1):R10.
5. Perrone G, Gaeta LM, Zagami M, et al. In situ identification of CD44+/CD24- cancer cells in primary human breast carcinomas. *PLoS One.* 2012;7(9):e43110.
6. Wang LB, He YQ, Wu LG, Chen DM, Fan H, Jia W. Isolation and characterization of human breast tumor stem cells. *Xi Bao Yu Fen Zi Mian Yi Xue Za Zhi.* 2012;28 (12):1261–1264.
7. Gotte M, Yip GW. Heparanase, hyaluronan, and CD44 in cancers: a breast carcinoma perspective. *Cancer Res.* 2006;66(21):10233–10237.
8. Herrera-Gayol A, Jothy S. Adhesion proteins in the biology of breast cancer: contribution of CD44. *Exp Mol Pathol.* 1999;66(2):149–156.
9. Rangaswami H, Bulbule A, Kundu GC. Osteopontin: role in cell signaling and cancer progression. *Trends Cell Biol.* 2006;16(2):79–87.
10. Schabath H, Runz S, Joumaa S, Altevogt P. CD24 affects CXCR4 function in pre-B lymphocytes and breast carcinoma cells. *J Cell Sci.* 2006;119(Pt 2):314–325.
11. Ginestier C, Hur MH, Charafe-Jauffret E, et al. ALDH1 is a marker of normal and malignant human mammary stem cells and a predictor of poor clinical outcome. *Cell Stem Cell.* 2007;1(5):555–567.
12. Balicki D. Moving forward in human mammary stem cell biology and breast cancer prognostication using ALDH1. *Cell Stem Cell.* 2007;1(5):485–487.
13. Chute JP, Muramoto GG, Whitesides J, et al. Inhibition of aldehyde dehydrogenase and retinoid signaling induces the expansion of human hematopoietic stem cells. *Proc Natl Acad Sci USA.* 2006;103(31):11707–11712.
14. Meyer MJ, Fleming JM, Lin AF, Hussnain SA, Ginsburg E, Vonderhaar BK. CD44posCD49fhiCD133/2hi defines xenograft-initiating cells in estrogen receptor-negative breast cancer. *Cancer Res.* 2010;70(11):4624–4633.
15. Cariati M, Naderi A, Brown JP, et al. Alpha-6 integrin is necessary for the tumourigenicity of a stem cell-like subpopulation within the MCF7 breast cancer cell line. *Int J Cancer.* 2008;122(2):298–304.
16. Vaillant F, Asselin-Labat ML, Shackleton M, Forrest NC, Lindeman GJ, Visvader JE. The mammary progenitor marker CD61/beta3 integrin identifies CSCs in mouse models of mammary tumorigenesis. *Cancer Res.* 2008;68(19):7711–7717.
17. Peitzsch C, Kurth I, Kunz-Schughart L, Baumann M, Dubrovska A. Discovery of the cancer stem cell related determinants of radioresistance. *Radiother Oncol.* 2013;108 (3):378–387.
18. Karamboulas C, Ailles L. Developmental signaling pathways in CSCs of solid tumors. *Biochim Biophys Acta.* 2013;1830(2):2481–2495.
19. Munoz P, Iliou MS, Esteller M. Epigenetic alterations involved in cancer stem cell reprogramming. *Mol Oncol.* 2012;6(6):620–636.
20. Ling L, Nurcombe V, Cool SM. Wnt signaling controls the fate of mesenchymal stem cells. *Gene.* 2009;433(1–2):1–7.
21. Liu S, Dontu G, Mantle ID, et al. Hedgehog signaling and bmi-1 regulate self-renewal of normal and malignant human mammary stem cells. *Cancer Res.* 2006;66(12): 6063–6071.
22. Rizzo P, Osipo C, Foreman K, Golde T, Osborne B, Miele L. Rational targeting of notch signaling in cancer. *Oncogene.* 2008;27(38):5124–5131.
23. Farnie G, Clarke RB. Mammary stem cells and breast cancer—role of notch signalling. *Stem Cell Rev.* 2007;3(2):169–175.

24. Yamanaka S. Induction of pluripotent stem cells from mouse fibroblasts by four transcription factors. *Cell Prolif.* 2008;41(suppl 1):51–56.
25. Rodriguez-Pinilla SM, Sarrio D, Moreno-Bueno G, et al. Sox2: a possible driver of the basal-like phenotype in sporadic breast cancer. *Mod Pathol.* 2007;20(4): 474–481.
26. Zhang J, Liang Q, Lei Y, et al. SOX4 induces epithelial-mesenchymal transition and contributes to breast cancer progression. *Cancer Res.* 2012;72(17):4597–4608.
27. Baccelli I, Schneeweiss A, Riethdorf S, et al. Identification of a population of blood circulating tumor cells from breast cancer patients that initiates metastasis in a xenograft assay. *Nat Biotechnol.* 2013;31(6):539–544.
28. Magbanua MJ, Sosa EV, Roy R, et al. Genomic profiling of isolated circulating tumor cells from metastatic breast cancer patients. *Cancer Res.* 2013;73(1):30–40.
29. Molnar B, Ladanyi A, Tanko L, Sreter L, Tulassay Z. Circulating tumor cell clusters in the peripheral blood of colorectal cancer patients. *Clin Cancer Res.* 2001;7(12): 4080–4085.
30. Allard WJ, Matera J, Miller MC, et al. Tumor cells circulate in the peripheral blood of all major carcinomas but not in healthy subjects or patients with nonmalignant diseases. *Clin Cancer Res.* 2004;10(20):6897–6904.
31. Theodoropoulos PA, Polioudaki H, Agelaki S, et al. Circulating tumor cells with a putative stem cell phenotype in peripheral blood of patients with breast cancer. *Cancer Lett.* 2010;288(1):99–106.
32. Sundar SS, Ganesan TS. Role of lymphangiogenesis in cancer. *J Clin Oncol.* 2007;25 (27):4298–4307.
33. Gilkes DM, Semenza GL. Role of hypoxia-inducible factors in breast cancer metastasis. *Future Oncol.* 2013;9(11):1623–1636.
34. Mimeault M, Batra SK. Hypoxia-inducing factors as master regulators of stemness properties and altered metabolism of cancer- and metastasis-initiating cells. *J Cell Mol Med.* 2013;17(1):30–54.
35. Pandit TS, Kennette W, Mackenzie L, et al. Lymphatic metastasis of breast cancer cells is associated with differential gene expression profiles that predict cancer stem cell-like properties and the ability to survive, establish and grow in a foreign environment. *Int J Oncol.* 2009;35(2):297–308.
36. Hay ED. The mesenchymal cell, its role in the embryo, and the remarkable signaling mechanisms that create it. *Dev Dyn.* 2005;233(3):706–720.
37. Thiery JP, Sleeman JP. Complex networks orchestrate epithelial-mesenchymal transitions. *Nat Rev Mol Cell Biol.* 2006;7(2):131–142.
38. Kalluri R. EMT: when epithelial cells decide to become mesenchymal-like cells. *J Clin Invest.* 2009;119(6):1417–1419.
39. Kalluri R, Weinberg RA. The basics of epithelial-mesenchymal transition. *J Clin Invest.* 2009;119(6):1420–1428.
40. Zeisberg M, Neilson EG. Biomarkers for epithelial-mesenchymal transitions. *J Clin Invest.* 2009;119(6):1429–1437.
41. Huang RY, Guilford P, Thiery JP. Early events in cell adhesion and polarity during epithelial-mesenchymal transition. *J Cell Sci.* 2012;125(Pt 19):4417–4422.
42. Yilmaz M, Christofori G. EMT, the cytoskeleton, and cancer cell invasion. *Cancer Metastasis Rev.* 2009;28(1–2):15–33.
43. Kourtidis A, Ngok SP, Anastasiadis PZ. P120 catenin: an essential regulator of cadherin stability, adhesion-induced signaling, and cancer progression. *Prog Mol Biol Transl Sci.* 2013;116:409–432.
44. Bax NA, Pijnappels DA, van Oorschot AA, et al. Epithelial-to-mesenchymal transformation alters electrical conductivity of human epicardial cells. *J Cell Mol Med.* 2011;15 (12):2675–2683.

45. Zhou BP, Deng J, Xia W, et al. Dual regulation of snail by GSK-3beta-mediated phosphorylation in control of epithelial-mesenchymal transition. *Nat Cell Biol.* 2004;6(10):931–940.
46. Yook JI, Li XY, Ota I, et al. A wnt-Axin2-GSK3beta cascade regulates Snail1 activity in breast cancer cells. *Nat Cell Biol.* 2006;8(12):1398–1406.
47. Qiao M, Sheng S, Pardee AB. Metastasis and AKT activation. *Cell Cycle.* 2008;7(19):2991–2996.
48. Sahlgren C, Gustafsson MV, Jin S, Poellinger L, Lendahl U. Notch signaling mediates hypoxia-induced tumor cell migration and invasion. *Proc Natl Acad Sci USA.* 2008;105(17):6392–6397.
49. Wu Y, Deng J, Rychahou PG, Qiu S, Evers BM, Zhou BP. Stabilization of snail by NF-kappaB is required for inflammation-induced cell migration and invasion. *Cancer Cell.* 2009;15(5):416–428.
50. Wang SP, Wang WL, Chang YL, et al. p53 controls cancer cell invasion by inducing the MDM2-mediated degradation of slug. *Nat Cell Biol.* 2009;11(6):694–704.
51. Jorda M, Olmeda D, Vinyals A, et al. Upregulation of MMP-9 in MDCK epithelial cell line in response to expression of the snail transcription factor. *J Cell Sci.* 2005;118(Pt. 15):3371–3385.
52. Peinado H, Olmeda D, Cano A. Snail, zeb and bHLH factors in tumour progression: an alliance against the epithelial phenotype? *Nat Rev Cancer.* 2007;7(6):415–428.
53. De Craene B, Berx G. Regulatory networks defining EMT during cancer initiation and progression. *Nat Rev Cancer.* 2013;13(2):97–110.
54. Xu J, Lamouille S, Derynck R. TGF-beta-induced epithelial to mesenchymal transition. *Cell Res.* 2009;19(2):156–172.
55. Yang MH, Hsu DS, Wang HW, et al. Bmi1 is essential in Twist1-induced epithelial-mesenchymal transition. *Nat Cell Biol.* 2010;12(10):982–992.
56. Yang F, Sun L, Li Q, et al. SET8 promotes epithelial-mesenchymal transition and confers TWIST dual transcriptional activities. *EMBO J.* 2012;31(1):110–123.
57. Sanchez-Tillo E, Lazaro A, Torrent R, et al. ZEB1 represses E-cadherin and induces an EMT by recruiting the SWI/SNF chromatin-remodeling protein BRG1. *Oncogene.* 2010;29(24):3490–3500.
58. Wang J, Scully K, Zhu X, et al. Opposing LSD1 complexes function in developmental gene activation and repression programmes. *Nature.* 2007;446(7138):882–887.
59. Postigo AA, Depp JL, Taylor JJ, Kroll KL. Regulation of smad signaling through a differential recruitment of coactivators and corepressors by ZEB proteins. *EMBO J.* 2003;22(10):2453–2462.
60. Wells A, Yates C, Shepard CR. E-cadherin as an indicator of mesenchymal to epithelial reverting transitions during the metastatic seeding of disseminated carcinomas. *Clin Exp Metastasis.* 2008;25(6):621–628.
61. Dykxhoorn DM, Wu Y, Xie H, et al. miR-200 enhances mouse breast cancer cell colonization to form distant metastases. *PLoS One.* 2009;4(9):e7181.
62. Chao YL, Shepard CR, Wells A. Breast carcinoma cells re-express E-cadherin during mesenchymal to epithelial reverting transition. *Mol Cancer.* 2010;9:179.
63. Chaffer CL, Dopheide B, Savagner P, Thompson EW, Williams ED. Aberrant fibroblast growth factor receptor signaling in bladder and other cancers. *Differentiation.* 2007;75(9):831–842.
64. Baraniak AP, Lasda EL, Wagner EJ, Garcia-Blanco MA. A stem structure in fibroblast growth factor receptor 2 transcripts mediates cell-type-specific splicing by approximating intronic control elements. *Mol Cell Biol.* 2003;23(24):9327–9337.
65. Baraniak AP, Chen JR, Garcia-Blanco MA. Fox-2 mediates epithelial cell-specific fibroblast growth factor receptor 2 exon choice. *Mol Cell Biol.* 2006;26(4):1209–1222.

66. Yates CC, Shepard CR, Stolz DB, Wells A. Co-culturing human prostate carcinoma cells with hepatocytes leads to increased expression of E-cadherin. *Br J Cancer.* 2007;96 (8):1246–1252.
67. Yates C, Wells A, Turner T. Luteinising hormone-releasing hormone analogue reverses the cell adhesion profile of EGFR overexpressing DU-145 human prostate carcinoma subline. *Br J Cancer.* 2005;92(2):366–375.
68. Zhou XD, Agazie YM. Inhibition of SHP2 leads to mesenchymal to epithelial transition in breast cancer cells. *Cell Death Differ.* 2008;15(6):988–996.
69. Kang MH, Kang HN, Kim JL, Kim JS, Oh SC, Yoo YA. Inhibition of PI3 kinase/akt pathway is required for BMP2-induced EMT and invasion. *Oncol Rep.* 2009;22 (3):525–534.
70. Brabletz T, Jung A, Reu S, et al. Variable beta-catenin expression in colorectal cancers indicates tumor progression driven by the tumor environment. *Proc Natl Acad Sci USA.* 2001;98(18):10356–10361.
71. Yee DS, Tang Y, Li X, et al. The wnt inhibitory factor 1 restoration in prostate cancer cells was associated with reduced tumor growth, decreased capacity of cell migration and invasion and a reversal of epithelial to mesenchymal transition. *Mol Cancer.* 2010;9:162.
72. Wissmann C, Wild PJ, Kaiser S, et al. WIF1, a component of the wnt pathway, is down-regulated in prostate, breast, lung, and bladder cancer. *J Pathol.* 2003;201(2):204–212.
73. Gupta S, Iljin K, Sara H, et al. FZD4 as a mediator of ERG oncogene-induced WNT signaling and epithelial-to-mesenchymal transition in human prostate cancer cells. *Cancer Res.* 2010;70(17):6735–6745.
74. Campbell JJ, Hedrick J, Zlotnik A, Siani MA, Thompson DA, Butcher EC. Chemokines and the arrest of lymphocytes rolling under flow conditions. *Science.* 1998;279(5349):381–384.
75. Kuang Y, Wu Y, Jiang H, Wu D. Selective G protein coupling by C-C chemokine receptors. *J Biol Chem.* 1996;271(8):3975–3978.
76. Porter DA, Krop IE, Nasser S, et al. A SAGE (serial analysis of gene expression) view of breast tumor progression. *Cancer Res.* 2001;61(15):5697–5702.
77. Greene GF, Kitadai Y, Pettaway CA, von Eschenbach AC, Bucana CD, Fidler IJ. Correlation of metastasis-related gene expression with metastatic potential in human prostate carcinoma cells implanted in nude mice using an in situ messenger RNA hybridization technique. *Am J Pathol.* 1997;150(5):1571–1582.
78. Bieche I, Chavey C, Andrieu C, et al. CXC chemokines located in the 4q21 region are up-regulated in breast cancer. *Endocr Relat Cancer.* 2007;14(4):1039–1052.
79. Kozlowski L, Zakrzewska I, Tokajuk P, Wojtukiewicz MZ. Concentration of interleukin-6 (IL-6), interleukin-8 (IL-8) and interleukin-10 (IL-10) in blood serum of breast cancer patients. *Rocz Akad Med Bialymst.* 2003;48:82–84.
80. Benoy IH, Salgado R, Van Dam P, et al. Increased serum interleukin-8 in patients with early and metastatic breast cancer correlates with early dissemination and survival. *Clin Cancer Res.* 2004;10(21):7157–7162.
81. Freund A, Chauveau C, Brouillet JP, et al. IL-8 expression and its possible relationship with estrogen-receptor-negative status of breast cancer cells. *Oncogene.* 2003;22 (2):256–265.
82. Lin Y, Huang R, Chen L, et al. Identification of interleukin-8 as estrogen receptor-regulated factor involved in breast cancer invasion and angiogenesis by protein arrays. *Int J Cancer.* 2004;109(4):507–515.
83. Muller A, Homey B, Soto H, et al. Involvement of chemokine receptors in breast cancer metastasis. *Nature.* 2001;410(6824):50–56.
84. Kucia M, Reca R, Miekus K, et al. Trafficking of normal stem cells and metastasis of CSCs involve similar mechanisms: pivotal role of the SDF-1-CXCR4 axis. *Stem Cells.* 2005;23(7):879–894.

85. Nagasawa T, Nakajima T, Tachibana K, et al. Molecular cloning and characterization of a murine pre-B-cell growth-stimulating factor/stromal cell-derived factor 1 receptor, a murine homolog of the human immunodeficiency virus 1 entry coreceptor fusin. *Proc Natl Acad Sci USA*. 1996;93(25):14726–14729.
86. Ma Q, Jones D, Springer TA. The chemokine receptor CXCR4 is required for the retention of B lineage and granulocytic precursors within the bone marrow microenvironment. *Immunity*. 1999;10(4):463–471.
87. Bagri A, Gurney T, He X, et al. The chemokine SDF1 regulates migration of dentate granule cells. *Development*. 2002;129(18):4249–4260.
88. Lazarini F, Tham TN, Casanova P, Arenzana-Seisdedos F, Dubois-Dalcq M. Role of the alpha-chemokine stromal cell-derived factor (SDF-1) in the developing and mature central nervous system. *Glia*. 2003;42(2):139–148.
89. Zou YR, Kottmann AH, Kuroda M, Taniuchi I, Littman DR. Function of the chemokine receptor CXCR4 in haematopoiesis and in cerebellar development. *Nature*. 1998;393(6685):595–599.
90. Tachibana K, Hirota S, Iizasa H, et al. The chemokine receptor CXCR4 is essential for vascularization of the gastrointestinal tract. *Nature*. 1998;393(6685):591–594.
91. Sun YX, Wang J, Shelburne CE, et al. Expression of CXCR4 and CXCL12 (SDF-1) in human prostate cancers (PCa) in vivo. *J Cell Biochem*. 2003;89(3):462–473.
92. Geminder H, Sagi-Assif O, Goldberg L, et al. A possible role for CXCR4 and its ligand, the CXC chemokine stromal cell-derived factor-1, in the development of bone marrow metastases in neuroblastoma. *J Immunol*. 2001;167(8):4747–4757.
93. Porcile C, Bajetto A, Barbero S, Pirani P, Schettini G. CXCR4 activation induces epidermal growth factor receptor transactivation in an ovarian cancer cell line. *Ann NY Acad Sci*. 2004;1030:162–169.
94. Hall JM, Korach KS. Stromal cell-derived factor 1, a novel target of estrogen receptor action, mediates the mitogenic effects of estradiol in ovarian and breast cancer cells. *Mol Endocrinol*. 2003;17(5):792–803.
95. Libura J, Drukala J, Majka M, et al. CXCR4-SDF-1 signaling is active in rhabdomyosarcoma cells and regulates locomotion, chemotaxis, and adhesion. *Blood*. 2002;100 (7):2597–2606.
96. Jankowski K, Kucia M, Wysoczynski M, et al. Both hepatocyte growth factor (HGF) and stromal-derived factor-1 regulate the metastatic behavior of human rhabdomyosarcoma cells, but only HGF enhances their resistance to radiochemotherapy. *Cancer Res*. 2003;63(22):7926–7935.
97. Schioppa T, Uranchimeg B, Saccani A, et al. Regulation of the chemokine receptor CXCR4 by hypoxia. *J Exp Med*. 2003;198(9):1391–1402.
98. Helbig G, Christopherson 2nd KW, Bhat-Nakshatri P, et al. NF-kappaB promotes breast cancer cell migration and metastasis by inducing the expression of the chemokine receptor CXCR4. *J Biol Chem*. 2003;278(24):21631–21638.
99. Curnow SJ, Wloka K, Faint JM, et al. Topical glucocorticoid therapy directly induces up-regulation of functional CXCR4 on primed T lymphocytes in the aqueous humor of patients with uveitis. *J Immunol*. 2004;172(11):7154–7161.
100. Franitza S, Kollet O, Brill A, et al. TGF-beta1 enhances SDF-1alpha-induced chemotaxis and homing of naive T cells by up-regulating CXCR4 expression and downstream cytoskeletal effector molecules. *Eur J Immunol*. 2002;32(1):193–202.
101. Han KH, Hong KH, Ko J, et al. Lysophosphatidylcholine up-regulates CXCR4 chemokine receptor expression in human CD4 T cells. *J Leukoc Biol*. 2004;76 (1):195–202.
102. Bachelder RE, Wendt MA, Mercurio AM. Vascular endothelial growth factor promotes breast carcinoma invasion in an autocrine manner by regulating the chemokine receptor CXCR4. *Cancer Res*. 2002;62(24):7203–7206.

103. Yonezawa A, Morita R, Takaori-Kondo A, et al. Natural alpha interferon-producing cells respond to human immunodeficiency virus type 1 with alpha interferon production and maturation into dendritic cells. *J Virol.* 2003;77(6):3777–3784.
104. Iikura M, Miyamasu M, Yamaguchi M, et al. Chemokine receptors in human basophils: inducible expression of functional CXCR4. *J Leukoc Biol.* 2001;70(1):113–120.
105. Jourdan P, Vendrell JP, Huguet MF, et al. Cytokines and cell surface molecules independently induce CXCR4 expression on CD4+ CCR7+ human memory T cells. *J Immunol.* 2000;165(2):716–724.
106. Peled A, Grabovsky V, Habler L, et al. The chemokine SDF-1 stimulates integrin-mediated arrest of CD34(+) cells on vascular endothelium under shear flow. *J Clin Invest.* 1999;104(9):1199–1211.
107. Aiuti A, Webb IJ, Bleul C, Springer T, Gutierrez-Ramos JC. The chemokine SDF-1 is a chemoattractant for human CD34+ hematopoietic progenitor cells and provides a new mechanism to explain the mobilization of CD34+ progenitors to peripheral blood. *J Exp Med.* 1997;185(1):111–120.
108. Ma Q, Jones D, Borghesani PR, et al. Impaired B-lymphopoiesis, myelopoiesis, and derailed cerebellar neuron migration in CXCR4- and SDF-1-deficient mice. *Proc Natl Acad Sci USA.* 1998;95(16):9448–9453.
109. Melgarejo E, Medina MA, Sanchez-Jimenez F, Urdiales JL. Monocyte chemoattractant protein-1: a key mediator in inflammatory processes. *Int J Biochem Cell Biol.* 2009;41(5):998–1001.
110. Soria G, Ben-Baruch A. The inflammatory chemokines CCL2 and CCL5 in breast cancer. *Cancer Lett.* 2008;267(2):271–285.
111. Tsuyada A, Chow A, Wu J, et al. CCL2 mediates cross-talk between cancer cells and stromal fibroblasts that regulates breast CSCs. *Cancer Res.* 2012;72(11):2768–2779.
112. Ueno T, Toi M, Saji H, et al. Significance of macrophage chemoattractant protein-1 in macrophage recruitment, angiogenesis, and survival in human breast cancer. *Clin Cancer Res.* 2000;6(8):3282–3289.
113. Saji H, Koike M, Yamori T, et al. Significant correlation of monocyte chemoattractant protein-1 expression with neovascularization and progression of breast carcinoma. *Cancer.* 2001;92(5):1085–1091.
114. Chin AR, Wang SE. Cytokines driving breast cancer stemness. *Mol Cell Endocrinol.* 2014;382(1):598–602.
115. Culig Z. Cytokine disbalance in common human cancers. *Biochim Biophys Acta.* 2011;1813(2):308–314.
116. Derynck R, Zhang YE. Smad-dependent and smad-independent pathways in TGF-beta family signalling. *Nature.* 2003;425(6958):577–584.
117. Frey RS, Mulder KM. Involvement of extracellular signal-regulated kinase 2 and stress-activated protein kinase/jun N-terminal kinase activation by transforming growth factor beta in the negative growth control of breast cancer cells. *Cancer Res.* 1997;57(4):628–633.
118. Inman GJ. Switching TGFbeta from a tumor suppressor to a tumor promoter. *Curr Opin Genet Dev.* 2011;21(1):93–99.
119. Joshi A, Cao D. TGF-beta signaling, tumor microenvironment and tumor progression: the butterfly effect. *Front Biosci (Landmark Ed).* 2010;15:180–194.
120. Meulmeester E, Ten Dijke P. The dynamic roles of TGF-beta in cancer. *J Pathol.* 2011;223(2):205–218.
121. Vincent T, Neve EP, Johnson JR, et al. A SNAIL1-SMAD3/4 transcriptional repressor complex promotes TGF-beta mediated epithelial-mesenchymal transition. *Nat Cell Biol.* 2009;11(8):943–950.
122. Voulgari A, Pintzas A. Epithelial-mesenchymal transition in cancer metastasis: mechanisms, markers and strategies to overcome drug resistance in the clinic. *Biochim Biophys Acta.* 2009;1796(2):75–90.

123. Wendt MK, Allington TM, Schiemann WP. Mechanisms of the epithelial-mesenchymal transition by TGF-beta. *Future Oncol.* 2009;5(8):1145–1168.
124. Jain P, Alahari SK. Breast CSCs: a new challenge for breast cancer treatment. *Front Biosci (Landmark Ed).* 2011;16:1824–1832.
125. Taube JH, Herschkowitz JI, Komurov K, et al. Core epithelial-to-mesenchymal transition interactome gene-expression signature is associated with claudin-low and metaplastic breast cancer subtypes. *Proc Natl Acad Sci USA.* 2010;107(35):15449–15454.
126. Shipitsin M, Campbell LL, Argani P, et al. Molecular definition of breast tumor heterogeneity. *Cancer Cell.* 2007;11(3):259–273.
127. Korkaya H, Kim GI, Davis A, et al. Activation of an IL6 inflammatory loop mediates trastuzumab resistance in HER2+ breast cancer by expanding the cancer stem cell population. *Mol Cell.* 2012;47(4):570–584.
128. Karnoub AE, Dash AB, Vo AP, et al. Mesenchymal stem cells within tumour stroma promote breast cancer metastasis. *Nature.* 2007;449(7162):557–563.
129. Marotta LL, Almendro V, Marusyk A, et al. The JAK2/STAT3 signaling pathway is required for growth of CD44(+)CD24(-) stem cell-like breast cancer cells in human tumors. *J Clin Invest.* 2011;121(7):2723–2735.
130. Waugh DJ, Wilson C. The interleukin-8 pathway in cancer. *Clin Cancer Res.* 2008;14(21):6735–6741.
131. Charafe-Jauffret E, Ginestier C, Iovino F, et al. Breast cancer cell lines contain functional CSCs with metastatic capacity and a distinct molecular signature. *Cancer Res.* 2009;69(4):1302–1313.
132. Ginestier C, Liu S, Diebel ME, et al. CXCR1 blockade selectively targets human breast CSCs in vitro and in xenografts. *J Clin Invest.* 2010;120(2):485–497.
133. Wirtz D. Particle-tracking microrheology of living cells: Principles and applications. *Annu Rev Biophys.* 2009;38:301–326.
134. Tseng Y, Lee JS, Kole TP, Jiang I, Wirtz D. Micro-organization and visco-elasticity of the interphase nucleus revealed by particle nanotracking. *J Cell Sci.* 2004;117(Pt. 10):2159–2167.
135. Dahl KN, Kahn SM, Wilson KL, Discher DE. The nuclear envelope lamina network has elasticity and a compressibility limit suggestive of a molecular shock absorber. *J Cell Sci.* 2004;117(Pt. 20):4779–4786.
136. Gerlitz G, Bustin M. Efficient cell migration requires global chromatin condensation. *J Cell Sci.* 2010;123(Pt. 13):2207–2217.
137. Crisp M, Liu Q, Roux K, et al. Coupling of the nucleus and cytoplasm: role of the LINC complex. *J Cell Biol.* 2006;172(1):41–53.
138. Stewart-Hutchinson PJ, Hale CM, Wirtz D, Hodzic D. Structural requirements for the assembly of LINC complexes and their function in cellular mechanical stiffness. *Exp Cell Res.* 2008;314(8):1892–1905.
139. Hale CM, Shrestha AL, Khatau SB, et al. Dysfunctional connections between the nucleus and the actin and microtubule networks in laminopathic models. *Biophys J.* 2008;95(11):5462–5475.
140. Wood LD, Parsons DW, Jones S, et al. The genomic landscapes of human breast and colorectal cancers. *Science.* 2007;318(5853):1108–1113.
141. Cross SE, Jin YS, Rao J, Gimzewski JK. Nanomechanical analysis of cells from cancer patients. *Nat Nanotechnol.* 2007;2(12):780–783.
142. Guck J, Schinkinger S, Lincoln B, et al. Optical deformability as an inherent cell marker for testing malignant transformation and metastatic competence. *Biophys J.* 2005;88(5):3689–3698.
143. Kienast Y, von Baumgarten L, Fuhrmann M, et al. Real-time imaging reveals the single steps of brain metastasis formation. *Nat Med.* 2010;16(1):116–122.

144. Joyce JA, Pollard JW. Microenvironmental regulation of metastasis. *Nat Rev Cancer.* 2009;9(4):239–252.
145. Hanahan D, Coussens LM. Accessories to the crime: functions of cells recruited to the tumor microenvironment. *Cancer Cell.* 2012;21(3):309–322.
146. Hanahan D, Weinberg RA. Hallmarks of cancer: the next generation. *Cell.* 2011;144(5):646–674.
147. Qian BZ, Pollard JW. Macrophage diversity enhances tumor progression and metastasis. *Cell.* 2010;141(1):39–51.
148. Condeelis J, Pollard JW. Macrophages: obligate partners for tumor cell migration, invasion, and metastasis. *Cell.* 2006;124(2):263–266.
149. Goswami S, Sahai E, Wyckoff JB, et al. Macrophages promote the invasion of breast carcinoma cells via a colony-stimulating factor-1/epidermal growth factor paracrine loop. *Cancer Res.* 2005;65(12):5278–5283.
150. Coniglio SJ, Eugenin E, Dobrenis K, et al. Microglial stimulation of glioblastoma invasion involves epidermal growth factor receptor (EGFR) and colony stimulating factor 1 receptor (CSF-1R) signaling. *Mol Med.* 2012;18:519–527.
151. Joyce JA, Baruch A, Chehade K, et al. Cathepsin cysteine proteases are effectors of invasive growth and angiogenesis during multistage tumorigenesis. *Cancer Cell.* 2004;5(5):443–453.
152. Gocheva V, Wang HW, Gadea BB, et al. IL-4 induces cathepsin protease activity in tumor-associated macrophages to promote cancer growth and invasion. *Genes Dev.* 2010;24(3):241–255.
153. Shree T, Olson OC, Elie BT, et al. Macrophages and cathepsin proteases blunt chemotherapeutic response in breast cancer. *Genes Dev.* 2011;25(23):2465–2479.
154. Biswas SK, Mantovani A. Macrophage plasticity and interaction with lymphocyte subsets: cancer as a paradigm. *Nat Immunol.* 2010;11(10):889–896.
155. Lu H, Clauser KR, Tam WL, et al. A breast cancer stem cell niche supported by juxtacrine signalling from monocytes and macrophages. *Nat Cell Biol.* 2014;16(11):1105–1117.
156. Jinushi M, Chiba S, Yoshiyama H, et al. Tumor-associated macrophages regulate tumorigenicity and anticancer drug responses of cancer stem/initiating cells. *Proc Natl Acad Sci USA.* 2011;108(30):12425–12430.
157. Yang J, Liao D, Chen C, et al. Tumor-associated macrophages regulate murine breast CSCs through a novel paracrine EGFR/Stat3/sox-2 signaling pathway. *Stem Cells.* 2013;31(2):248–258.
158. Mao Y, Keller ET, Garfield DH, Shen K, Wang J. Stromal cells in tumor microenvironment and breast cancer. *Cancer Metastasis Rev.* 2013;32(1–2):303–315.
159. Adams EF, Coldham NG, James VH. Steroidal regulation of oestradiol-17 beta dehydrogenase activity of the human breast cancer cell line MCF-7. *J Endocrinol.* 1988;118(1):149–154.
160. Stuelten CH, Busch JI, Tang B, et al. Transient tumor-fibroblast interactions increase tumor cell malignancy by a TGF-beta mediated mechanism in a mouse xenograft model of breast cancer. *PLoS One.* 2010;5(3):e9832.
161. Hugo HJ, Lebret S, Tomaskovic-Crook E, et al. Contribution of fibroblast and mast cell (afferent) and tumor (efferent) IL-6 effects within the tumor microenvironment. *Cancer Microenviron.* 2012;5(1):83–93.
162. Hu M, Peluffo G, Chen H, Gelman R, Schnitt S, Polyak K. Role of COX-2 in epithelial-stromal cell interactions and progression of ductal carcinoma in situ of the breast. *Proc Natl Acad Sci USA.* 2009;106(9):3372–3377.
163. Duda DG, Cohen KS, Kozin SV, et al. Evidence for incorporation of bone marrow-derived endothelial cells into perfused blood vessels in tumors. *Blood.* 2006;107(7):2774–2776.

164. Lyden D, Hattori K, Dias S, et al. Impaired recruitment of bone-marrow-derived endothelial and hematopoietic precursor cells blocks tumor angiogenesis and growth. *Nat Med.* 2001;7(11):1194–1201.
165. Grunewald M, Avraham I, Dor Y, et al. VEGF-induced adult neovascularization: recruitment, retention, and role of accessory cells. *Cell.* 2006;124(1):175–189.
166. Spring H, Schuler T, Arnold B, Hammerling GJ, Ganss R. Chemokines direct endothelial progenitors into tumor neovessels. *Proc Natl Acad Sci USA.* 2005;102(50): 18111–18116.
167. Ghiabi P, Jiang J, Pasquier J, et al. Endothelial cells provide a notch-dependent protumoral niche for enhancing breast cancer survival, stemness and pro-metastatic properties. *PLoS One.* 2014;9(11):e112424.
168. Wiseman BS, Werb Z. Stromal effects on mammary gland development and breast cancer. *Science.* 2002;296(5570):1046–1049.
169. Subbaramaiah K, Howe LR, Bhardwaj P, et al. Obesity is associated with inflammation and elevated aromatase expression in the mouse mammary gland. *CancerPrev Res(Phila).* 2011;4(3):329–346.
170. Park J, Euhus DM, Scherer PE. Paracrine and endocrine effects of adipose tissue on cancer development and progression. *Endocr Rev.* 2011;32(4):550–570.
171. Celis JE, Moreira JM, Cabezon T, et al. Identification of extracellular and intracellular signaling components of the mammary adipose tissue and its interstitial fluid in high risk breast cancer patients: toward dissecting the molecular circuitry of epithelial-adipocyte stromal cell interactions. *Mol Cell Proteomics.* 2005;4(4):492–522.
172. Dirat B, Bochet L, Dabek M, et al. Cancer-associated adipocytes exhibit an activated phenotype and contribute to breast cancer invasion. *Cancer Res.* 2011;71(7): 2455–2465.
173. D'Esposito V, Passaretti F, Hammarstedt A, et al. Adipocyte-released insulin-like growth factor-1 is regulated by glucose and fatty acids and controls breast cancer cell growth in vitro. *Diabetologia.* 2012;55(10):2811–2822.
174. Miyoshi Y, Funahashi T, Tanaka S, et al. High expression of leptin receptor mRNA in breast cancer tissue predicts poor prognosis for patients with high, but not low, serum leptin levels. *Int J Cancer.* 2006;118(6):1414–1419.
175. Schaffler A, Scholmerich J, Buechler C. Mechanisms of disease: adipokines and breast cancer—endocrine and paracrine mechanisms that connect adiposity and breast cancer. *Nat Clin Pract Endocrinol Metab.* 2007;3(4):345–354.
176. Dalmas E, Tordjman J, Guerre-Millo M, Clement K. Adipose tissue, a new playground for immune cells. *Med Sci (Paris).* 2011;27(11):993–999.
177. Whittaker CA, Bergeron KF, Whittle J, Brandhorst BP, Burke RD, Hynes RO. The echinoderm adhesome. *Dev Biol.* 2006;300(1):252–266.
178. Ozbek S, Balasubramanian PG, Chiquet-Ehrismann R, Tucker RP, Adams JC. The evolution of extracellular matrix. *Mol Biol Cell.* 2010;21(24):4300–4305.
179. Egeblad M, Rasch MG, Weaver VM. Dynamic interplay between the collagen scaffold and tumor evolution. *Curr Opin Cell Biol.* 2010;22(5):697–706.
180. O'Brien J, Lyons T, Monks J, et al. Alternatively activated macrophages and collagen remodeling characterize the postpartum involuting mammary gland across species. *Am J Pathol.* 2010;176(3):1241–1255.
181. Gudjonsson T, Ronnov-Jessen L, Villadsen R, Rank F, Bissell MJ, Petersen OW. Normal and tumor-derived myoepithelial cells differ in their ability to interact with luminal breast epithelial cells for polarity and basement membrane deposition. *J Cell Sci.* 2002;115(Pt. 1):39–50.
182. Hamilton SR, Fard SF, Paiwand FF, et al. The hyaluronan receptors CD44 and rhamm (CD168) form complexes with ERK1,2 that sustain high basal motility in breast cancer cells. *J Biol Chem.* 2007;282(22):16667–16680.

183. Auvinen P, Tammi R, Kosma VM, et al. Increased hyaluronan content and stromal cell CD44 associate with HER2 positivity and poor prognosis in human breast cancer. *Int J Cancer.* 2013;132(3):531–539.
184. Wozniak MA, Desai R, Solski PA, Der CJ, Keely PJ. ROCK-generated contractility regulates breast epithelial cell differentiation in response to the physical properties of a three-dimensional collagen matrix. *J Cell Biol.* 2003;163(3):583–595.
185. Paszek MJ, Zahir N, Johnson KR, et al. Tensional homeostasis and the malignant phenotype. *Cancer Cell.* 2005;8(3):241–254.
186. Mott JD, Werb Z. Regulation of matrix biology by matrix metalloproteinases. *Curr Opin Cell Biol.* 2004;16(5):558–564.
187. Yamashita YM, Fuller MT, Jones DL. Signaling in stem cell niches: lessons from the drosophila germline. *J Cell Sci.* 2005;118(Pt 4):665–672.
188. Tanentzapf G, Devenport D, Godt D, Brown NH. Integrin-dependent anchoring of a stem-cell niche. *Nat Cell Biol.* 2007;9(12):1413–1418.
189. Kollet O, Dar A, Shivtiel S, et al. Osteoclasts degrade endosteal components and promote mobilization of hematopoietic progenitor cells. *Nat Med.* 2006;12(6):657–664.
190. Lymperi S, Ferraro F, Scadden DT. The HSC niche concept has turned 31. Has our knowledge matured? *Ann NY Acad Sci.* 2010;1192:12–18.
191. Garcion E, Halilagic A, Faissner A, ffrench-Constant C. Generation of an environmental niche for neural stem cell development by the extracellular matrix molecule tenascin C. *Development.* 2004;131(14):3423–3432.
192. Bi Y, Ehirchiou D, Kilts TM, et al. Identification of tendon stem/progenitor cells and the role of the extracellular matrix in their niche. *Nat Med.* 2007;13(10):1219–1227.
193. Li N, Zhang Y, Naylor MJ, et al. Beta1 integrins regulate mammary gland proliferation and maintain the integrity of mammary alveoli. *EMBO J.* 2005;24(11):1942–1953.
194. Lambert JF, Colvin GA, Zhong S, et al. H2-mismatched transplantation with repetitive cell infusions and CD40 ligand antibody infusions without myeloablation. *Br J Haematol.* 2002;119(1):155–163.
195. Fuchs E, Tumbar T, Guasch G. Socializing with the neighbors: stem cells and their niche. *Cell.* 2004;116(6):769–778.
196. Lechler T, Fuchs E. Asymmetric cell divisions promote stratification and differentiation of mammalian skin. *Nature.* 2005;437(7056):275–280.
197. Yamashita YM. Selfish stem cells compete with each other. *Cell Stem Cell.* 2008;2(1):3–4.
198. Tsunoda T, Inada H, Kalembeyi I, et al. Involvement of large tenascin-C splice variants in breast cancer progression. *Am J Pathol.* 2003;162(6):1857–1867.
199. Oskarsson T, Acharyya S, Zhang XH, et al. Breast cancer cells produce tenascin C as a metastatic niche component to colonize the lungs. *Nat Med.* 2011;17(7):867–874.
200. Maruhashi T, Kii I, Saito M, Kudo A. Interaction between periostin and BMP-1 promotes proteolytic activation of lysyl oxidase. *J Biol Chem.* 2010;285(17): 13294–13303.
201. Oskarsson T, Massague J. Extracellular matrix players in metastatic niches. *EMBO J.* 2012;31(2):254–256.
202. Wang GL, Semenza GL. Purification and characterization of hypoxia-inducible factor 1. *J Biol Chem.* 1995;270(3):1230–1237.
203. Wang GL, Jiang BH, Rue EA, Semenza GL. Hypoxia-inducible factor 1 is a basic-helix-loop-helix-PAS heterodimer regulated by cellular O2 tension. *Proc Natl Acad Sci USA.* 1995;92(12):5510–5514.
204. Semenza GL. HIF-1: upstream and downstream of cancer metabolism. *Curr Opin Genet Dev.* 2010;20(1):51–56.
205. Vaupel P. The role of hypoxia-induced factors in tumor progression. *Oncologist.* 2004;9(suppl. 5):10–17.

206. Maxwell PH, Wiesener MS, Chang GW, et al. The tumour suppressor protein VHL targets hypoxia-inducible factors for oxygen-dependent proteolysis. *Nature.* 1999;399 (6733):271–275.
207. Salceda S, Caro J. Hypoxia-inducible factor 1alpha (HIF-1alpha) protein is rapidly degraded by the ubiquitin-proteasome system under normoxic conditions. Its stabilization by hypoxia depends on redox-induced changes. *J Biol Chem.* 1997;272 (36):22642–22647.
208. Ratcliffe PJ. Oxygen sensing and hypoxia signalling pathways in animals: the implications of physiology for cancer. *J Physiol.* 2013;591(8):2027–2042.
209. Semenza GL. Oxygen sensing, hypoxia-inducible factors, and disease pathophysiology. *Annu Rev Pathol.* 2014;9:47–71.
210. Bennewith KL, Durand RE. Quantifying transient hypoxia in human tumor xenografts by flow cytometry. *Cancer Res.* 2004;64(17):6183–6189.
211. Brurberg KG, Thuen M, Ruud EB, Rofstad EK. Fluctuations in pO2 in irradiated human melanoma xenografts. *Radiat Res.* 2006;165(1):16–25.
212. Cabarcas SM, Mathews LA, Farrar WL. The cancer stem cell niche—there goes the neighborhood? *Int J Cancer.* 2011;129(10):2315–2327.
213. Iriondo O, Rabano M, Domenici G, et al. Distinct breast cancer stem/progenitor cell populations require either HIF1alpha or loss of PHD3 to expand under hypoxic conditions. *Oncotarget.* 2015;6(31):31721–31739.
214. Liu SP, Fu RH, Wu DC, et al. Mouse-induced pluripotent stem cells generated under hypoxic conditions in the absence of viral infection and oncogenic factors and used for ischemic stroke therapy. *Stem Cells Dev.* 2014;23(4):421–433.
215. Heddleston JM, Li Z, Lathia JD, Bao S, Hjelmeland AB, Rich JN. Hypoxia inducible factors in CSCs. *Br J Cancer.* 2010;102(5):789–795.
216. Yatabe N, Kyo S, Maida Y, et al. HIF-1-mediated activation of telomerase in cervical cancer cells. *Oncogene.* 2004;23(20):3708–3715.
217. Duncan R, Izzo L. Dendrimer biocompatibility and toxicity. *Adv Drug Deliv Rev.* 2005;57(15):2215–2237.
218. Nori A, Kopecek J. Intracellular targeting of polymer-bound drugs for cancer chemotherapy. *Adv Drug Deliv Rev.* 2005;57(4):609–636.
219. Kabanov AV, Alakhov VY. Pluronic block copolymers in drug delivery: from micellar nanocontainers to biological response modifiers. *Crit Rev Ther Drug Carrier Syst.* 2002;19 (1):1–72.
220. Batrakova EV, Kabanov AV. Pluronic block copolymers: evolution of drug delivery concept from inert nanocarriers to biological response modifiers. *J Control Release.* 2008;130(2):98–106.
221. Bae Y, Kataoka K. Intelligent polymeric micelles from functional poly(ethylene glycol)-poly(amino acid) block copolymers. *Adv Drug Deliv Rev.* 2009;61(10): 768–784.
222. Kakizawa Y, Kataoka K. Block copolymer micelles for delivery of gene and related compounds. *Adv Drug Deliv Rev.* 2002;54(2):203–222.
223. Torchilin VP. Recent advances with liposomes as pharmaceutical carriers. *Nat Rev Drug Discov.* 2005;4(2):145–160.
224. Rana S, Bajaj A, Mout R, Rotello VM. Monolayer coated gold nanoparticles for delivery applications. *Adv Drug Deliv Rev.* 2012;64(2):200–216.
225. Ghosh P, Han G, De M, Kim CK, Rotello VM. Gold nanoparticles in delivery applications. *Adv Drug Deliv Rev.* 2008;60(11):1307–1315.
226. Sun C, Qu R, Chen H, et al. Degradation behavior of chitosan chains in the "green" synthesis of gold nanoparticles. *Carbohydr Res.* 2008;343(15):2595–2599.
227. Constantinides PP, Chaubal MV, Shorr R. Advances in lipid nanodispersions for parenteral drug delivery and targeting. *Adv Drug Deliv Rev.* 2008;60(6):757–767.

228. Gener P, Gouveia LP, Sabat GR, et al. Fluorescent CSC models evidence that targeted nanomedicines improve treatment sensitivity of breast and colon CSCs. *Nanomedicine.* 2015;11(8):1883–1892.
229. Toole BP. Hyaluronan: from extracellular glue to pericellular cue. *Nat Rev Cancer.* 2004;4(7):528–539.
230. Lesley J, Hascall VC, Tammi M, Hyman R. Hyaluronan binding by cell surface CD44. *J Biol Chem.* 2000;275(35):26967–26975.
231. Lesley J, English N, Charles C, Hyman R. The role of the CD44 cytoplasmic and transmembrane domains in constitutive and inducible hyaluronan binding. *Eur J Immunol.* 2000;30(1):245–253.
232. Muntimadugu E, Kumar R, Saladi S, Rafeeqi TA, Khan W. CD44 targeted chemotherapy for co-eradication of breast CSCs and cancer cells using polymeric nanoparticles of salinomycin and paclitaxel. *Colloids Surf B Biointerfaces.* 2016;143:532–546.
233. Wei X, Senanayake TH, Warren G, Vinogradov SV. Hyaluronic acid-based nanogel-drug conjugates with enhanced anticancer activity designed for the targeting of CD44-positive and drug-resistant tumors. *Bioconjug Chem.* 2013;24(4):658–668.
234. Taipale J, Chen JK, Cooper MK, et al. Effects of oncogenic mutations in smoothened and patched can be reversed by cyclopamine. *Nature.* 2000;406(6799):1005–1009.
235. Bar EE, Chaudhry A, Lin A, et al. Cyclopamine-mediated hedgehog pathway inhibition depletes stem-like cancer cells in glioblastoma. *Stem Cells.* 2007;25(10):2524–2533.
236. Von Hoff DD, LoRusso PM, Rudin CM, et al. Inhibition of the hedgehog pathway in advanced basal-cell carcinoma. *N Engl J Med.* 2009;361(12):1164–1172.
237. Olive KP, Jacobetz MA, Davidson CJ, et al. Inhibition of hedgehog signaling enhances delivery of chemotherapy in a mouse model of pancreatic cancer. *Science.* 2009;324 (5933):1457–1461.
238. Bora-Singhal N, Perumal D, Nguyen J, Chellappan S. Gli1-mediated regulation of Sox2 facilitates self-renewal of stem-like cells and confers resistance to EGFR inhibitors in non-small cell lung cancer. *Neoplasia.* 2015;17(7):538–551.
239. Munchhof MJ, Li Q, Shavnya A, et al. Discovery of PF-04449913, a potent and orally bioavailable inhibitor of smoothened. *ACS Med Chem Lett.* 2011;3(2):106–111.
240. Hyman JM, Firestone AJ, Heine VM, et al. Small-molecule inhibitors reveal multiple strategies for hedgehog pathway blockade. *Proc Natl Acad Sci USA.* 2009;106(33): 14132–14137.
241. Lauth M, Bergstrom A, Shimokawa T, Toftgard R. Inhibition of GLI-mediated transcription and tumor cell growth by small-molecule antagonists. *Proc Natl Acad Sci USA.* 2007;104(20):8455–8460.
242. Jang GB, Hong IS, Kim RJ, et al. Wnt/beta-catenin small-molecule inhibitor CWP232228 preferentially inhibits the growth of breast cancer stem-like cells. *Cancer Res.* 2015;75(8):1691–1702.
243. Shim Y, Song JM. Quantum dot nanoprobe-based high-content monitoring of notch pathway inhibition of breast cancer stem cell by capsaicin. *Mol Cell Probes.* 2015;29 (6):376–381.
244. Tsai JH, Hsu LS, Lin CL, et al. 3,5,4′-trimethoxystilbene, a natural methoxylated analog of resveratrol, inhibits breast cancer cell invasiveness by downregulation of PI3K/akt and wnt/beta-catenin signaling cascades and reversal of epithelial-mesenchymal transition. *Toxicol Appl Pharmacol.* 2013;272(3):746–756.
245. Bidwell BN, Slaney CY, Withana NP, et al. Silencing of Irf7 pathways in breast cancer cells promotes bone metastasis through immune escape. *Nat Med.* 2012;18(8):1224–1231.
246. Wei J, Barr J, Kong LY, et al. Glioma-associated cancer-initiating cells induce immunosuppression. *Clin Cancer Res.* 2010;16(2):461–473.
247. Schatton T, Schutte U, Frank NY, et al. Modulation of T-cell activation by malignant melanoma initiating cells. *Cancer Res.* 2010;70(2):697–708.

248. Di Tomaso T, Mazzoleni S, Wang E, et al. Immunobiological characterization of CSCs isolated from glioblastoma patients. *Clin Cancer Res.* 2010;16(3):800–813.
249. Parmiani G, Russo V, Marrari A, et al. Universal and stemness-related tumor antigens: potential use in cancer immunotherapy. *Clin Cancer Res.* 2007;13(19):5675–5679.
250. Engelmann K, Shen H, Finn OJ. MCF7 side population cells with characteristics of cancer stem/progenitor cells express the tumor antigen MUC1. *Cancer Res.* 2008;68 (7):2419–2426.
251. Warrell Jr RP, Frankel SR, Miller Jr WH et al. Differentiation therapy of acute promyelocytic leukemia with tretinoin (all-trans-retinoic acid). *N Engl J Med.* 1991;324 (20):1385–1393.
252. Carey JO, Posekany KJ, deVente JE, Pettit GR, Ways DK. Phorbol ester-stimulated phosphorylation of PU.1: association with leukemic cell growth inhibition. *Blood.* 1996;87(10):4316–4324.
253. Wu HL, Li SF, Zeng BW, Yu RQ. Spectrophotometric determination of multicomponent vitamins by flexible tolerance simplex method. *Yao Xue Xue Bao.* 1991;26 (3):214–218.
254. Olsson I, Gullberg U, Ivhed I, Nilsson K. Induction of differentiation of the human histiocytic lymphoma cell line U-937 by 1 alpha, 25-dihydroxycholecalciferol. *Cancer Res.* 1983;43(12 Pt 1):5862–5867.
255. Meshorer E, Misteli T. Chromatin in pluripotent embryonic stem cells and differentiation. *Nat Rev Mol Cell Biol.* 2006;7(7):540–546.
256. Widschwendter M, Fiegl H, Egle D, et al. Epigenetic stem cell signature in cancer. *Nat Genet.* 2007;39(2):157–158.
257. Li E. Chromatin modification and epigenetic reprogramming in mammalian development. *Nat Rev Genet.* 2002;3(9):662–673.
258. Tam WL, Weinberg RA. The epigenetics of epithelial-mesenchymal plasticity in cancer. *Nat Med.* 2013;19(11):1438–1449.
259. Munster PN, Troso-Sandoval T, Rosen N, Rifkind R, Marks PA, Richon VM. The histone deacetylase inhibitor suberoylanilide hydroxamic acid induces differentiation of human breast cancer cells. *Cancer Res.* 2001;61(23):8492–8497.
260. Uchida H, Maruyama T, Nagashima T, Asada H, Yoshimura Y. Histone deacetylase inhibitors induce differentiation of human endometrial adenocarcinoma cells through up-regulation of glycodelin. *Endocrinology.* 2005;146(12):5365–5373.
261. Dimou A, Syrigos KN, Saif MW. Overcoming the stromal barrier: technologies to optimize drug delivery in pancreatic cancer. *Ther Adv Med Oncol.* 2012;4(5):271–279.

CHAPTER SIX

Therapeutic Use of Estrogen Receptor β Agonists in Prevention and Treatment of Endocrine Therapy Resistant Breast Cancers: Observations From Preclinical Models

Kumaraguruparan Ramasamy, Cathy Samayoa, Naveen Krishnegowda, Rajeshwar R. Tekmal[1]
The University of Texas Health Science Center at San Antonio, San Antonio, TX United States
[1] Corresponding author. E-mail address: tekmal@uthscsa.edu

Contents

Abstract

Breast cancer is one of the most common cancers in the world. The majority of breast cancers express estrogen receptor (ER)α, and endocrine therapy is the primary therapeutic approach to treat ER positive breast cancers. However, developing resistance and side effects are common events of these therapeutic strategies. Recent studies have evaluated the role of ERs sub types and demonstrated that ERα is a tumorigenic

Progress in Molecular Biology and Translational Science, Volume 151
ISSN 1877-1173
http://dx.doi.org/10.1016/bs.pmbts.2017.08.002

and ERβ functions as a tumor suppressor. In recent years, preclinical studies focused on the use of natural and synthetic ERβ agonists to treat wide varieties of cancers, including breast cancer. Successful studies conducted so far, have established that ERβ agonists are effective both alone and in combination with chemotherapeutic agents. These data have suggested that use of ERβ agonists in combination with endocrine therapy may be an effective treatment strategy in hormone receptor positive breast cancers. The present review focuses on the tumor suppressive role of ERβ, and the efficacy and mechanisms of several natural and synthetic ERβ agonists against breast cancer.

1. INTRODUCTION

Estrogen mediates its biological effects through the estrogen receptors (ERs). Approximately 70% of breast cancers are ERα positive at the time of diagnosis, urging the development of treatment strategies to target ERα.[1] Current treatment strategies for ERα-positive breast cancer focus on abrogating ERα signaling, because estrogen mediates its tumor promoting activity through ERα. These treatment strategies include endocrine therapies, such as antiestrogens, or aromatase inhibitors. The most commonly used endocrine (hormonal) therapies inhibit ER activity by either targeting the ER protein itself or degrading the receptor of its ligand.[2–4] Menopausal status has been reported to play an important role in the determination of the treatment modality in the ERα positive breast cancer. Younger women who have not undergone menopause are treated with antiestrogens, such as tamoxifen; whereas postmenopausal women will commonly receive aromatase inhibitors (AIs).[4] In premenopausal breast cancer patients, where the ovaries produce the majority of estrogen, the standard treatment is antiestrogens. Selective ER modulators (SERMs), for example, tamoxifen, compete with the estrogen ligand and have mixed agonistic/antagonistic capacities.[5–7] Selective ER degraders (SERDs), for example, fulvestrant, have antagonistic activity and induce ER protein degradation. SERDs do not have agonistic activity and may be useful in cases where patients cannot metabolize tamoxifen.[8–10] Aromatase inhibitors are used primarily in postmenopausal women to block estrogen synthesis occurring mostly in tissue other than the ovaries. AIs directly inhibit the estrogen biosynthesis enzyme aromatase, which is responsible for the conversion of androgens to estrogens in all tissue types, including in the breast tumors themselves.[11,12] Letrozole, an aromatase inhibitor, reduces estrogen levels and in turn also diminishes ER signaling.[13] In addition to indirectly abrogating ERα signaling by decreasing estrogen levels, AIs also impact carcinogenesis by reducing genotoxic estrogen and preventing DNA damage.[14] While endocrine therapies have proven to be very effective in both early and metastatic cancers, both de novo and acquired resistance to endocrine treatments remain a key clinical challenge. Approximately 30% of women with ERα-positive breast cancers have been reported to have de novo resistance to hormone therapy. Though the remaining women initially respond to hormone therapy, they develop secondary acquired resistance in later stages, which leads to endocrine resistance that eventually culminates in tumor progression and metastasis.[15] Clinical and experimental investigations have revealed that therapy resistances can be caused by: activation of growth factor pathways,[16] loss,[17,18] or mutations[19–21] in ERα, dysregulation of ER coregulators,[22] and ligand independent activation of ERα signaling.[23] While current endocrine therapies focus on ERα signaling, emerging studies highlight the importance of ERβ. Unlike ERα, β has been shown to have a tumor-suppressive function in various cancers, including breast cancer.[24–26] Recent studies have identified, synthesized, and tested the clinical safety of ERβ-selective agonists.[27–29] In this review, we summarize the recent evidence for the antitumoral activity of ERβ signaling in breast cancer progression and discuss the utility of using ERβ agonists in the treatment of breast cancer.

2. ESTROGEN RECEPTORS α AND β

The effects of estrogens are mediated by ERs. The ERs are ligand activated nuclear receptors that function as transcription factors. Two similar, but distinct ERs have been identified: ERα (ESR1) in 1985[30] and ERβ (ESR2) in 1996.[31] The human ERα gene is located on chromosome 6 and the ERβ gene is on chromosome 14. ERα is made up of 595 amino acids, and a molecular size of 66 kDa, while ERβ is made up of 530 amino acids and a molecular size of 54 kDa. ERα and β are structurally similar with six regions (A–F) in the primary amino acid sequence found in majority of nuclear receptor superfamily. Both contain three functional domains with varying levels of homology. (1) The N-terminal domain (NTD or AF1) contains a transactivation domain and is responsible for the recruitment of coactivators/corepressors. (2) The DNA-binding domain (DBD or C) recognizes DNA sequences known as estrogen response elements (ERE)

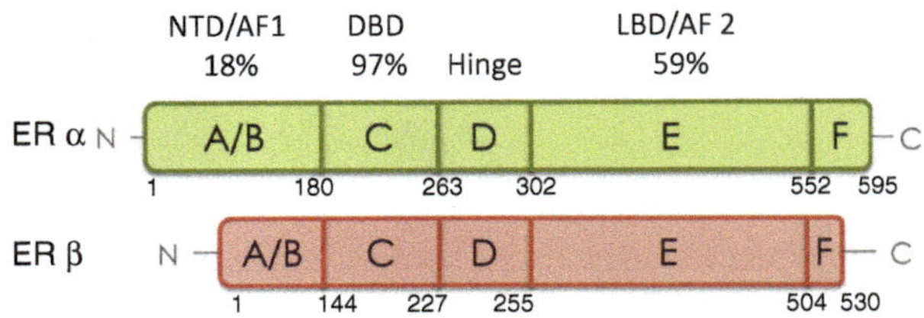

Fig. 1 Protein structure and degree of homology of estrogen receptor (ER)α and β. A schematic representation of structural domains of ERα and β. The ERs are composed of three main domains: the N-terminal domain (NTD), which contains the activation function1, the DNA-binding domain (DBD), and the ligand-binding domain (LBD), containing the activation function 2. The percentage amino acid homologies between ERα and β are also shown.

in the promoter region of target genes. (3) The ligand-binding domain (LBD or D/E/F) is responsible for binding to coregulatory and chaperone proteins. The two ERs share about 97% similarity in the DBD, 59% in LBD, and only 16% in their NTD domain (Fig. 1).[32,33] The differences in the LBD are responsible for the shape of ligand-binding pocket, and specific ligands for each receptor subtype have been designed and synthesized. Because of these structural differences, a particular ligand may have a different binding affinity for ERα and β subtypes. Although both activated by estrogen, their gene activation and tissue distribution patterns vary. Normal mammary gland tissue expresses both ERs, and in breast cancer, the frequency of ER expression varies.[34]

Both the receptors can dimerize and translocate to the nucleus after ligand-binding and regulate transcription of target genes through binding to estrogen response elements (ERE) in the DNA. The receptors can also interact with other transcription factors, such as activating protein-1 (AP-1), stimulating protein 1 (SP-1), and nuclear factor-κB (NF-κB).[35–37] Studies have also showed that ERs can modulate the enzyme activities (e.g., kinases and phosphatases), which are required for normal physiological function, by changing their phosphorylation status.[38] Although both the receptors share extensive homology, increasing evidences suggests that they have opposite functions in cancer biology and therapy.[39] Overall, completed studies have suggested that ERα contributes to tumorigenesis by inducing cell proliferation, while ERβ functions as a tumor suppressor with significant antitumoral activity.

3. ESTROGEN RECEPTOR β IN BREAST CANCER

Preclinical and clinical studies have shown an inverse relationship between ERβ expression and breast cancer.[40] The full-length ERβ protein, named as ERβ1, is generated from 530 amino acids with an estimated molecular weight of 59.2 kDa. Four ERβ isoforms exist as a result of alternative splicing mechanisms named as ERβ2, ERβ3, ERβ4, and ERβ5 (Fig. 2).[41–43] It has been reported that the ligands could not bind to these different variants because of C-terminal truncation. ERβ2, ERβ4, and ERβ5 do not necessarily form homodimers with themselves, but they can form a heterodimer with ERβ1 or with ERα, there by inhibiting their binding to ERE elements on DNA.[42,43] Several lines of evidence suggest that ERβ functions as a tumor suppressor in vitro models.[40] Exogenous expression of ERβ in ERα-positive breast cancer cells slowed down the mitogenic responses initiated by ERα, reduced cell motility and invasion, and inhibited tumor formation and angiogenesis in mouse xenografts.[44–47] The role of ERβ as a tumor suppressor is further supported by findings of several epidemiologic studies demonstrating a loss of ERβ in higher-grade versus lower-grade breast cancer tissue.[48] A wide variety of studies, including those from our laboratory, showed the antitumoral efficacy of ERβ in breast cancer.[44–47,49]

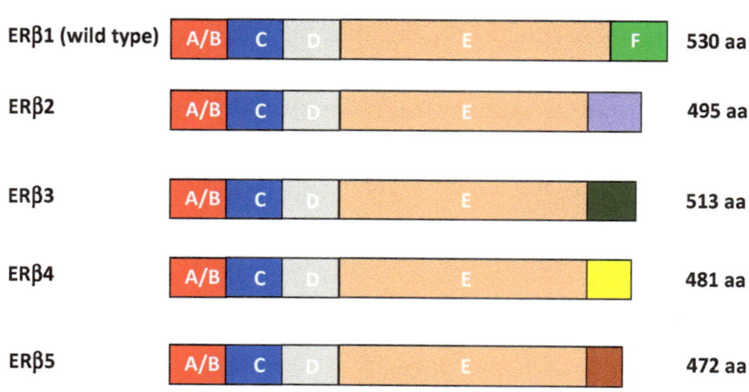

Fig. 2 Schematic representations of different isoforms of ERβ. ERβ variants are formed alternative splicing of the last coding exon. All ERβ variants have a truncated C-terminus and expressed in different type of cancers.

4. ESTROGEN RECEPTOR β AGONISTS IN BREAST CANCER

Estrogen is the endogenous ligand for both ERs. However, studies have identified plant derived and synthetic ERβ-agonists.[50,51] These compounds were reported to bind with a higher affinity to ERβ when compared to ERα. Upon binding, ERβ agonists activate ERβ, and can then function as a transcriptional regulator. In particular, phytoestrogens are natural, plant-derived compounds that have a high affinity to bind to ERβ.[52–55] Epidemiological studies investigating the protective effects of diet on breast cancer risks have suggested that a diet high in soy may be protective in the development of breast cancer. It has been reported that the women in Asia have lower incidence rates when compared to women in the United States and European countries. However, once they migrate to the United States, their rates increase to that of women in the United States.[56,57] This suggests that their protective effect is independent of genetics, and more influenced by their diet and lifestyle. Studies in soy consumption also highlight the differences in soy consumption between women in Asia and the United States. In the traditional Asian diet, the daily intake of soy protein is estimated at 20–30 g (100 mg isoflavones) compared to 1 g of soy protein in non-Asian diet.[56–58] The major isoflavones present in soy are genistein, daidzein, and S-equol; the active metabolite of daidzein, which are having structural and functional similarities like 17-β-estradiol.[59–61] They can bind to both ERα and β, but have a higher affinity for ERβ. In addition, flavonoids also exhibited anticancer efficacy in different type of cancers.

4.1 Genistein

Genistein, a plant phytoestrogen that resembles 17-β-estradiol structure, and possess weak estrogenic activity as it binds to ERs. Genistein showed a 20- to 30-fold higher affinity toward ERβ than for ERα. A wide variety of studies have shown anticancer efficacy of genistein and in a range of cancers, including breast cancers.[53,60] Genistein treatment inhibited breast cancer cells growth and proliferation and increased apoptosis in MCF-7 breast cancer cells.[62] Interestingly, genistein treatment inhibited breast cancer stem-like cells through downregulating Hedgehog–Gli1 Signaling Pathway both in vitro and in vivo.[63] Pons *et al.*[64] has reported that genistein modulates proliferation and mitochondrial functionality in breast cancer cells depending on the ERα/β ratio. They observed that genistein treatment induces cell cycle arrest and an improvement of mitochondrial functionality in T47D cells with a low ERα/β ratio, but not in MCF-7 (high ERα/β ratio) and MDA-MB-231 (ER-negative) cells. High concentrations of genistein have been reported to decrease cell proliferation in T47D cells by an ERK1/2-independent decrease in phosphorylated p90RSK. Interestingly, genistein inhibits both constitutive and EGF-stimulated invasion in MDA-MB-231 and MDA-MB-468, which are triple negative breast cancer cells.[65] Genistein inhibited the invasion in MCF-7 and MDA-MB-231 cells both in vitro *and* in vivo *tumorxenografts*. This inhibition was mainly due to the downregulation of matrix metalloproteinase (MMP)-9 and upregulation of tissue inhibitor of metalloproteinase-1. In particular, genistein also inhibited angiogenesis by decreasing vessel density and decreasing the levels of vascular endothelial growth factor, and transforming growth factor-β in MDA-MB-231 xenograft.[66] Genistein treatment significantly down regulated antiapoptotic protein BCl-2 and upregulated Bax, a proapoptoic protein thereby inducing apoptosis in MDA-MB-231 cells.[67] Mai *et al.*[68] have reported that genistein sensitizes the inhibitory effect of tamoxifen on the growth of ER-positive and HER2-overexpressing human breast cancer cells. Genistein also reduced mammary adipogenicity and enhanced expression of PTEN and E-cadherin in mice.[69] In addition, the compound has been shown to alter CSCs in the sera of adult mice by decreasing their ability to self-renew and form mammospheres, as well as target basal stem-like $CD44^+/CD24^-/ESA^+$ and the luminal progenitor $CD24^+$ subpopulations in MCF-7 and MDA-MB-231 cells.[70]

4.2 Daidzein

Daidzein, another isoflavone present in soy, showed fivefold increased affinity to ERβ. More recently, daidzein has been reported to inhibit cancer cell growth.[71] Daidzein treatment inhibited TNF-α induced cellular migration and invasion in ER-negative MCF10DCIS human breast cancer cells by downregulating hedgehog–Gli1 signaling.[72] Preclinical studies showed that daidzein inhibited the breast cancer progression by impeding cell proliferation and inducing apoptosis. Inhibition of tumor necrosis factor-α (TNF) induced migration and invasion was inhibited by daidzein.[72] In a recent study, Lecomte *et al.*[73] have shown that daidzein decreased the expression of CXCL12 in MCF-7 and T47D cells. CXCL12 is a key factor for proliferation when it is produced by breast cancer cells and is important for migration and metastasis when it is produced by target tissues.

Metabolomic reprogramming has been recognized as a one of the hallmarks of cancer. Metabolic reprogramming is required for both malignant transformation and tumor development, including invasion and metastasis. A recent study have shown that daidzein induced specific changes in alanine, cystathionine, hypotaurine, phosphoenolpyruvate, and succinate levels in MDA-MB-231 cells.[74] Amaral *et al.*[71] have recently investigated the soybean extract has stronger antiaromatase activity and antiproliferative activity in ER^+ aromatase-overexpressing breast cancer cells.

4.3 S-Equol

S-equol, an active metabolite of the soy isoflavone daidzein, has been found to be selective ERβ agonist. Soy metabolism can vary depending on the gut microbiota, and could affect how women benefit from soy.[75] About 50% of women in Asia can metabolize soy into S-equol, but only 25% of women in the United States are S-equol producers.[58,59] It has been reported that S-equol significantly decreased the methylation of the cytosine phosphate guanine (CpG) islands in the promoters of BRCA1 and BRCA2 in MCF-7 and MDA-MB-231 cells.[76] Later the authors also found increased expression of BRCA1 and BRCA2 proteins in the nuclei and cytoplasm to show that S-equol demethylating BRCA1 and BRCA2 genes.[76] S-equol inhibited MDA-MB-231 and T47D cell growth and sensitized these cells to irradiation. This effect was accompanied by increased radiation-induced DNA damage and apoptosis.[77] Magee *et al.*[78] investigated the role of S-equol on the migration and invasion of MDA-MB-231 breast cancer cells. In this study, the authors demonstrated that S-equol inhibited the invasion of breast cancer cells by down regulating MMP-2.[78] A study from our group has demonstrated that the ERβ phosphorylation switch is important for its antitumoral activity. We showed that phosphorylation status of tyrosine residue (Y36) is important for its antitumoral activity. This leads to transcriptional activation of ERβ-target specific genes thereby inhibiting the tumor cell growth.[79] Recently Yuan *et al.*[80] demonstrated that S-equol treatment did not affect the growth of ERβ KO cells, which further validated that ERβ phosphorylation status. To confirm this, authors treated the MDA-MB-231 cells with S-equol, and showed that S-equol increased the phosphorylation of ERβ. Further, they validated the growth inhibitory effect of S-equol in MDA-MB-231 cells xenograft model. S-equol inhibited the MDA-MB-231 cells growth and also decreased Ki-67 positive cells, which is accompanied by significant increase in ERβ phosphorylation status.[80] Our preliminary studies have shown that S-equol inhibited the short- and long-term growth of both endocrine therapy sensitive and resistant breast cancer cells (Tekmal, unpublished observations). As we discussed earlier, endocrine therapies target ERα in two ways, by using antiestrogen, such as tamoxifen, works by binding to and inhibiting the ERα and aromatase inhibitors, on the other hand, work by reducing the amount of estrogen being produced, which in turn decreases the amount of ERα activation and signaling. A woman's menopausal status will be the key to determine the endocrine therapy. In premenopausal women, tamoxifen, an antiestrogen, is typically recommended whereas in the postmenopausal women, letrozole, an aromatase inhibitor, is typically recommended. Since breast tumors that are treated with endocrine therapy commonly develop therapy resistance, we recently used aromatase overexpressing cells (MCF7aro and ZR75aro) and therapy resistance cells (letrozole resistance cell line—LTLT) to study the mechanism of therapy resistance. S-equol significantly inhibited the growth of both sensitive and resistance cells (Tekmal, unpublished observations). In addition to breast cancer, the anticancer efficacy of S-equol has also been demonstrated in other cancer models.[75,81]

5. LIQUIRITIGENIN

For centuries in traditional Chinese medicine, licorice root has been used to relieve the symptoms associated with menopause. MF101 is used to treat the menopausal symptoms, including hot flashes. Cvoro *et al.*[83] has showed that MF101, which is composed of 22 individual plants, induced the conformational change of ERβ, which allows ERβ to bind to ERE and recruit coactivators of target genes. Liquiritigenin, is a plant derived compound, isolated from the root of *Glycyrrhizaeuralensis* Fisch, a plant present in MF101 formulation. Previous studies investigating the use of low-dose liquiritigenin for menopause, demonstrated its safety, as there were neither any effects on uterine weight, nor growth of breast cancer cells.[82–84] Studies have shown that liquiritigenin inhibited the growth of breast cancer cells in different pathways.[75] It has been recently reported that liquiritigenin decreased the invasion of MDA-MB-231 cells and HS578T cells.[85] Hinsche *etal.*[86] have shown that treatment with selective ERβ agonist liquiritigenin resulted in a significant decrease of CXCR4 protein expression, an important prometastatic factor, in TNBC cell lines HCC1806 and HCC1937. A preliminary study from our laboratory demonstrate that liquiritigenin can inhibit the growth of not only therapy sensitive breast cancer cells, but also who have developed resistance to conventional therapies. To explore the translational significance of liquiritigenin using MCF7Aro cells as a model of postmenopausal breast cancer, we established MCF7aro xenograft mouse model. Our in vivo data showed that liquiritigenin inhibits the growth of MCF7aro tumors by modulating the transcription factors, which are needed for tumors to grow (Tekmal, unpublished observations).

6. LY500307

Other than naturally occurring ERβ agonists, the synthetic compounds are also available to target ERβ in different cancer models. One of the compounds is LY500307, a selective ERβ agonist with a 12-fold higher affinity for ERβ than ERα, and 32-fold functional potency. LY500307 in prostate cancer demonstrated a decrease in prostate weight, and decreased benign prostate hyperplasia (BPH).[87] Based on these studies, a clinical trial was designed with the primary outcome of evaluating the utility of ERβ agonists on BPH in men. Results from that study showed no treatment efficacy in BPH. Additionally, at the dose administered, there were no adverse effects. This study demonstrated that the ERβ agonist, LY500307, was safe to use, but did not significantly decrease the incidence of BPH. LY500307 was also shown to decrease tumor volume in a mouse model of glioblastoma.[88] Currently ERβ agonists, LY500307, and S-equol, are in clinical trials for the treatment of schizophrenia (NCT01874756) and for the reduction of polyps (NCT01402648), and participants are being recruited to evaluate ERβ agonists in triple negative breast cancers that express ERβ (NCT02352025).

The other synthetic ERβ agonists ERB-041 showed a 200-fold selectivity and WAY-200070 showed a 68-fold higher selectivity showed growth inhibitory effects when combined with tamoxifen in MCF7 and T47-D cells.[89] Ruddy *et al.*[90] found that Way-200070 had a marked reduction in G2–M phase correlated with effects on cyclin D1 and cyclin E expression in a SERM/SERD-resistant breast cancer cell line. Diarylprepionitrile, a selective ERβ agonist, exhibited a 70-fold higher relative binding affinity and 70-fold higher relative potency in transcription assays with ERβ than with ERα. Using a xenograft model, we have also demonstrated that diarylprepionitrile, an ERβ agonist, restored sensitivity to aromatase inhibition in letrozole

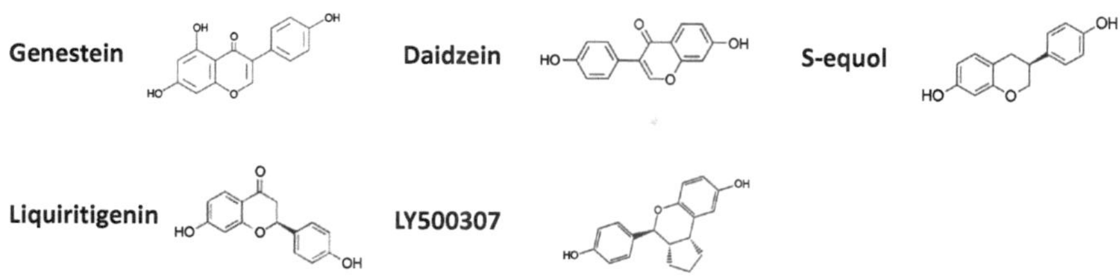

Fig. 3 Structures of natural and synthetic ERβ agonists.

resistant breast cancer cell lines, and also decreased the ERα/β ratio.[49] Fig. 3 presents the chemical structures of ERβ agonists.

7. ERβ AGONISTS ON TRANSGENIC MOUSE MODELS OF BREAST CANCER

Transgenic mouse models of breast cancer have proven useful to study mammary gland development, including carcinogenesis. The mouse mammary tumor virus (MMTV) is essential in generating mouse models to study mammary gland specific effects. Transgenic mouse models of breast cancer can exhibit differences in latency and incidence, and serve as a system to study phenotypes and signaling pathways affected by oncogenes.[91]

Studies in MMTV-aromatase mice, which have local estrogen production in the mammary gland, have demonstrated an increase in mammary gland epithelial cells proliferation, which eventually leads to premalignant lesions.[92] This model is useful to study intratumoral estrogen production that is often seen in the microenvironment of the human mammary gland harboring a tumor.[93] Administration of the aromatase inhibitor, letrozole, was able to reverse the changes caused by aromatase overexpression in the mammary gland. In efforts to study the contribution of aromatase in HER2/neu driven oncogenesis, our group generated of a double transgenic mouse model; MMTV-aromatase × MMTV-HER2/neu. MMTV-HER2/neu mice contain a valine to glutamic acid substitution at amino acid 664, which results in an activated for of HER2/neu. This mimics the amplification and overexpression seen in approximately 10%–20% of all breast cancer cases.[94] MMTV-HER2 mice form mammary gland tumors at about 7 months of age with a tumor incidence of 65%.[95] Early in development, these mice exhibit premalignant lesions and hyperbranching, which makes them good models to study therapeutic agents, which may affect this phenotype. In the MMTV-

HER2/neu mice, increased HER2 activity alone is enough to drive oncogenesis in the mammary gland. This highlights the significance of MMTV-HER2/neu overexpression and amplification in human tumors as drivers of tumorigenesis.[96] It was hypothesized that this double transgenic mice would have a shorter latency and increased tumor frequency than then MMTV-HER2/neu mice. Surprisingly, increased estrogenic signaling in the MMTV-HER2/neu mice led to increased tumor free survival, which was mediated through ERβ.[97] This study provided insights into the protective role of ERβ signaling for the prevention of breast cancer, and suggested that administration of an ERβ-selective agonist may prevent premalignant lesions and the development of carcinomas. Recently we used MMTV-HER2/neu transgenic mouse model to demonstrate the anticancer efficacy of S-equol and LY500307. Both S-equol and LY500307 treatment reduced significantly the mammary gland branching, with no significant changes in body weight and also able to decrease the number of hyperplastic lesions (Tekmal, unpublished observations).

8. CONCLUSIONS AND FUTURE PERSPECTIVES

Endocrine therapy has been demonstrated as effective approach to treat patients with ERα-positive breast cancer. However, treatment duration, side effects, and relapse are critical factors to consider. Endocrine therapy has been traditionally administered for 5 years, but in more recent studies the benefit of treating for 10 years has been investigated. One significant concern is the side effects and therapy resistance that accompany long-term endocrine therapy. In a vast majority of patients, therapy resistance emerges and leads to relapse, metastasis and mortality. Therefore, resistance to endocrine therapy continues to be a subject of great importance. Preclinical and clinical studies have showed that ERβ is an important and promising receptor and target in breast cancer management.[25,49] Therefore, ERβ can play a major role in prevention, treatment, and prognosis of breast cancer. Studies described here and elsewhere summarized the anticancer efficacy of both natural and synthetic ERβ agonists in various in vitro and in vivo breast cancer models. The ongoing clinical studies have demonstrated that ERβ agonists show no side effects, which encourages their use in breast cancer treatment. Additional clinical research is warranted to evaluate further the chemotherapeutic efficacy of ERβ agonists in various human cancers. A complete understanding of specific molecular targets of ERβ agonists may lead to better translation of their potential anticancer effects in both preclinical, as well as clinical settings in the future.

REFERENCES

1. Clark GM, Osborne CK, McGuire WL. Correlations between estrogen receptor, progesterone receptor, and patient characteristics in human breast cancer. *J Clin Oncol.* 1984;2:1102–1109.
2. Prat A, Adamo B, Cheang MC, Anders CK, Carey LA, Perou CM. Molecular characterization of basal-like and non-basal-like triple-negative breast cancer. *Oncologist.* 2013;18:123–133.
3. Goss PE. Addressing the future: combination with targeted therapies, adjuvant setting and beyond. *Anticancer Drugs.* 2008;19(suppl 2):S3–S5.
4. Di Cosimo S, Baselga J. Targeted therapies in breast cancer: where are we now? *Eur J Cancer.* 2008;44:2781–2790.
5. Frasor J, Stossi F, Danes JM, Komm B, Lyttle CR, Katzenellenbogen BS. Selective estrogen receptor modulators: discrimination of agonistic versus antagonistic activities by gene expression profiling in breast cancer cells. *Cancer Res.* 2004;64:1522–1533.
6. Jordan VC. The science of selective estrogen receptor modulators: concept to clinical practice. *Clin Cancer Res.* 2006;12:5010–5013.
7. Osborne CK. Tamoxifen in the treatment of breast cancer. *N Engl J Med.* 1998;339:1609–1618.
8. McDonnell DP, Wardell SE, Norris JD. Oral selective estrogen receptor downregulators (SERDs), a breakthrough endocrine therapy for breast cancer. *J Med Chem.* 2015;58:4883–4887.
9. Howell SJ, Johnston SR, Howell A. The use of selective estrogen receptor modulators and selective estrogen receptor down-regulators in breast cancer. *Best Pract Res Clin Endocrinol Metab.* 2004;18:47–66.
10. Jin Y, Desta Z, Stearns V, Ward B, Ho H, Lee KH, Skaar T, Storniolo AM, Li L, Araba A, et al. CYP2D6 genotype, antidepressant use, and tamoxifen metabolism during adjuvant breast cancer treatment. *J Natl Cancer Inst.* 2005;97:30–39.
11. Winer EP, Hudis C, Burstein HJ, Wolff AC, Pritchard KI, Ingle JN, Chlebowski RT, Gelber R, Edge SB, Gralow J, et al. American Society of Clinical Oncology technology assessment on the use of aromatase inhibitors as adjuvant therapy for postmenopausal women with hormone receptor-positive breast cancer: status report 2004. *J Clin Oncol.* 2005;23:619–629.
12. Goss PE, Qi S, Cheung AM, Hu H, Mendes M, Pritzker KP. Effects of the steroidal aromatase inhibitor exemestane and the nonsteroidal aromatase inhibitor letrozole on bone and lipid metabolism in ovariectomized rats. *Clin Cancer Res.* 2004;10:5717–5723.
13. Santen RJ, Brodie H, Simpson ER, Siiteri PK, Brodie A. History of aromatase: saga of an important biological mediator and therapeutic target. *Endocr Rev.* 2009;30:343–375.
14. Miller WR, Jackson J. The therapeutic potential of aromatase inhibitors. *Expert Opin Investig Drugs.* 2003;12:337–351.
15. Musgrove EA, Sutherland RL. Biological determinants of endocrine resistance in breast cancer. *Nat Rev Cancer.* 2009;9:631–643.
16. Meng S, Tripathy D, Shete S, Ashfaq R, Haley B, Perkins S, Beitsch P, Khan A, Euhus D, Osborne C, et al. HER-2 gene amplification can be acquired as breast cancer progresses. *Proc Natl Acad Sci USA.* 2004;101:9393–9398.
17. Gutierrez MC, Detre S, Johnston S, Mohsin SK, Shou J, Allred DC, Schiff R, Osborne CK, Dowsett M. Molecular changes in tamoxifen-resistant breast cancer: relationship between estrogen receptor, HER-2, and p38 mitogen-activated protein kinase. *J Clin Oncol.* 2005;23:2469–2476.

18. Encarnacion CA, Ciocca DR, McGuire WL, Clark GM, Fuqua SA, Osborne CK. Measurement of steroid hormone receptors in breast cancer patients on tamoxifen. *Breast Cancer Res Treat.* 1993;26:237–246.
19. Robinson DR, Wu YM, Vats P, Su F, Lonigro RJ, Cao X, Kalyana-Sundaram S, Wang R, Ning Y, Hodges L, et al. Activating ESR1 mutations in hormone-resistant metastatic breast cancer. *Nat Genet.* 2013;45:1446–1451.
20. Toy W, Shen Y, Won H, Green B, Sakr RA, Will M, Li Z, Gala K, Fanning S, King TA, et al. ESR1 ligand-binding domain mutations in hormone-resistant breast cancer. *Nat Genet.* 2013;45:1439–1445.
21. Zhang QX, Borg A, Wolf DM, Oesterreich S, Fuqua SA. An estrogen receptor mutant with strong hormone-independent activity from a metastatic breast cancer. *Cancer Res.* 1997;57:1244–1249.
22. Osborne CK, Bardou V, Hopp TA, Chamness GC, Hilsenbeck SG, Fuqua SA, Wong J, Allred DC, Clark GM, Schiff R. Role of the estrogen receptor coactivator AIB1 (SRC-3) and HER-2/neu in tamoxifen resistance in breast cancer. *J Natl Cancer Inst.* 2003;95:353–361.
23. Carascossa S, Dudek P, Cenni B, Briand PA, Picard D. CARM1 mediates the ligand-independent and tamoxifen-resistant activation of the estrogen receptor alpha by cAMP. *Genes Dev.* 2010;24:708–719.
24. Bossard C, Busson M, Vindrieux D, Gaudin F, Machelon V, Brigitte M, Jacquard C, Pillon A, Balaguer P, Balabanian K, Lazennec G. Potential role of estrogen receptor beta as a tumor suppressor of epithelial ovarian cancer. *PLoS One.* 2012;7:e44787.
25. Sareddy GR, Nair BC, Gonugunta VK, Zhang QG, Brenner A, Brann DW, Tekmal RR, Vadlamudi RK. Therapeutic significance of estrogen receptor beta agonists in gliomas. *Mol Cancer Ther.* 2012;11:1174–1182.
26. Strom A, Hartman J, Foster JS, Kietz S, Wimalasena J, Gustafsson JA. Estrogen receptor beta inhibits 17beta-estradiol-stimulated proliferation of the breast cancer cell line T47D. *Proc Natl Acad Sci USA.* 2004;101:1566–1571.
27. Kuiper GG, Lemmen JG, Carlsson B, Corton JC, Safe SH, van der Saag PT, van der Burg B, Gustafsson JA. Interaction of estrogenic chemicals and phytoestrogens with estrogen receptor beta. *Endocrinology.* 1998;139:4252–4263.
28. Muthyala RS, Ju YH, Sheng S, Williams LD, Doerge DR, Katzenellenbogen BS, Helferich WG, Katzenellenbogen JA. Equol, a natural estrogenic metabolite from soy isoflavones: convenient preparation and resolution of R- and S-equols and their differing binding and biological activity through estrogen receptors alpha and beta. *Bioorg Med Chem.* 2004;12:1559–1567.
29. Paruthiyil S, Cvoro A, Zhao X, Wu Z, Sui Y, Staub RE, Baggett S, Herber CB, Griffin C, Tagliaferri M, et al. Drug and cell type-specific regulation of genes with different classes of estrogen receptor beta-selective agonists. *PLoS One.* 2009;4:e6271.
30. Walter P, Green S, Greene G, Krust A, Bornert JM, Jeltsch JM, Staub A, Jensen E, Scrace G, Waterfield M, et al. Cloning of the human estrogen receptor cDNA. *Proc Natl Acad Sci USA.* 1985;82:7889–7893.
31. Kuiper GG, Enmark E, Pelto-Huikko M, Nilsson S, Gustafsson JA. Cloning of a novel receptor expressed in rat prostate and ovary. *Proc Natl Acad Sci USA.* 1996;93:5925–5930.
32. Jia M, Dalhman-Wright K, Gustafsson JA. Estrogen receptor alpha and beta in health and disease. *Best Pract Res Clin Endocrinol Metab.* 2015;29:557–5568.
33. Pettersson K, Gustafsson JA. Role of estrogen receptor beta in estrogen action. *Annu Rev Physiol.* 2001;63:165–192.
34. Cowley SM, Hoare S, Mosselman S, Parker MG. Estrogen receptors alpha and beta form heterodimers on DNA. *J Biol Chem.* 1997;272:19858–19862.

35. Paech K1. Webb P, Kuiper GG, Nilsson S, Gustafsson J, Kushner PJ, Scanlan TS. Differential ligand activation of estrogen receptors ERalpha and ERbeta at AP1 sites. *Science.* 1997;277:1508–1510.
36. Porter W1. Saville B, Hoivik D, Safe S. Functional synergy between the transcription factor Sp1 and the estrogen receptor. *Mol Endocrinol.* 1997;11:1569–1580.
37. Frasor J1. Weaver A, Pradhan M, Dai Y, Miller LD, Lin CY, Stanculescu A. Positive cross-talk between estrogen receptor and NF-kappaB in breast cancer. *Cancer Res.* 2009;69: 8918–8925.
38. Levin ER[1]. Integration of the extranuclear and nuclear actions of estrogen. *Mol Endocrinol.* 2005;19:1951–1959.
39. Thomas C, Gustafsson JA. The different roles of ER subtypes in cancer biology and therapy. *Nat Rev Cancer.* 2011;11:597–608.
40. Haldosén LA, Zhao C, Dahlman-Wright K. Estrogen receptor beta in breast cancer. *Mol Cell Endocrinol.* 2014;382:665–672.
41. Moore JT, McKee DD, Slentz-Kesler K, Moore LB, Jones SA, Horne EL, et al. Cloning and characterization of human estrogen receptor beta isoforms. *Biochem Biophys Res Commun.* 1998;247:75–78.
42. Ogawa S, Inoue S, Watanabe T, Orimo A, Hosoi T, Ouchi Y, et al. Molecular cloning and characterization of human estrogen receptor betacx: a potential inhibitor ofestrogen action in human. *Nucleic Acids Res.* 1998;26:3505–3512.
43. Leung YK, Mak P, Hassan S, Ho SM. Estrogen receptor (ER)-beta isoforms: a key to understanding ER-beta signaling. *Proc Natl Acad Sci USA.* 2006;103:13162–13167.
44. Paruthiyil S, Parmar H, Kerekatte V, Cunha GR, Firestone GL, Leitman DC. Estrogen receptor beta inhibits human breast cancer cell proliferation and tumor formation by causing a G2 cell cycle arrest. *Cancer Res.* 2004;64:423–428.
45. Roger P, Sahla ME, Makela S, Gustafsson JA, Baldet P, Rochefort H. Decreased expression of estrogen receptor beta protein in proliferative preinvasive mammary tumors. *Cancer Res.* 2001;61:2537–2541.
46. Lazennec G, Bresson D, Lucas A, Chauveau C, Vignon F. ER beta inhibits proliferation and invasion of breast cancer cells. *Endocrinology.* 2001;142:4120–4130.
47. Hartman J, Lindberg K, Morani A, Inzunza J, Strom A, Gustafsson JA. Estrogen receptor beta inhibits angiogenesis and growth of T47D breast cancer xenografts. *Cancer Res.* 2006;66:11207–11213.
48. Murphy LC, Leygue E. The role of estrogen receptor-β in breast cancer. *Semin Reprod Med.* 2012;30:5–13.
49. Nair HB, Kirma NB, Ganapathy M, Vadlamudi RK, Tekmal RR. Estrogen receptor-beta activation in combination with letrozole blocks the growth of breast cancer tumors resistant to letrozole therapy. *Steroids.* 2011;76:792–796.
50. Kuiper GG, Carlsson B, Grandien K, Enmark E, Haggblad J, Nilsson S, Gustafsson JA. Comparison of the ligand binding specificity and transcript tissue distribution of estrogen receptors alpha and beta. *Endocrinology.* 1997;138:863–870.
51. Paruthiyil S, Cvoro A, Zhao X, Wu Z, Sui Y, Staub RE, et al. Drug and cell type-specific regulation of genes with different classes of estrogen receptor beta-selective agonists. *PLoS One.* 2009;4:e6271.
52. Bak MJ1. Das Gupta S1. Wahler J1. Suh N[2]. Role of dietary bioactive natural products in estrogen receptor-positive breast cancer. *Semin Cancer Biol.* 2016;40–41:170–191.
53. Marzagalli M, Montagnani Marelli M, Casati L, Fontana F, Moretti RM, Limonta P. Estrogen receptor β in melanoma: from molecular insights to potential clinical utility. *Front Endocrinol (Lausanne).* 2016;7:140.
54. Patisaul HB, Jefferson W. The pros and cons of phytoestrogens. *Front Neuroendocrinol.* 2010;31:400–419.

55. Kamangar F, Dores GM, Anderson WF. Patterns of cancer incidence, mortality, and prevalence across five continents: defining priorities to reduce cancer disparities in different geographic regions of the world. *J Clin Oncol.* 2006;24:2137–2150.
56. Ziegler RG, Hoover RN, Pike MC, Hildesheim A, Nomura AM, West DW, Wu-Williams AH, Kolonel LN, Horn-Ross PL, Rosenthal JF, et al. Migration patterns and breast cancer risk in Asian–American women. *J Natl Cancer Inst.* 1993;85:1819–1827.
57. Shu XO, Zheng Y, Cai H, Gu K, Chen Z, Zheng W, Lu W. Food intake and breast cancer survival. *JAMA.* 2009;302:2437–2443.
58. Messina M, Nagata C, Wu AH. Estimated Asian adult soy protein and isoflavone intakes. *Nutr Cancer.* 2006;55:1–12.
59. Horn-Ross PL, John EM, Lee M, Stewart SL, Koo J, Sakoda LC, Shiau AC, Goldstein J, Davis P, Perez-Stable EJ. Phytoestrogen consumption and breast cancer risk in a multiethnic population: The Bay Area Breast Cancer Study. *Am J Epidemiol.* 2001;154: 434–441.
60. Russo M, Russo GL, Daglia M, Kasi PD, Ravi S, Nabavi SF, et al. Understanding genistein in cancer: the "good" and the "bad" effects: a review. *Food Chem.* 2016;196: 589–600.
61. Rowland IR, Wiseman H, Sanders TA, Adlercreutz H, Bowey EA. Interindividual variation in metabolism of soy isoflavones and lignans: influence of habitual diet on equol production by the gut microflora. *Nutr Cancer.* 2000;36:27–32.
62. Uifălean A, Schneider S, Ionescu C, Lalk M, Iuga CA. Soy isoflavones and breast cancer cell lines: molecular mechanisms and future perspectives. *Molecules.* 2015;21:E13.
63. Fan P, Fan S, Wang H, Mao J, Shi Y, Ibrahim MM, Ma W, Yu X, Hou Z, Wang B, Li L. Genistein decreases the breast cancer stem-like cell population through Hedgehog pathway. *Stem Cell ResTher.* 2013;4:146.
64. Pons DG1. Nadal-Serrano M, Blanquer-Rossello MM, Sastre-Serra J, Oliver J, Roca P. Genistein modulates proliferation and mitochondrial functionality in breast cancer cells depending on ERalpha/ERbeta ratio. *J Cell Biochem.* 2014;115:949–958.
65. Shao ZM, Wu J, Shen ZZ, Barsky SH. Genistein inhibits both constitutive and EGF-stimulated invasion in ER-negative human breast carcinoma cell lines. *Anticancer Res.* 1998;18:1435–1439.
66. Shao ZM, Wu J, Shen ZZ, Barsky SH. Genistein exerts multiple suppressive effects on human breast carcinoma cells. *Cancer Res.* 1998;58:4851–4857.
67. Li Y, Upadhyay S, Bhuiyan M, Sarkar FH. Induction of apoptosis in breast cancer cells MDA-MB-231 by genistein. *Oncogene.* 1999;18:3166–3172.
68. Mai Z, Blackburn GL, Zhou JR. Genistein sensitizes inhibitory effect of tamoxifen on the growth of estrogen receptor-positive and HER2-overexpressing human breast cancer cells. *Mol Carcinog.* 2007;46:534–542.
69. Montales MT1. Rahal OM, Nakatani H, Matsuda T, Simmen RC. Repression of mammary adipogenesis by genistein limits mammosphere formation of human MCF-7 cells. *J Endocrinol.* 2013;218:135–149.
70. Montales MT1. Rahal OM, Kang J, Rogers TJ, Prior RL, Wu X, Simmen RC. Repression of mammosphere formation of human breast cancer cells by soy isoflavone genistein and blueberry polyphenolic acids suggests diet-mediated targeting of cancer stem-like/progenitor cells. *Carcinogenesis.* 2012;33:652–660.
71. Amaral C, Toloi MRT, Vasconcelos LD, Fonseca MJV, Correia-da-Silva G, Teixeira N. The role of soybean extracts and isoflavones in hormone-dependent breast cancer: aromatase activity and biological effects. *Food Funct.* 2017.
72. Bao C, Namgung H, Lee J, Park HC, Ko J, Moon H, Ko HW, Lee HJ. Daidzein suppresses tumor necrosis factor-α induced migration and invasion by inhibiting hedgehog/Gli1 signaling in human breast cancer cells. *J Agric Food Chem.* 2014;62:3759–3767.

73. Lecomte S, Lelong M, Bourgine G, Efstathiou T, Saligaut C, Pakdel F. Assessment of the potential activity of major dietary compounds as selective estrogen receptor modulators in two distinct cell models for proliferation and differentiation. *Toxicol Appl Pharmacol.* 2017;325:61–70.
74. Uifălean A, Schneider S, Gierok P, Ionescu C, Iuga CA, Lalk M. The impact of soy isoflavones on MCF-7 and MDA-MB-231 breast cancer cells using a global metabolomic approach. *Int J Mol Sci.* 2016;17:E1443.
75. Sareddy GR, Vadlamudi RK. Cancer therapy using natural ligands that target estrogen receptor beta. *Chin J Nat Med.* 2015;13:801–807.
76. Bosviel R, Durif J, Déchelotte P, Bignon YJ, Bernard-Gallon D. Epigenetic modulation of BRCA1 and BRCA2 gene expression by equol in breast cancer cell lines. *Br J Nutr.* 2012;108:1187–1193.
77. Taghizadeh B1. Ghavami L, Nikoofar A, Goliaei B. Equol as a potent radiosensitizer in estrogen receptor-positive and -negative human breast cancer cell lines. *Breast Cancer.* 2015;22:382–390.
78. Magee PJ, Allsopp P, Samaletdin A, Rowland IR. Daidzein, R-(+)equol and S- (-)equol inhibit the invasion of MDA-MB-231 breast cancer cells potentially via the down-regulation of matrix metalloproteinase-2. *Eur J Nutr.* 2014;53:345–350.
79. Yuan B, Cheng L, Chiang HC, Xu X, Han Y, Su H, Wang L, Zhang B, Lin J, Li X, Xie X, Wang T, Tekmal RR, Curiel TJ, Yuan ZM, Elledge R, Hu Y, Ye Q, Li R. A phosphotyrosine switch determines the antitumor activity of ER β. *J Clin Invest.* 2014;124: 3378–3390.
80. Yuan B, Cheng L, Gupta K, Chiang HC, Gupta HB, Sareddy GR, Wang D, Lathrop K, Elledge R, Wang P, McHardy S, Vadlamudi R, Curiel TJ, Hu Y, Ye Q, Li R. Tyrosine phosphorylation regulates ER β ubiquitination, protein turnover, and inhibition of breast cancer. *Oncotarget.* 2016;7:42585–42597.
81. Liu J, Viswanadhapalli S, Garcia L, Zhou M, Nair BC, Kost E, Tekmal RR, Li R, Rao MK, Curiel T, Vadlamudi RK, Sareddy GR. Therapeutic utility of natural estrogen receptor beta agonists on ovarian cancer. *Oncotarget.* 2017;8:50002–50014.
82. Grady D1. Sawaya GF, Johnson KC, Koltun W, Hess R, Vittinghoff E, et al. MF101, a selective estrogen receptor beta modulator for the treatment of menopausal hot flushes: a phase II clinical trial. *Menopause.* 2009;16:458–465.
83. Cvoro A, Paruthiyil S, Jones J, Tzagarakis-Foster C, Clegg NJ, et al. Selective activation of estrogen receptor-beta transcriptional pathways by an herbal extract. *Endocrinology.* 2007;148:538–547.
84. Stoval. l DW, Pinkerton JV. MF-101, an estrogen receptor beta agonist for the treatment of vasomotor symptoms in peri- and postmenopausal women. *Curr Opin Investig Drugs.* 2009;10:365–371.
85. Schüler-Toprak S, Häring J, Inwald EC, Moehle C, Ortmann O, Treeck O. Agonists and knockdown of estrogen receptor β differentially affect invasion of triple-negative breast cancer cells in vitro. *BMC Cancer.* 2016;16:951.
86. Hinsche O, Girgert R, Emons G, Gründker C. Estrogen receptor β selective agonists reduce invasiveness of triple-negative breast cancer cells. *Int J Oncol.* 2015;46:878–884.
87. Roehrborn CG, Spann ME, Myers SL, Serviss CR, Hu L, Jin Y. Estrogen receptor beta agonist LY500307 fails to improve symptoms in men with enlarged prostate secondary to benign prostatic hypertrophy. *Prostate Cancer Prostatic Dis.* 2015;18:43–48.
88. Sareddy GR, Li X, Liu J, Viswanadhapalli S, Garcia L, Gruslova A, Cavazos D, Garcia M, Strom AM, Gustafsson JA, Tekmal RR, Brenner A, Vadlamudi RK. Selective estrogen receptor β agonist LY500307 as a novel therapeutic agent for glioblastoma. *Sci Rep.* 2016;6:24185.

89. Lattrich C, Schüler S, Häring J, Skrzypczak M, Ortmann O, Treeck O. Effects of a combined treatment with tamoxifen and estrogen receptor β agonists on human breast cancer cell lines. *Arch Gynecol Obstet.* 2014;289:163–171.
90. Ruddy SC, Lau R, Cabrita MA, McGregor C, McKay BC, Murphy LC, Wright JS, Durst T, Pratt MA. Preferential estrogen receptor β ligands reduce Bcl-2 expression in hormone-resistant breast cancer cells to increase autophagy. *Mol CancerTher.* 2014;13: 1882–1893.
91. Kirma NB1. Tekmal RR. Transgenic mouse models of hormonal mammary carcinogenesis: advantages and limitations. *J Steroid Biochem Mol Biol.* 2012;131:76–82.
92. Kirma N, Gill K, Mandava U, Tekmal RR. Overexpression of aromatase leads to hyperplasia and changes in the expression of genes involved in apoptosis, cell cycle, growth, and tumor suppressor functions in the mammary glands of transgenic mice. *Cancer Res.* 2001;61:1910–1918.
93. Sasano H, Harada N. Intratumoral aromatase in human breast, endometrial, and ovarian malignancies. *Endocr Rev.* 1998;19:593–607.
94. Press MF, Slamon DJ, Flom KJ, Park J, Zhou JY, Bernstein L. Evaluation of HER-2/neu gene amplification and overexpression: comparison of frequently used assay methods in a molecularly characterized cohort of breast cancer specimens. *J Clin Oncol.* 2002;20:3095–3105.
95. Tekmal RR, Nair HB, Perla RP, Kirma N. HER-2/neu x aromatase double transgenic mice model: the effects of aromatase overexpression on mammary tumorigenesis. *J Steroid Biochem Mol Biol.* 2007;106:111–118.
96. Ursini-Siegel J, Schade B, Cardiff RD, Muller WJ. Insights from transgenic mouse models of ERBB2-induced breast cancer. *Nat Rev Cancer.* 2007;7:389–397.
97. Nair HB, Perla RP, Kirma NB, Krishnegowda NK, Ganapathy M, Rajhans R, Nair SS, Saikumar P, Vadlamudi RK, Tekmal RR. Estrogen receptor-beta mediates the protective effects of aromatase induction in the MMTV-Her-2/neu x aromatase double transgenic mice. *Horm Cancer.* 2012;3:26–36.

FURTHER READING

Setchell KD, Brown NM, Lydeking-Olsen E. The clinical importance of the metabolite equol-a clue to the effectiveness of soy and its isoflavones. *J Nutr.* 2002;132:3577–3584.

CHAPTER SEVEN

Experimental Therapy of Advanced Breast Cancer: Targeting NFAT1–MDM2–p53 Pathway

Jiang-Jiang Qin[*,†], Wei Wang[*,†], Ruiwen Zhang[*,†,1]
[*] University of Houston, Houston, TX, United States
[†] Texas Tech University Health Sciences Center, Amarillo, TX, United States
[1] Corresponding author. E-mail address: ruiwen.zhang@ttuhsc.edu

Contents

Abstract

Advanced breast cancer, especially advanced triple-negative breast cancer, is typically more aggressive and more difficult to treat than other breast cancer phenotypes. There is currently no curable option for breast cancer patients with advanced diseases, highlighting the urgent need for novel treatment strategies. We have recently discovered that the nuclear factor of activated T cells 1 (NFAT1) activates the murine double minute 2 (*MDM2*) oncogene. Both MDM2 and NFAT1 are overexpressed and

Progress in Molecular Biology and Translational Science, Volume 151
ISSN 1877-1173
http://dx.doi.org/10.1016/bs.pmbts.2017.07.005

constitutively activated in breast cancer, particularly in advanced breast cancer, and contribute to its initiation, progression, and metastasis. MDM2 regulates cancer cell proliferation, cell cycle progression, apoptosis, migration, and invasion through both p53-dependent and -independent mechanisms. We have proposed to target the NFAT1–MDM2–p53 pathway for the treatment of human cancers, especially breast cancer. We have recently identified NFAT1 and MDM2 dual inhibitors that have shown excellent in vitro and in vivo activities against breast cancer, including triple-negative breast cancer. Herein, we summarize recent advances made in the understanding of the oncogenic functions of MDM2 and NFAT1 in breast cancer, as well as current targeting strategies and representative inhibitors. We also propose several strategies for inhibiting the NFAT1–MDM2–p53 pathway, which could be useful for developing more specific and effective inhibitors for breast cancer therapy.

1. INTRODUCTION

Breast cancer remains the most common cancer and the principal cause of cancer-related death among women worldwide.[1] Among various breast cancer phenotypes, triple-negative breast cancer (TNBC), which lacks the expression of estrogen receptors (ER), progesterone receptors (PR), and human epidermal growth factor receptor 2 (HER2), is the most aggressive form and represents approximately 15% of breast cancer.[2–4] Advanced breast cancer, comprising the most serious stages, stage 3 (locally advanced disease) and stage 4 (metastatic stage) of breast cancer, is generally considered as a treatable but still incurable disease because of its high rates of recurrence and resistance to chemotherapy.[5,6] Due to recent progress in understanding, diagnosing, and treating advanced breast cancer, particularly the HER2-positive and Luminal-like subtypes, overall survival for patients with advanced diseases has improved.[7,8] However, long-term survival for patients with advanced breast cancer, especially for patients with advanced TNBC remains poor,[2,3,8] emphasizing the unmet need for novel treatment strategies for these advanced diseases.

The murine double minute 2 (*MDM2*) oncogene has been identified as a master regulator of the p53 tumor suppressor.[9,10] The deregulation of MDM2–p53 pathway owing to the amplification and overexpression of MDM2 and the mutation and deletion of p53 has been implicated in the initiation, progression, and metastasis of breast cancer.[11,12] We have recently discovered that the nuclear factor of activated T cells 1 (NFAT1) transcription factor activates the transcription of *MDM2*.[13] NFAT1 is also highly expressed and constitutively activated in breast cancer and contributes to the onset and metastatic progression of this disease.[14] The activation of the NFAT1–MDM2–p53 pathway has been further observed in breast cancer, including TNBC; this pathway may be a promising molecular target for developing effective treatments for advanced breast cancer.[13] In this article, we review recent progress in our understanding of the functions of MDM2 and NFAT1 in breast cancer at different stages, the current targeting approaches and inhibitors of the NFAT1–MDM2–p53 pathway, and their potentials for advanced breast cancer treatment.

2. THE MDM2 ONCOGENE IN BREAST CANCER

2.1 The MDM2 Oncogene

More than 2 decades ago, MDM2 was identified as an oncogene, as its overexpression induces tumorigenicity in mouse 3T3 cells.[15] MDM2 amplification and overexpression are highly prevalent in human cancers,[16–18] including breast cancer,[19] and play pivotal roles in cancer onset, progression, metastasis, and response to therapy via both p53-dependent and -independent mechanisms.[20,21] Due to significant advances in our understanding of MDM2 biology, MDM2 has been found to be a complicated and highly regulated protein. MDM2 consists of multiple functional domains, including the N-terminal p53-binding domain, the centrally located acidic domain, the zinc finger domain, and the C-terminal RING finger domain.[20,22] The p53-binding domain is essential for the binding of MDM2 to p53, as well as MDM2's inhibitory effect on p53 transcriptional activity.[23] This domain is also involved in the interaction between MDM2 and other proteins,[22] for example, Ring1- and YY1-binding protein (RYBP).[24,25] The acidic domain of MDM2 has been demonstrated to play an important role in p53 ubiquitination and degradation.[26,27] This domain also aids in interactions with ribosomal proteins S7 and S25.[28,29] The linker region between the p53-binding domain and the acidic domain contains a nuclear localization signal (NLS) and a nuclear export signal (NES), which are essential for the subcellular localization of MDM2.[30] There is also a cryptic nucleolar localization signal in the C-terminal domain of MDM2.[31] The zinc finger domain plays a vital role in the binding of MDM2 with various proteins, including the tumor suppressor $p14^{ARF}$ [32,33] and several ribosomal proteins.[34] The RING finger domain is critical for MDM2's E3 ligase activity, which controls the ubiquitination of p53 and MDM2 itself.[35] In addition, the RING finger domain contributes to MDM2 homodimerization and MDM2–MDMX heterodimerization.[36]

2.2 The MDM2–p53 Feedback Loop in Breast Cancer

MDM2 is a major negative regulator of p53 in normal cells and cancer cells.[9,10,23] MDM2 forms a tight complex with p53 and promotes p53 ubiquitination through its E3 ligase activity,[9,10] enabling the recognition and interaction of p53 with the proteasome.[37] Subsequently, the ubiquitinated p53 is degraded by the proteasome. In addition, MDM2 also promotes the interaction of the SUMO E3 ligase PIASy with p53 and induces the sumoylation and nuclear export of p53,[38,39] consequently inhibiting the binding of p53 to transcriptional coactivators and decreasing p53-mediated transcription of the downstream target genes, including *MDM2*.[40] On the other hand, p53 directly binds to the p53-responsive P2 promoter of *MDM2* and mediates transcription of the *MDM2* gene.[41] The increased MDM2 protein, in turn, binds to p53 and promotes its degradation and inhibits its transcriptional activity, therefore forming the MDM2–p53 negative feedback loop (Fig. 1). This feedback loop is critical for controlling the cellular levels of p53 in normal cells because the high expression level of p53 is harmful for normal cell growth and differentiation. However, the amplification and

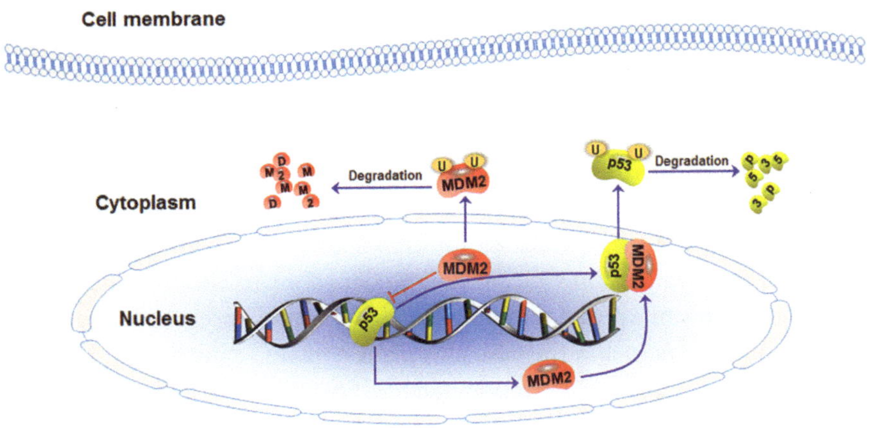

Fig. 1 The schematic diagram illustrates the murine double minute 2 (MDM2)–p53 feedback loop in breast cancer. MDM2 oncoprotein forms a tight complex with the p53 tumor suppressor and shuttles p53 out of the nucleus, inhibiting p53-mediated transcription and promoting p53 ubiquitination and proteasomal degradation. On the other hand, p53 binds to the *MDM2* P2 promoter and activates *MDM2* transcription. In addition, MDM2 also targets itself for ubiquitination and degradation.

overexpression of MDM2 are commonly observed in cancer cells due to deregulation of the MDM2–p53 pathway.[16–19] In addition, several modulatory proteins have been reported to regulate the MDM2–p53 interaction, such as RYBP,[24,25] ribosomal protein S7 and S25,[28,29,34] and PA28γ,[42] which have been reviewed in the recent papers.[22,30]

2.3 p53-Independent Functions of MDM2 in Breast Cancer

MDM2 is amplified and overexpressed in human cancer with p53 mutations and deletions, which promotes cancer initiation and progression through p53-independent mechanisms.[11,12,43,44] It has been observed that the overexpression of MDM2 splice variants without the binding ability to p53 still can induce tumor formation in transgenic mice via a p53-independent mechanism.[45,46] In breast cancer cells, estrogen was found to promote cell proliferation by activating MDM2, independent of p53.[47] Furthermore, MDM2 promotes cell cycle progression and inhibits cell apoptosis by regulating several important mediators in a p53-independent manner. Previously, we discovered that MDM2 inhibition by an antisense antihuman MDM2 mixed-backbone oligonucleotide increased the expression levels of p21, Bax, and hypophosphorylated retinoblastoma (pRb) and reduced the levels of hyperphosphorylated pRb (ppRb) and E2F1 in p53-*null* cancer cells.[48] Based on these observations, MDM2 has been further characterized as a negative regulator of the cyclin-dependent kinase inhibitor p21 by two laboratories independently.[49–51] It has been found that MDM2 interacts with p21 and promotes its degradation in an ubiquitin-independent proteasomal degradation pathway. MDM2 also induces the proteasomal degradation of the tumor suppressor retinoblastoma (Rb) in both ubiquitin-dependent and -independent manners.[52,53] We have found that MDM2 stabilizes the transcription factor E2F1 by inhibiting its ubiquitination,[54] thereby activating the transcription of E2F1's target genes related to cell cycle progression, for example, cyclin E.[55] MDM2 also interacts with the Forkhead box O (FOXO) transcription factors, including FOXO3a and FOXO4, and promotes their degradation via an MDM2-mediated ubiquitin–proteasome pathway.[56,57] In addition, MDM2 increases the translation of X-linked inhibitor of apoptosis protein (XIAP) by interacting with the internal ribosome entry segment (IRES) of the XIAP 5′-UTR and upregulating XIAP IRES activity.[58]

MDM2 has also been demonstrated to have a p53-independent role in regulating the epithelial–mesenchymal transition (EMT). In breast cancer cells, MDM2 increases cell motility and invasiveness by interacting with

E-cadherin (one of the EMT markers) and promoting its ubiquitination and degradation via an endosome pathway.[59] This negative regulation of E-cadherin by MDM2 has been further validated using tumor specimens from breast cancer patients with lymph node metastasis.[59] In addition, MDM2 plays a crucial role in DNA replication and DNA repair. The MDM2 expression has been implicated in the chromosome instability, which has been discussed in a recent review paper.[44] All these p53-independent functions of MDM2 further strengthen the potential of inhibiting MDM2 for cancer prevention and therapy.

2.4 Inhibiting MDM2 for Advanced Breast Cancer Therapy

The *MDM2* oncogene has been demonstrated as a promising molecular target for cancer therapy.[22,30,60,61] Several MDM2-targeting approaches have been developed: (1) inhibiting MDM2–p53 interaction, (2) blocking MDM2 E3 ligase activity, (3) destabilizing MDM2 protein, and (4) down-regulating MDM2 expression. Currently, the majority of MDM2 inhibitors have been developed to inhibit the MDM2–p53 interaction, releasing p53 from MDM2-mediated degradation and activating the transcription of p53 target genes.[62] Several MDM2–p53 binding inhibitors have shown potent efficacy against human cancers harboring wild-type p53, and are undergoing further evaluation in clinical trials, for example, RG7112,[63] HDM201,[64] AMG 232,[65] and NVP-CGM097.[66] However, due to frequent mutation and deletion of p53 in advanced breast cancer, especially advanced TNBC, the MDM2–p53 binding inhibitors often show limited efficacy.[12] Many MDM2 inhibitors have also been developed to block the E3 ligase activity of MDM2, activating the wild-type p53, such as HLI98C[67] and MEL23.[68] JNJ-26854165 (serdemetan), a novel inhibitor of the MDM2 E3 ligase activity, has been evaluated in a phase I trial performed in patients with advanced solid tumors, including seven patients with advanced breast cancer.[69] A dose- and exposure-dependent p53 induction was observed in some of the patients. However, only one of seven patients with advanced breast cancer showed a partial response to JNJ-26854165. Recent studies have indicated that JNJ-26854165 also exerts its anticancer activity in p53-mutant cancer cells through p53-independent mechanisms.[70]

We have been focusing on the MDM2-targeting therapeutics that can directly inhibit MDM2 and induce cancer cell death, regardless of p53 status, and have developed several MDM2 inhibitors, including antisense oligonucleotides and siRNA,[48,71–76] genistein,[77] curcumin,[78] ginsenosides,[79–81] and makaluvamine analogs.[82,83] These MDM2 inhibitors have shown excellent efficacy against human cancers, including breast cancer, regardless of the p53 status in the cancer cells. We have recently proposed the development of small-molecule inhibitors that can directly bind to MDM2 and induce its degradation, resulting in the discovery of a novel MDM2 inhibitor, termed SP141.[84,85] SP141 can directly bind to the p53-binding domain of MDM2, enhancing MDM2 autoubiquitination and proteasomal degradation.[84] In human breast cancer cell lines and animal models, SP141 inhibits cancer cell growth and proliferation and induces cell cycle arrest and apoptosis in vitro and suppresses xenograft tumor growth in vivo, independent of p53.[84] Furthermore, SP141 also inhibits breast cancer cell migration in vitro and prevents lung metastasis of breast cancer in vivo by modulating the expression of EMT markers.[84] The pharmacokinetic profile of SP141 has also been investigated in our recent studies[86,87]; although further studies should be performed using more clinically relevant models of advanced breast cancer.

MDM2's oncogenic functions, MDM2-targeting strategies, and the MDM2 inhibitors under development have been comprehensively discussed in recent review papers.[22,30,60,61] Despite the significant advances in this field, challenges remain in the pursuit of clinically applicable MDM2 inhibitors for cancer therapy, including low specificity, poor bioavailability, limited efficacy, and potential toxicity. As MDM2 overexpression can trigger both p53-dependent and -independent pathways, inhibiting one of the MDM2 pathways could only have limited efficacy in human cancers. For cancers containing mutant p53, inhibitors that target MDM2–p53 binding may not be as effective as those that directly inhibit MDM2 expression. Moreover, p53 activation may lead to unexpected adverse effects to normal tissues. Thus, strategies to specifically and effectively inhibit MDM2 and prevent unnecessary toxicity are critically needed.

3. NFAT1, AN ONCOGENIC TRANSCRIPTION FACTOR IN BREAST CANCER

3.1 The NFAT, a Family of Transcription Factors

The NFAT family of transcription factors consist of five members, including four calcium-regulated isoforms, NFAT1 (NFATc2 or NFATp), NFAT2 (NFATc1 or NFATc), NFAT3 (NFATc4), and NFAT4 (NFATc3 or NFATx) and a tonicity-responsive enhancer-binding protein (TonEBP or NFAT5).[14,88] All NFAT proteins have an N-terminal NFAT homology domain (NHD), a highly conserved Rel homology domain (RHD), and a C-terminal domain, except for NFAT5.[14,88] All the calcium-responsive NFAT members, NFAT1–NFAT4 contain the NHD, which possesses a potent transactivation domain and a calcineurin docking site.[89] However, NFAT5 lacks the calcineurin docking site and is completely insensitive to calcium and calcineurin.[90] Indeed, NFAT5 can be activated in response to osmotic stress.[91] The NHD also contains NLS1 and 2 and NES, which control NFAT subcellular localization through phosphorylation and dephosphorylation of the NFAT proteins at the serine-rich regions (SRR).[92,93] The RHD is also known as the DNA-binding domain (DBD) of NFAT proteins, which shows a structural similarity to the DBD of nuclear factor-κB (NF-κB).[94] The calcineurin–NFAT signaling pathway, the posttranslational modifications of NFAT proteins, including phosphorylation by various kinases, as well as the transcriptional and posttranscriptional regulation of NFAT isoforms have been comprehensively discussed in the recent review papers.[14,88] Despite significant progress in the understanding of NFAT biology, the role of these domains in the functions of NFAT proteins has not yet been completely understood.

3.2 NFAT1 as an Oncogenic Transcription Factor in Breast Cancer

Since the discovery of NFAT, as an inducible transcription factor that is capable of transactivating interleukin-2 (IL-2) during early T-cell activation,[95,96] its expression and functions have been implicated in numerous aspects of vertebrate development.[97,98] NFAT proteins have also been reported to play critical roles in malignant phenotypes and tumor development and progression.[14,99,100] The major NFAT isoform, NFAT1 is highly expressed and aberrantly activated in various human cancers, including breast,[101–104] pancreatic,[105–108] lung,[109–111] cervical,[112] and colon cancer,[113] as well as melanoma,[114] and contributes to the initiation, progression, and metastasis to these diseases. This section will focus on the role of NFAT1 in breast cancer, especially advanced breast cancer.

Jauliac *et al.* have found that NFAT1 is highly expressed in tumor tissues from patients with grade III invasive ductal carcinoma and may play a pivotal role in invasive and metastatic properties of breast cancer.[101] A tissue microarray study has been recently performed and indicated a significant increase of NFAT1 expression in invasive ductal carcinoma with lymph node metastasis, in comparison to normal adjacent tissue and primary tumor.[102] In a recent paper, Kaunisto *et al.* have further reported that NFAT1 expression level is significantly higher in tumor-invasive front epithelium than in the central epithelium or normal breast epithelium.[103] In addition, the stromal NFAT1 is also expressed at higher levels in the central and invasive front tumor stroma than in normal breast stroma.[103] More importantly, in comparison with other subtypes of breast cancer, TNBC shows a higher activation rate of the calcineurin/NFAT1 pathway, which is essential to tumorigenesis and metastasis.[104]

NFAT1 regulates cell cycle progression, apoptosis, migration, invasion, angiogenesis, and resistance to chemotherapy by transactivating its downstream target genes.[14,99] Specific to breast cancer, NFAT1 has been discovered to promote breast cancer cell migration and invasion through the induction of cyclooxygenase-2 (COX-2), Glypican-6 (GPC-6), and autotaxin, as well as the synthesis of prostaglandins.[115–117] NFAT1 stabilizes SnoN and mediates transforming growth factor-β (TGF-β)–induced EMT in breast cancer cells by upregulating N-cadherin expression and downregulating E-cadherin expression.[118] NFAT1 also cooperates with Smad2 and promotes breast cancer progression by upregulating the cyclin-dependent kinase 2 (CDK2), cyclin-dependent kinase 4 (CDK4), and cyclin E.[119] The NFAT1-induced expression of IL-8 promotes intratumoral neutrophil infiltration in breast cancer.[103] In addition, the negative cross talk between NFAT1 and signal transducer and activator of transcription 5 (Stat5) signaling cascades has been implicated in the growth and metastasis of breast cancer.[102]

Several mediators have been involved in the regulation of NFAT1 in breast cancer cells. The α6β4 integrin promotes breast carcinoma invasion by upregulating and activating NFAT1.[101,117] The tumor suppressor FOXP1 binds to NFAT1 on DNA and suppresses NFAT1 transcriptional activity, resulting in enhanced breast cancer cell migration.[120] Plasma membrane calcium ATPase isoform 4 (PMCA4) interacts with calcineurin and inhibits the calcineurin/NFAT1 pathway, suppressing vascular endothelial growth factor (VEGF)–mediated angiogenesis,[121] while the PMCA2–calcineurin binding inhibits the calcineurin/NFAT1 pathway, leading to the inhibition of breast cancer cell apoptosis.[122] Akt blocks breast cancer cell migration and invasion by inducing MDM2-mediated NFAT1 ubiquitination and proteasomal degradation,[123,124] whereas GM3 synthase promotes breast cancer lung metastasis by inhibiting phosphoinositide-3 kinase (PI3K)/Akt pathway and activating NFAT1.[125] In addition, Wnt5a and Secreted frizzle-related protein 2 (SFRP2) have also been found to induce breast cancer metastasis and angiogenesis by activating NFAT1.[126–128]

3.3 Inhibiting NFAT1 for Advanced Breast Cancer Therapy

Pharmacological inhibition of NFAT1 signaling has been demonstrated as a promising therapeutic strategy for cancer therapy.[14] Several inhibitors of calcineurin/NFAT1 pathway, such as cyclosporine A (CsA) have also shown promising efficacy against human cancer in vitro and in vivo.[129] A calcineurin inhibitor, tacrolimus (also known as FK506), directly binds to the immunophilin FKBPB12 and inhibits the calcineurin's phosphatase activity, leading to dephosphorylation of NFAT proteins, has shown potent antibreast cancer activity in vitro and in vivo.[127,128] Tacrolimus has been found to inhibit SFRP2-induced angiogenesis and the growth of angiosarcoma xenografts in mice in vivo.[128] In the MMTV-neu transgenic mice, treatment with tacrolimus markedly decreases the tumor growth rate by reducing the tumor microvascular density.[127] Quercetin, a well-known anticancer flavonoid, inhibits tumor growth, induces tumor necrosis, and suppresses oncocyte proliferation in mice bearing breast cancer xenograft tumors.[130] Mechanistically, quercetin suppresses the calcineurin/NFAT pathway, resulting in the inhibition of the expression of VEGF and VEGFR2.[130] Moreover, the human herpesvirus 6B (HHV-6B) U54 tegument protein is reported to inhibit breast cancer cell proliferation through the inhibition of NFAT activation.[131]

As NFAT1 transcription factor plays a crucial role in the immune response through the regulation the transcription of cytokine genes, the long-term use of NFAT1 inhibitors may result in a reduction of immunosurveillance in tumor microenvironments, subsequently causing an increase in cancer development.[132,133] Although the potential adverse effects of the current NFAT1 inhibitors are yet to be determined, novel targeting approaches for specific inhibition of oncogenic NFAT1 are still urgently needed.

4. TARGETING NFAT1–MDM2–P53 PATHWAY FOR ADVANCED BREAST CANCER THERAPY

4.1 The NFAT1–MDM2–p53 Pathway

In our continued efforts to develop specific and potent MDM2 inhibitors, genistein has been found to inhibit MDM2 at both transcriptional and posttranslational levels.[77] The inhibition of *MDM2* transcription by genistein (**1**) requires the NFAT transcription site in the *MDM2* P2 promoter. This is an intriguing observation with an obvious explanation that NFAT is

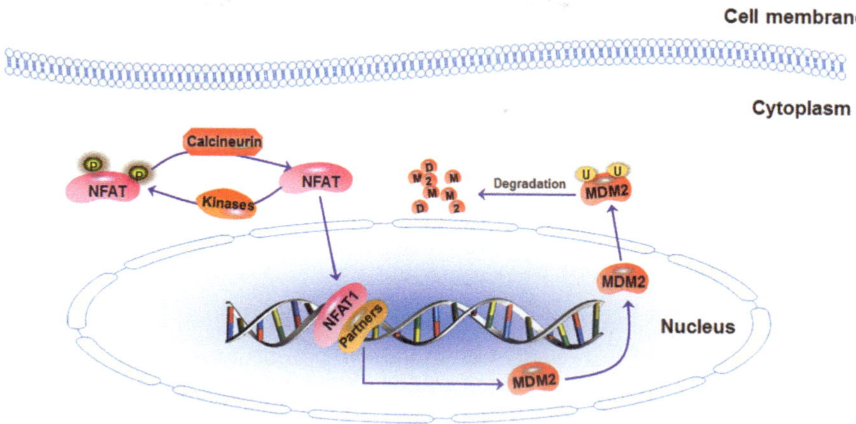

Fig. 2 The schematic diagram illustrates the nuclear factor of activated T cells 1 (NFAT1)–MDM2 pathway in breast cancer. In the cytoplasm, calcineurin dephosphorylates NFAT1 and induces its nuclear translocation. NFAT1 can also be rephosphorylated by various kinases. In the nucleus, the dephosphorylated NFAT1 interacts with multiple transcriptional partners to activate the transcription of *MDM2*. MDM2 targets itself for ubiquitination and degradation.

involved in regulating *MDM2* transcription. In this regard, our recent findings have demonstrated that NFAT1 is the predominant NFAT isoform regulating *MDM2* transcription. It binds to the *MDM2* P2 promoter and activates MDM2 expression in a p53-independent manner (Fig. 2).[13] The NFAT1–MDM2–p53 signaling pathway is subsequently validated in human cancer cells, including breast cancer cells in vitro and in human carcinoma tissues in vivo.[13] Considering these findings, it is important to ask whether targeting NFAT1–MDM2–p53 pathway could be the–MDM2–p53 pathway could be a useful approach for cancer therapy.

We first wanted to know if the known NFAT1 inhibitors could be used to target the NFAT1–MDM2–p53 pathway. Despite the structural similarities of NFAT isoforms, there is increasing evidence that these NFAT isoforms are not functionally equivalent but promote distinct signaling outputs in human cancers.[14] Of note, NFAT1 and NFAT2 are frequently overexpressed and activated in human cancers, contributing to cancer development, progression, and metastasis,[134–136] whereas NFAT3 functions as a tumor suppressor in breast cancer by inhibiting Lipocalin 2 expression and reducing cell motility.[137] However, the majority of NFAT inhibitors, such as CsA and tacrolimus, have been developed to target the calcineurin–NFAT pathway or other upstream regulators, leading to an unwanted impact on all NFAT isoforms.[14] Therefore, these NFAT inhibitors may not be useful in inhibiting the NFAT1–MDM2–p53 pathway for cancer therapy. Although genistein (**1**) has shown significant inhibitory effects on NFAT1–MDM2 interaction, more specific small-molecule inhibitors need to be developed for cancer therapy.

4.2 Representative Small-Molecule Inhibitors of NFAT1–MDM2–p53 Pathway

We have recently identified several small molecules (Fig. 3) that can effectively inhibit NFAT1 and MDM2 and show potent anticancer activity in vitro and in vivo. Two novel ginsenosides, 25-OCH_3-PPD (**2**) and 25-OH-PPD (**3**) were initially identified as MDM2 inhibitors,[79,80] and exhibited potent anticancer activity against various types of human cancer, including prostate,[138–140] pancreatic,[141] lung,[142] and breast cancer.[143] Both compounds have been found to inhibit cell viability and proliferation and induce cell cycle arrest and apoptosis in vitro. They also suppress xenograft tumor growth and sensitize tumors to chemotherapy in vivo. It has been further discovered that 25-OCH_3-PPD inhibits breast cancer cell migration in vitro and prevents the lung metastasis of breast cancer in vivo.[143] These activities have been attributed to its inhibitory effects on MDM2 expression and the modulatory effects on the expression of proteins related to cell growth,

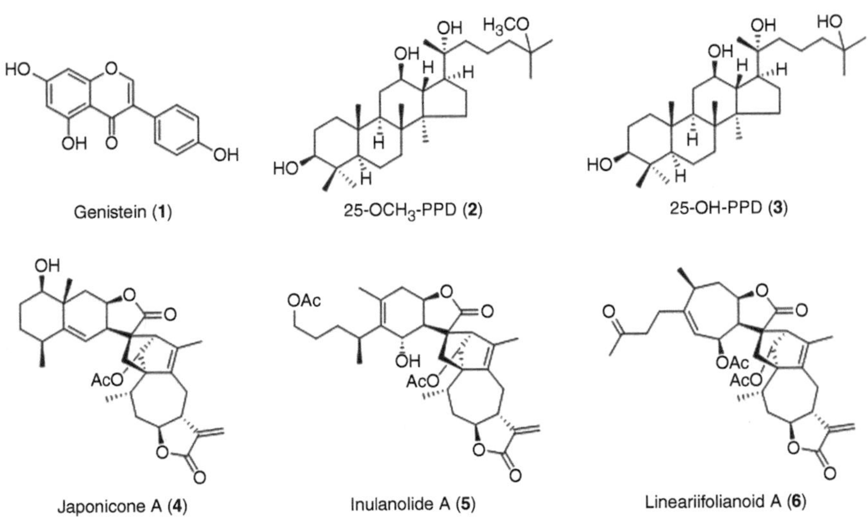

Fig. 3 Structures of representative small-molecule inhibitors of the NFAT1–MDM2–p53 pathway.

cell cycle progression, apoptosis, and EMT. Indeed, 25-OCH$_3$-PPD (**2**) inhibits MDM2 at both transcriptional and posttranslational levels.[143] The treatment of 25-OCH$_3$-PPD enhances the MDM2 ubiquitination and degradation in breast cancer cells, independent of p53. Furthermore, 25-OCH$_3$-PPD suppresses *MDM2* transcription by inhibiting NFAT1 expression (unpublished data).

In our recent studies, we shed new light on the development of NFAT1–MDM2–p53 inhibitors. In a computational structure-based screening, a novel class of natural product MDM2 inhibitors, including japonicone A (JapA, **4**), inulanolide A (InuA, **5**), and lineariifolianoid A (LinA, **6**) have been identified.[144] Our results have demonstrated that JapA exerts its antibreast cancer activity via a dual-targeting mechanism. JapA directly binds to the p53-binding domain of the MDM2 protein and promotes MDM2 autoubiquitination and degradation, independent of p53.[144] This compound also inhibits NFAT1 and NFAT1-mediated *MDM2* transcription by suppressing NFAT1 nuclear localization, its binding to the MDM2 P2 promoter, and the induction NFAT1 ubiquitination and degradation.[145] Our in vitro and in vivo studies have shown the potent antibreast cancer activity and safety profile of JapA, indicating that targeting the NFAT1–MDM2–p53 pathway is a promising approach for treating breast cancer, especially advanced breast cancer. More importantly, our data highlight the importance of developing dual NFAT1–MDM2 inhibitors instead of "mono" inhibitors for breast cancer therapy. It has been further discovered that both InuA and LinA suppress breast cancer growth and metastasis in vitro and in vivo by dually inhibiting NFAT1 and MDM2.[146,147] Based on these results, we have proposed several targeting strategies for developing NFAT1–MDM2 inhibitors for the treatment of breast cancer, especially advanced breast cancer (Fig. 4).

5. CONCLUSIONS

The overexpression and constitutive activation of MDM2 and NFAT1 are common events in breast cancer and have been linked with the progression and metastasis of advanced breast cancer, especially advanced TNBC. Previous studies have shown that the "mono" small-molecule inhibitors of MDM2 and NFAT1 can inhibit breast tumor growth and metastasis, but may not be able to cause complete tumor regression due to inherent redundancy or acquired resistance. We have recently demonstrated that NFAT1

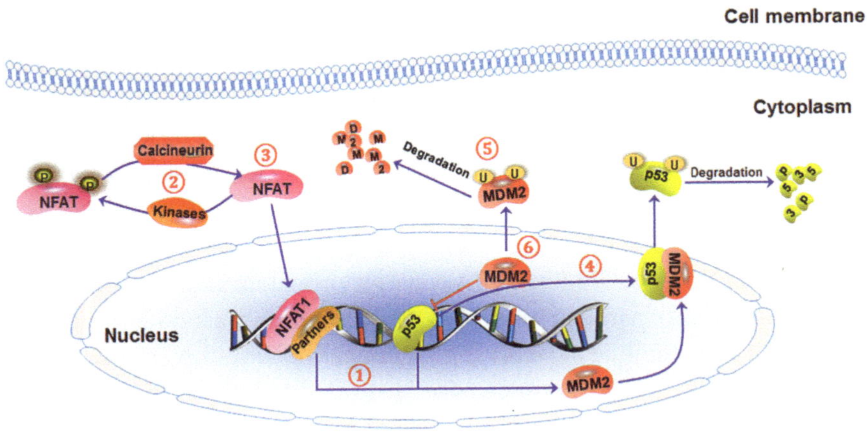

Fig. 4 The schematic diagram illustrates targeting strategies for the NFAT1–MDM2–p53 pathway. To effectively inhibit NFAT1's expression and activity, three targeting strategies have been proposed: (1) blocking NFAT1-*MDM2* DNA binding, (2) promoting NFAT1 phosphorylation and/or inhibiting NFAT1 dephosphorylation, and (3) directly destabilizing NFAT1 protein. To effectively inhibit MDM2's expression and activity, three targeting strategies have been proposed: (4) blocking MDM2–p53 interaction, (5) promoting MDM2 autoubiquitination and proteasomal degradation, and (6) directly inhibiting MDM2 expression. Small molecules that can dually inhibit NFAT1–MDM2 may be developed using these targeting strategies.

activates MDM2 in a p53-independent manner, and validated that the NFAT1–MDM2–p53 pathway is activated in breast cancer cell lines and carcinoma tissues. We, therefore, have proposed that targeting NFAT1–MDM2–p53 pathway is a promising strategy for developing novel and effective anticancer therapeutics, leading to the discovery of several small-molecule inhibitors, as well as their evaluation in preclinical models of breast cancer, including TNBC in vitro and in vivo.

Although the dual inhibitors of NFAT1 and MDM2 show great promise in our preliminary studies, concerns regarding its potential immune toxicity should be addressed before clinical trials are performed. Furthermore, the clinical relevance of the NFAT1–MDM2–p53 pathway in advanced breast cancer, especially advanced TNBC should be further investigated. It will be tempting to search for more specific NFAT1–MDM2 dual inhibitors. However, two questions should be answered. (1) How should specific and potent inhibitors for both NFAT1 and MDM2 be identified? (2) How should the adverse effects of these inhibitors on normal functioning NFAT1 be minimized? Both questions need to be addressed in future studies.

ACKNOWLEDGMENTS

This work was supported by the National Institutes of Health (NIH) grant R01 CA186662 (to R.Z.). The content is solely the responsibility of the authors, and do not necessarily represent the official views of the National Institutes of Health. W.W. was supported by the American Cancer Society (ACS) grant RSG-15-009-01-CDD.

REFERENCES

1. Brenner DR, Brockton NT, Kotsopoulos J, et al. Breast cancer survival among young women: a review of the role of modifiable lifestyle factors. *Cancer Causes Control.* 2016;27:459–472.
2. Szekely B, Silber AL, Pusztai L. New therapeutic strategies for triple-negative breast cancer. *Oncology.* 2017;31:130–137.
3. Denkert C, Liedtke C, Tutt A, von Minckwitz G. Molecular alterations in triple-negative breast cancer-the road to new treatment strategies. *Lancet.* 2017;389:2430–2442.
4. Bianchini G, Balko JM, Mayer IA, Sanders ME, Gianni L. Triple-negative breast cancer: challenges and opportunities of a heterogeneous disease. *Nat Rev Clin Oncol.* 2016;13:674–690.
5. Liedtke C, Kolberg HC. Systemic therapy of advanced/metastatic breast cancer—current evidence and future concepts. *Breast Care.* 2016;11:275–281.
6. Sonnenblick A, Ponde N, Piccart M. Metastatic breast cancer: the Odyssey of personalization. *Mol Oncol.* 2016;10:1147–1159.
7. Cardoso F, Costa A, Norton L, et al. ESO-ESMO 2nd international consensus guidelines for advanced breast cancer (advanced breast cancer2). *Ann Oncol.* 2014;25: 1871–1888.
8. Puliti D, Bucchi L, Mancini S, et al. Advanced breast cancer rates in the epoch of service screening: the 400,000 women cohort study from Italy. *Eur J Cancer.* 2017;75:109–116.
9. Haupt Y, Maya R, Kazaz A, Oren M. Mdm2 promotes the rapid degradation of p53. *Nature.* 1997;387:296–299.
10. Kubbutat MH, Jones SN, Vousden KH. Regulation of p53 stability by Mdm2. *Nature.* 1997;387:299–303.
11. Haupt S, Vijayakumaran R, Panimaya J, Burgess A, Lim E, Haupt Y. The role of MDM2 and MDM4 in breast cancer development and prevention. *J Mol Cell Biol.* 2017;9:53–61.
12. Turner N, Moretti E, Siclari O, et al. Targeting triple negative breast cancer: is p53 the answer? *Cancer Treat Rev.* 2013;39:541–550.
13. Zhang X, Zhang Z, Cheng J, et al. Transcription factor NFAT1 activates the mdm2 oncogene independent of p53. *J Biol Chem.* 2012;287:30468–30476.
14. Qin JJ, Nag S, Wang W, et al. NFAT as cancer target: mission possible? *Biochim Biophys Acta.* 2014;1846:297–311.
15. Cahilly-Snyder L, Yang-Feng T, Francke U, George DL. Molecular analysis and chromosomal mapping of amplified genes isolated from a transformed mouse 3T3 cell line. *Somat Cell Mol Genet.* 1987;13:235–244.
16. Ware PL, Snow AN, Gvalani M, Pettenati MJ, Qasem SA. MDM2 copy numbers in well-differentiated and dedifferentiated liposarcoma: characterizing progression to high-grade tumors. *Am J Clin Pathol.* 2014;141:334–341.
17. Momand J, Jung D, Wilczynski S, Niland J. The MDM2 gene amplification database. *Nucleic Acids Res.* 1998;26:3453–3459.
18. Onel K, Cordon-Cardo C. MDM2 and prognosis. *Mol Cancer Res.* 2004;2:1–8.
19. Yu Q, Li Y, Mu K, et al. Amplification of Mdmx and overexpression of MDM2 contribute to mammary carcinogenesis by substituting for p53 mutations. *Diagn Pathol.* 2014;9:71.

20. Karni-Schmidt O, Lokshin M, Prives C. The roles of MDM2 and MDMX in cancer. *Annu Rev Pathol.* 2016;11:617–644.
21. Wade M, Li YC, Wahl GM. MDM2, MDMX and p53 in oncogenesis and cancer therapy. *Nat Rev Cancer.* 2013;13:83–96.
22. Nag S, Qin J, Srivenugopal KS, Wang M, Zhang R. The MDM2-p53 pathway revisited. *J Biomed Res.* 2013;27:254–271.
23. Momand J, Zambetti GP, Olson DC, George D, Levine AJ. The mdm-2 oncogene product forms a complex with the p53 protein and inhibits p53-mediated transactivation. *Cell.* 1992;69:1237–1245.
24. Chen D, Zhang J, Li M, Rayburn ER, Wang H, Zhang R. RYBP stabilizes p53 by modulating MDM2. *EMBO Rep.* 2009;10:166–172.
25. Wang W, Qin JJ, Voruganti S, Nag S, Zhou J, Zhang R. Polycomb group (PcG) proteins and human cancers: multifaceted functions and therapeutic implications. *Med Res Rev.* 2015;35:1220–1267.
26. Kawai H, Wiederschain D, Yuan ZM. Critical contribution of the MDM2 acidic domain to p53 ubiquitination. *Mol Cell Biol.* 2003;23:4939–4947.
27. Meulmeester E, Frenk R, Stad R, et al. Critical role for a central part of Mdm2 in the ubiquitylation of p53. *Mol Cell Biol.* 2003;23:4929–4938.
28. Chen D, Zhang Z, Li M, et al. Ribosomal protein S7 as a novel modulator of p53-MDM2 interaction: binding to MDM2, stabilization of p53 protein, and activation of p53 function. *Oncogene.* 2007;26:5029–5037.
29. Zhang X, Wang W, Wang H, Wang MH, Xu W, Zhang R. Identification of ribosomal protein S25 (RPS25)-MDM2-p53 regulatory feedback loop. *Oncogene.* 2013;32:2782–2791.
30. Qin JJ, Nag S, Voruganti S, Wang W, Zhang R. Natural product MDM2 inhibitors: anticancer activity and mechanisms of action. *Curr Med Chem.* 2012;19:5705–5725.
31. Lohrum MA, Ashcroft M, Kubbutat MH, Vousden KH. Identification of a cryptic nucleolar-localization signal in MDM2. *Nat Cell Biol.* 2000;2:179–181.
32. Kamijo T, Weber JD, Zambetti G, Zindy F, Roussel MF, Sherr CJ. Functional and physical interactions of the ARF tumor suppressor with p53 and Mdm2. *Proc Natl Acad Sci USA.* 1998;95:8292–8297.
33. Stott FJ, Bates S, James MC, et al. The alternative product from the human CDKN2A locus, p14(ARF), participates in a regulatory feedback loop with p53 and MDM2. *EMBO J.* 1998;17:5001–5014.
34. Wang W, Nag S, Zhang X, et al. Ribosomal proteins and human diseases: pathogenesis, molecular mechanisms, and therapeutic implications. *Med Res Rev.* 2015;35:225–285.
35. Fang S, Jensen JP, Ludwig RL, Vousden KH, Weissman AM. Mdm2 is a RING finger-dependent ubiquitin protein ligase for itself and p53. *J Biol Chem.* 2000;275:8945–8951.
36. Leslie PL, Ke H, Zhang Y. The MDM2 RING domain and central acidic domain play distinct roles in MDM2 protein homodimerization and MDM2-MDMX protein heterodimerization. *J Biol Chem.* 2015;290:12941–12950.
37. Kulikov R, Letienne J, Kaur M, Grossman SR, Arts J, Blattner C. Mdm2 facilitates the association of p53 with the proteasome. *Proc Natl Acad Sci USA.* 2010;107:10038–10043.
38. Carter S, Bischof O, Dejean A, Vousden KH. C-terminal modifications regulate MDM2 dissociation and nuclear export of p53. *Nat Cell Biol.* 2007;9:428–435.
39. Geyer RK, Yu ZK, Maki CG. The MDM2 RING-finger domain is required to promote p53 nuclear export. *Nat Cell Biol.* 2000;2:569–573.
40. Oliner JD, Pietenpol JA, Thiagalingam S, Gyuris J, Kinzler KW, Vogelstein B. Oncoprotein MDM2 conceals the activation domain of tumour suppressor p53. *Nature.* 1993;362:857–860.

41. Barak Y, Gottlieb E, Juven-Gershon T, Oren M. Regulation of mdm2 expression by p53: alternative promoters produce transcripts with nonidentical translation potential. *Genes Dev.* 1994;8:1739–1749.
42. Zhang Z, Zhang R. Proteasome activator PA28 gamma regulates p53 by enhancing its MDM2-mediated degradation. *EMBO J.* 2008;27:852–864.
43. Zhang Z, Zhang R. p53-independent activities of MDM2 and their relevance to cancer therapy. *Curr Cancer Drug Targets.* 2005;5:9–20.
44. Bouska A, Eischen CM. Murine double minute 2: p53-independent roads lead to genome instability or death. *Trends Biochem Sci.* 2009;34:279–286.
45. Fridman JS, Hernando E, Hemann MT, de Stanchina E, Cordon-Cardo C, Lowe SW. Tumor promotion by Mdm2 splice variants unable to bind p53. *Cancer Res.* 2003;63:5703–5706.
46. Steinman HA, Burstein E, Lengner C, et al. An alternative splice form of Mdm2 induces p53-independent cell growth and tumorigenesis. *J Biol Chem.* 2004;279: 4877–4886.
47. Brekman A, Singh KE, Polotskaia A, Kundu N, Bargonetti J. A p53-independent role of Mdm2 in estrogen-mediated activation of breast cancer cell proliferation. *Breast Cancer Res.* 2011;13:R3.
48. Zhang Z, Li M, Wang H, Agrawal S, Zhang R. Antisense therapy targeting MDM2 oncogene in prostate cancer: effects on proliferation, apoptosis, multiple gene expression, and chemotherapy. *Proc Natl Acad Sci USA.* 2003;100:11636–11641.
49. Zhang Z, Wang H, Li M, Agrawal S, Chen X, Zhang R. MDM2 is a negative regulator of p21WAF1/CIP1, independent of p53. *J Biol Chem.* 2004;279:16000–16006.
50. Jin Y, Lee H, Zeng SX, Dai MS, Lu H. MDM2 promotes p21waf1/cip1 proteasomal turnover independently of ubiquitylation. *EMBO J.* 2003;22:6365–6377.
51. Xu H, Zhang Z, Li M, Zhang R. MDM2 promotes proteasomal degradation of p21Waf1 via a conformation change. *J Biol Chem.* 2010;285:18407–18414.
52. Uchida C, Miwa S, Kitagawa K, et al. Enhanced Mdm2 activity inhibits pRB function via ubiquitin-dependent degradation. *EMBO J.* 2005;24:160–169.
53. Sdek P, Ying H, Chang DL, et al. MDM2 promotes proteasome-dependent ubiquitin-independent degradation of retinoblastoma protein. *Mol Cell.* 2005;20:699–708.
54. Zhang Z, Wang H, Li M, Rayburn ER, Agrawal S, Zhang R. Stabilization of E2F1 protein by MDM2 through the E2F1 ubiquitination pathway. *Oncogene.* 2005;24: 7238–7247.
55. Ohtani K, DeGregori J, Nevins JR. Regulation of the cyclin E gene by transcription factor E2F1. *Proc Natl Acad Sci USA.* 1995;92:12146–12150.
56. Yang JY, Zong CS, Xia W, et al. ERK promotes tumorigenesis by inhibiting FOXO3a via MDM2-mediated degradation. *Nat Cell Biol.* 2008;10:138–148.
57. Brenkman AB, de Keizer PL, van den Broek NJ, Jochemsen AG, Burgering BM. Mdm2 induces mono-ubiquitination of FOXO4. *PLoS One.* 2008;3:e2819.
58. Gu L, Zhu N, Zhang H, Durden DL, Feng Y, Zhou M. Regulation of XIAP translation and induction by MDM2 following irradiation. *Cancer Cell.* 2009;15:363–375.
59. Yang JY, Zong CS, Xia W, et al. MDM2 promotes cell motility and invasiveness by regulating E-cadherin degradation. *Mol Cell Biol.* 2006;26:7269–7282.
60. Burgess A, Chia KM, Haupt S, Thomas D, Haupt Y, Lim E. Clinical overview of MDM2/X-targeted therapies. *Front Oncol.* 2016;6:7.
61. Lemos A, Leao M, Soares J, et al. Medicinal chemistry strategies to disrupt the p53-MDM2/MDMX interaction. *Med Res Rev.* 2016;36:789–844.
62. Zhang B, Golding BT, Hardcastle IR. Small-molecule MDM2-p53 inhibitors: recent advances. *Future Med Chem.* 2015;7:631–645.
63. Andreeff M, Kelly KR, Yee K, et al. Results of the phase I trial of RG7112, a small-molecule MDM2 antagonist in leukemia. *Clin Cancer Res.* 2016;22:868–876.

64. Furet P, Masuya K, Kallen J, et al. Discovery of a novel class of highly potent inhibitors of the p53-MDM2 interaction by structure-based design starting from a conformational argument. *Bioorg Med Chem Lett.* 2016;26:4837–4841.
65. Rew Y, Sun D. Discovery of a small molecule MDM2 inhibitor (AMG 232) for treating cancer. *J Med Chem.* 2014;57:6332–6341.
66. Reuther C, Heinzle V, Nolting S, et al. The HDM2 (MDM2) inhibitor NVP-CGM097 inhibits tumor cell proliferation and shows additive effects with 5-Fluorouracil on the p53-p21-Rb-E2F1 cascade in the p53wildtype neuroendocrine tumor cell line GOT1. *Neuroendocrinology.* 2016. [Nov 21].
67. Yang Y, Ludwig RL, Jensen JP, et al. Small molecule inhibitors of HDM2 ubiquitin ligase activity stabilize and activate p53 in cells. *Cancer Cell.* 2005;7:547–559.
68. Herman AG, Hayano M, Poyurovsky MV, et al. Discovery of Mdm2-MdmX E3 ligase inhibitors using a cell-based ubiquitination assay. *Cancer Discov.* 2011;1:312–325.
69. Tabernero J, Dirix L, Schoffski P, et al. A phase I first-in-human pharmacokinetic and pharmacodynamic study of serdemetan in patients with advanced solid tumors. *Clin Cancer Res.* 2011;17:6313–6321.
70. You L, Liu H, Huang J, et al. The novel anticancer agent JNJ-26854165 is active in chronic myeloid leukemic cells with unmutated BCR/ABL and T315I mutant BCR/ABL through promoting proteosomal degradation of BCR/ABL proteins. *Oncotarget.* 2017;8:7777–7790.
71. Chen L, Agrawal S, Zhou W, Zhang R, Chen J. Synergistic activation of p53 by inhibition of MDM2 expression and DNA damage. *Proc Natl Acad Sci USA.* 1998;95: 195–200.
72. Chen L, Lu W, Agrawal S, Zhou W, Zhang R, Chen J. Ubiquitous induction of p53 in tumor cells by antisense inhibition of MDM2 expression. *Mol Med.* 1999;5:21–34.
73. Wang H, Nan L, Yu D, Agrawal S, Zhang R. Antisense anti-MDM2 oligonucleotides as a novel therapeutic approach to human breast cancer: in vitro and in vivo activities and mechanisms. *Clin Cancer Res.* 2001;7:3613–3624.
74. Wang H, Wang S, Nan L, Yu D, Agrawal S, Zhang R. Antisense anti-MDM2 mixed-backbone oligonucleotides enhance therapeutic efficacy of topoisomerase I inhibitor irinotecan in nude mice bearing human cancer xenografts: in vivo activity and mechanisms. *Int J Oncol.* 2002;20:745–752.
75. Wang H, Nan L, Yu D, Lindsey JR, Agrawal S, Zhang R. Anti-tumor efficacy of a novel antisense anti-MDM2 mixed-backbone oligonucleotide in human colon cancer models: p53-dependent and p53-independent mechanisms. *Mol Med.* 2002;8:185–199.
76. Wang H, Yu D, Agrawal S, Zhang R. Experimental therapy of human prostate cancer by inhibiting MDM2 expression with novel mixed-backbone antisense oligonucleotides: in vitro and in vivo activities and mechanisms. *Prostate.* 2003;54:194–205.
77. Li M, Zhang Z, Hill DL, Chen X, Wang H, Zhang R. Genistein, a dietary isoflavone, down-regulates the MDM2 oncogene at both transcriptional and posttranslational levels. *Cancer Res.* 2005;65:8200–8208.
78. Li M, Zhang Z, Hill DL, Wang H, Zhang R. Curcumin, a dietary component, has anticancer, chemosensitization, and radiosensitization effects by down-regulating the MDM2 oncogene through the PI3K/mTOR/ETS2 pathway. *Cancer Res.* 2007;67: 1988–1996.
79. Zhao Y, Wang W, Han L, et al. Isolation, structural determination, and evaluation of the biological activity of 20(S)-25-methoxyl-dammarane-3beta, 12beta, 20-triol [20(S)-25-OCH3-PPD], a novel natural product from *Panax notoginseng. Med Chem.* 2007;3:51–60.
80. Wang W, Zhao Y, Rayburn ER, Hill DL, Wang H, Zhang R. In vitro anti-cancer activity and structure-activity relationships of natural products isolated from fruits of Panax ginseng. *Cancer Chemother Pharmacol.* 2007;59:589–601.

81. Nag SA, Qin JJ, Wang W, Wang MH, Wang H, Zhang R. Ginsenosides as anticancer agents: in vitro and in vivo activities, structure-activity relationships, and molecular mechanisms of action. *Front Pharmacol.* 2012;3:25.
82. Wang W, Rayburn ER, Velu SE, Nadkarni DH, Murugesan S, Zhang R. In vitro and in vivo anticancer activity of novel synthetic makaluvamine analogues. *Clin Cancer Res.* 2009;15:3511–3518.
83. Wang W, Rayburn ER, Velu SE, et al. A novel synthetic iminoquinone, BA-TPQ, as an anti-breast cancer agent: in vitro and in vivo activity and mechanisms of action. *Breast Cancer Res Treat.* 2010;123:321–331.
84. Wang W, Qin JJ, Voruganti S, et al. The pyrido[b]indole MDM2 inhibitor SP-141 exerts potent therapeutic effects in breast cancer models. *Nat Commun.* 2014;5:5086.
85. Wang W, Qin JJ, Voruganti S, et al. Identification of a new class of MDM2 inhibitor that inhibits growth of orthotopic pancreatic tumors in mice. *Gastroenterology.* 2014;147: 893–902e892.
86. Nag S, Qin JJ, Patil S, et al. A quantitative LC-MS/MS method for determination of SP-141, a novel pyrido[b]indole anticancer agent, and its application to a mouse PK study. *J Chromatogr B.* 2014;969:235–240.
87. Nag S, Qin JJ, Voruganti S, et al. Development and validation of a rapid HPLC method for quantitation of SP-141, a novel pyrido[b]indole anticancer agent, and an initial pharmacokinetic study in mice. *Biomed Chromatogr.* 2015;29:654–663.
88. Pan MG, Xiong Y, Chen F. NFAT gene family in inflammation and cancer. *Curr Mol Med.* 2013;13:543–554.
89. Luo C, Shaw KT, Raghavan A, et al. Interaction of calcineurin with a domain of the transcription factor NFAT1 that controls nuclear import. *Proc Natl Acad Sci USA.* 1996;93:8907–8912.
90. Lopez-Rodriguez C, Aramburu J, Rakeman AS, Rao A. NFAT5, a constitutively nuclear NFAT protein that does not cooperate with Fos and Jun. *Proc Natl Acad Sci USA.* 1999;96:7214–7219.
91. Miyakawa H, Woo SK, Dahl SC, Handler JS, Kwon HM. Tonicity-responsive enhancer binding protein, a rel-like protein that stimulates transcription in response to hypertonicity. *Proc Natl Acad Sci USA.* 1999;96:2538–2542.
92. Okamura H, Aramburu J, Garcia-Rodriguez C, et al. Concerted dephosphorylation of the transcription factor NFAT1 induces a conformational switch that regulates transcriptional activity. *Mol Cell.* 2000;6:539–550.
93. Beals CR, Clipstone NA, Ho SN, Crabtree GR. Nuclear localization of NF-ATc by a calcineurin-dependent, cyclosporin-sensitive intramolecular interaction. *Genes Dev.* 1997;11:824–834.
94. Nolan GP. NF-AT-AP-1 and Rel-bZIP: hybrid vigor and binding under the influence. *Cell.* 1994;77:795–798.
95. Shaw JP, Utz PJ, Durand DB, Toole JJ, Emmel EA, Crabtree GR. Identification of a putative regulator of early T cell activation genes. *Science.* 1988;241:202–205.
96. Durand DB, Shaw JP, Bush MR, Replogle RE, Belagaje R, Crabtree GR. Characterization of antigen receptor response elements within the interleukin-2 enhancer. *Mol Cell Biol.* 1988;8:1715–1724.
97. Minami T. Calcineurin-NFAT activation and DSCR-1 auto-inhibitory loop: how is homoeostasis regulated? *J Biochem.* 2014;155:217–226.
98. Fric J, Zelante T, Wong AY, Mertes A, Yu HB, Ricciardi-Castagnoli P. NFAT control of innate immunity. *Blood.* 2012;120:1380–1389.
99. Mognol GP, Carneiro FR, Robbs BK, Faget DV, Viola JP. Cell cycle and apoptosis regulation by NFAT transcription factors: new roles for an old player. *Cell Death Dis.* 2016;7:e2199.

100. Shou J, Jing J, Xie J, et al. Nuclear factor of activated T cells in cancer development and treatment. *Cancer Lett.* 2015;361:174–184.
101. Jauliac S, Lopez-Rodriguez C, Shaw LM, Brown LF, Rao A, Toker A. The role of NFAT transcription factors in integrin-mediated carcinoma invasion. *Nat Cell Biol.* 2002;4:540–544.
102. Zheng J, Fang F, Zeng X, Medler TR, Fiorillo AA, Clevenger CV. Negative cross talk between NFAT1 and Stat5 signaling in breast cancer. *Mol Endocrinol.* 2011;25: 2054–2064.
103. Kaunisto A, Henry WS, Montaser-Kouhsari L, et al. NFAT1 promotes intratumoral neutrophil infiltration by regulating IL8 expression in breast cancer. *Mol Oncol.* 2015;9:1140–1154.
104. Quang CT, Leboucher S, Passaro D, et al. The calcineurin/NFAT pathway is activated in diagnostic breast cancer cases and is essential to survival and metastasis of mammary cancer cells. *Cell Death Dis.* 2015;6:e1658.
105. Holzmann K, Kohlhammer H, Schwaenen C, et al. Genomic DNA-chip hybridization reveals a higher incidence of genomic amplifications in pancreatic cancer than conventional comparative genomic hybridization and leads to the identification of novel candidate genes. *Cancer Res.* 2004;64:4428–4433.
106. Baumgart S, Glesel E, Singh G, et al. Restricted heterochromatin formation links NFATc2 repressor activity with growth promotion in pancreatic cancer. *Gastroenterology.* 2012;142:388–398.
107. Griesmann H, Ripka S, Pralle M, et al. WNT5A-NFAT signaling mediates resistance to apoptosis in pancreatic cancer. *Neoplasia.* 2013;15:11–22.
108. Baumgart S, Chen NM, Zhang JS, et al. GSK-3beta governs inflammation-induced NFATc2 signaling hubs to promote pancreatic cancer progression. *Mol Cancer Ther.* 2016;15:491–502.
109. Zhang K, Li N, Chen Z, et al. High expression of nuclear factor of activated T cells in Chinese primary non-small cell lung cancer tissues. *Int J Biol Markers.* 2007;22:221–225.
110. Chen ZL, Zhao SH, Wang Z, et al. Expression and unique functions of four nuclear factor of activated T cells isoforms in non-small cell lung cancer. *Chin J Cancer.* 2011;30:62–68.
111. Liu JF, Zhao SH, Wu SS. Depleting NFAT1 expression inhibits the ability of invasion and migration of human lung cancer cells. *Cancer Cell Int.* 2013;13:41.
112. Noordhuis MG, Fehrmann RS, Wisman GB, et al. Involvement of the TGF-beta and beta-catenin pathways in pelvic lymph node metastasis in early-stage cervical cancer. *Clin Cancer Res.* 2011;17:1317–1330.
113. Gerlach K, Daniel C, Lehr HA, et al. Transcription factor NFATc2 controls the emergence of colon cancer associated with IL-6-dependent colitis. *Cancer Res.* 2012;72:4340–4350.
114. Shoshan E, Braeuer RR, Kamiya T, et al. NFAT1 directly regulates IL8 and MMP3 to promote melanoma tumor growth and metastasis. *Cancer Res.* 2016;76:3145–3155.
115. Yiu GK, Toker A. NFAT induces breast cancer cell invasion by promoting the induction of cyclooxygenase-2. *J Biol Chem.* 2006;281:12210–12217.
116. Yiu GK, Kaunisto A, Chin YR, Toker A. NFAT promotes carcinoma invasive migration through glypican-6. *Biochem J.* 2011;440:157–166.
117. Chen M, O'Connor KL. Integrin alpha6beta4 promotes expression of autotaxin/ENPP2 autocrine motility factor in breast carcinoma cells. *Oncogene.* 2005;24: 5125–5130.
118. Sengupta S, Jana S, Biswas S, Mandal PK, Bhattacharyya A. Cooperative involvement of NFAT and SnoN mediates transforming growth factor-beta (TGF-beta) induced EMT in metastatic breast cancer (MDA-MB 231) cells. *Clin Exp Metastasis.* 2013;30: 1019–1031.

119. Sengupta S, Jana S, Bhattacharyya A. TGF-beta-Smad2 dependent activation of CDC 25A plays an important role in cell proliferation through NFAT activation in metastatic breast cancer cells. *Cell Signal.* 2014;26:240–252.
120. Oskay Halacli S. FOXP1 enhances tumor cell migration by repression of NFAT1 transcriptional activity in MDA-MB-231 cells. *Cell Biol Int.* 2017;41:102–110.
121. Baggott RR, Alfranca A, Lopez-Maderuelo D, et al. Plasma membrane calcium ATPase isoform 4 inhibits vascular endothelial growth factor-mediated angiogenesis through interaction with calcineurin. *Arterioscler Thromb Vasc Biol.* 2014;34:2310–2320.
122. Baggott RR, Mohamed TM, Oceandy D, et al. Disruption of the interaction between PMCA2 and calcineurin triggers apoptosis and enhances paclitaxel-induced cytotoxicity in breast cancer cells. *Carcinogenesis.* 2012;33:2362–2368.
123. Yoeli-Lerner M, Yiu GK, Rabinovitz I, Erhardt P, Jauliac S, Toker A. Akt blocks breast cancer cell motility and invasion through the transcription factor NFAT. *Mol Cell.* 2005;20:539–550.
124. Yoeli-Lerner M, Chin YR, Hansen CK, Toker A. Akt/protein kinase b and glycogen synthase kinase-3beta signaling pathway regulates cell migration through the NFAT1 transcription factor. *Mol Cancer Res.* 2009;7:425–432.
125. Gu Y, Zhang J, Mi W, et al. Silencing of GM3 synthase suppresses lung metastasis of murine breast cancer cells. *Breast Cancer Res.* 2008;10:R1.
126. Dejmek J, Safholm A, Kamp Nielsen C, Andersson T, Leandersson K. Wnt-5a/Ca^{2+}-induced NFAT activity is counteracted by Wnt-5a/Yes-Cdc42-casein kinase 1alpha signaling in human mammary epithelial cells. *Mol Cell Biol.* 2006;26: 6024–6036.
127. Siamakpour-Reihani S, Caster J, Bandhu Nepal D, et al. The role of calcineurin/NFAT in SFRP2 induced angiogenesis—a rationale for breast cancer treatment with the calcineurin inhibitor tacrolimus. *PLoS One.* 2011;6:e20412.
128. Courtwright A, Siamakpour-Reihani S, Arbiser JL, et al. Secreted frizzle-related protein 2 stimulates angiogenesis via a calcineurin/NFAT signaling pathway. *Cancer Res.* 2009;69:4621–4628.
129. Lee CR, Chun JN, Kim SY, et al. Cyclosporin A suppresses prostate cancer cell growth through CaMKKbeta/AMPK-mediated inhibition of mTORC1 signaling. *Biochem Pharmacol.* 2012;84:425–431.
130. Zhao X, Wang Q, Yang S, et al. Quercetin inhibits angiogenesis by targeting calcineurin in the xenograft model of human breast cancer. *Eur J Pharmacol.* 2016;781:60–68.
131. Iampietro M, Gravel A, Flamand L. Inhibition of breast cancer cell proliferation through disturbance of the calcineurin/NFAT pathway by human herpesvirus 6B U54 tegument protein. *J Virol.* 2014;88:12910–12914.
132. Hogan PG. Calcium-NFAT transcriptional signalling in T cell activation and T cell exhaustion. *Cell Calcium.* 2017;63:66–69.
133. Muller MR, Rao A. NFAT, immunity and cancer: a transcription factor comes of age. *Nat Rev Immunol.* 2010;10:645–656.
134. Tie X, Han S, Meng L, Wang Y, Wu A. NFAT1 is highly expressed in, and regulates the invasion of, glioblastoma multiforme cells. *PLoS One.* 2013;8:e66008.
135. Baumgart S, Chen NM, Siveke JT, et al. Inflammation-induced NFATc1-STAT3 transcription complex promotes pancreatic cancer initiation by KrasG12D. *Cancer Discov.* 2014;4:688–701.
136. Oikawa T, Nakamura A, Onishi N, Yamada T, Matsuo K, Saya H. Acquired expression of NFATc1 downregulates E-cadherin and promotes cancer cell invasion. *Cancer Res.* 2013;73:5100–5109.
137. Fougere M, Gaudineau B, Barbier J, et al. NFAT3 transcription factor inhibits breast cancer cell motility by targeting the Lipocalin 2 gene. *Oncogene.* 2010;29: 2292–2301.

138. Wang W, Wang H, Rayburn ER, Zhao Y, Hill DL, Zhang R. 20(S)-25-methoxyl-dammarane-3beta, 12beta, 20-triol, a novel natural product for prostate cancer therapy: activity in vitro and in vivo and mechanisms of action. *Br J Cancer.* 2008;98:792–802.
139. Wang W, Rayburn ER, Hao M, et al. Experimental therapy of prostate cancer with novel natural product anti-cancer ginsenosides. *Prostate.* 2008;68:809–819.
140. Voruganti S, Qin JJ, Sarkar S, et al. Oral nano-delivery of anticancer ginsenoside 25-OCH3-PPD, a natural inhibitor of the MDM2 oncogene: nanoparticle preparation, characterization, in vitro and in vivo anti-prostate cancer activity, and mechanisms of action. *Oncotarget.* 2015;6:21379–21394.
141. Wang W, Rayburn ER, Zhao Y, Wang H, Zhang R. Novel ginsenosides 25-OH-PPD and 25-OCH3-PPD as experimental therapy for pancreatic cancer: anticancer activity and mechanisms of action. *Cancer Lett.* 2009;278:241–248.
142. Wang W, Rayburn ER, Hang J, Zhao Y, Wang H, Zhang R. Anti-lung cancer effects of novel ginsenoside 25-OCH(3)-PPD. *Lung Cancer.* 2009;65:306–311.
143. Wang W, Zhang X, Qin JJ, et al. Natural product ginsenoside 25-OCH3-PPD inhibits breast cancer growth and metastasis through down-regulating MDM2. *PLoS One.* 2012;7:e41586.
144. Qin JJ, Wang W, Voruganti S, Wang H, Zhang WD, Zhang R. Identification of a new class of natural product MDM2 inhibitor: in vitro and in vivo anti-breast cancer activities and target validation. *Oncotarget.* 2015;6:2623–2640.
145. Qin JJ, Wang W, Voruganti S, Wang H, Zhang WD, Zhang R. Inhibiting NFAT1 for breast cancer therapy: new insights into the mechanism of action of MDM2 inhibitor JapA. *Oncotarget.* 2015;6:33106–33119.
146. Qin JJ, Wang W, Sarkar S, Voruganti S, Agarwal R, Zhang R, Inulanolide A. as a new dual inhibitor of NFAT1-MDM2 pathway for breast cancer therapy. *Oncotarget.* 2016;7:32566–32578.
147. Qin JJ, Sarkar S, Voruganti S, Agarwal R, Wang W, Zhang R. Identification of lineariifolianoid A as a novel dual NFAT1 and MDM2 inhibitor for human cancer therapy. *J Biomed Res.* 2016;30:322–333.

CHAPTER EIGHT

Microbiota and Breast Cancer

Sridhar Mani[1]
Albert Einstein College of Medicine INC, Bronx, NY, United States
[1] Corresponding author. E-mail address: sridhar.mani@einstein.yu.edu

Contents

Abstract

Breast carcinogenesis and/or cancer growth and/or drug sensitivity has a multifactorial etiology—perhaps the least well-characterized aspect being that of the distant environmental influences, namely, the microbiota that inhabit humans. For the purposes of this chapter, and to keep the subject matter well defined, microbiota is defined as bacterial microbes only. In this chapter, the pathways that lead to priming of breast cancer and/or the maintenance of a malignant state and/or influences on drug sensitivity via bacterial influences will be identified and described. Specifically, major influence of bacteria found within breast tissue, skin, the oropharynx, and the gastrointestinal tract will be discussed as pertinent to published data. Finally, some preliminary conclusions of the important contributions and pathways of specific bacteria to breast carcinogenesis and/or cancer will be presented.

1. INTRODUCTION

Breast cancer evolves as a complex interaction between genes and environment; yet under the perfect conditions of uncommon genetic mutations and environmental exposure, only a fraction of cases develop as cancer.[1] Thus, there are other yet undescribed or poorly characterized pathways

Progress in Molecular Biology and Translational Science, Volume 151
ISSN 1877-1173
http://dx.doi.org/10.1016/bs.pmbts.2017.07.004

that lead to or promote the development of breast cancer. Foremost among them are over trillion bacterial forms that inhabit us as symbiotic hosts and of late have had their limelight in virtually every aspect of human health. Yet, in terms of breast carcinogenesis or malignancy, very little is known as to how bacteria from within us could potentiate or negate it. In fact, as of this chapter, there are no reviews that comprehensively cover this subject. Furthermore, many primary scientific papers on the subject are largely focused on associations rather than cause-and-effect. So a critical aspect of the subject matter is a theoretical model that needs to be frame worked, based on which, experimental evidence for or against it may be formulated. Indeed, a broad-based approach suggests that bacteria resident in certain compartments [breast tissue, skin, oropharynx, vaginal/reproductive tract, and the gastrointestinal (GI) tract] may directly and/or indirectly influence the development of breast cancer. The indirect influences include but are not limited to host tissue interactions with bacterial structural components (e.g., flagellin), bacterial metabolism of endobiotic or xenobiotic compounds (e.g., estrogens), and bacterial secretion products (e.g., quorum sensing peptides) (Fig. 1).

2. THE ESTRABOLOME AND BREAST CANCER

Breast cancer produces an overwhelming health burden—there are 232,670 new breast cancer cases and 40,000 deaths among women living with breast cancer in the United States in 2014.[2] These staggering statistics linger despite optimal therapies available to women in this decade. It is also known that breast cancer incidence begins to rise within the third decade of life, most likely a result of the effects of ovarian hormones on breast tissue.[3–5] There are many nonmodifiable risk factors for breast cancer (age, BRCA mutations, family and reproductive history, and high-dose radiation to the chest); however, of the modifiable forms (high-endogenous estrogens, hormone therapy, obesity, and alcohol consumption), there is a common theme that relates to endogenous estrogen balance and its subsequent effect on breast tissues (estrogen receptor expression/interactions).[6–9] It has been unequivocally established that endogenous estrogen burden is associated with breast cancer development in postmenopausal women.[10,11] There is a similar association in premenopausal women; however this has not been firmly established due to the small studies and greater variation in hormone levels due to the menstrual cycle.[10] Given the strong positive association

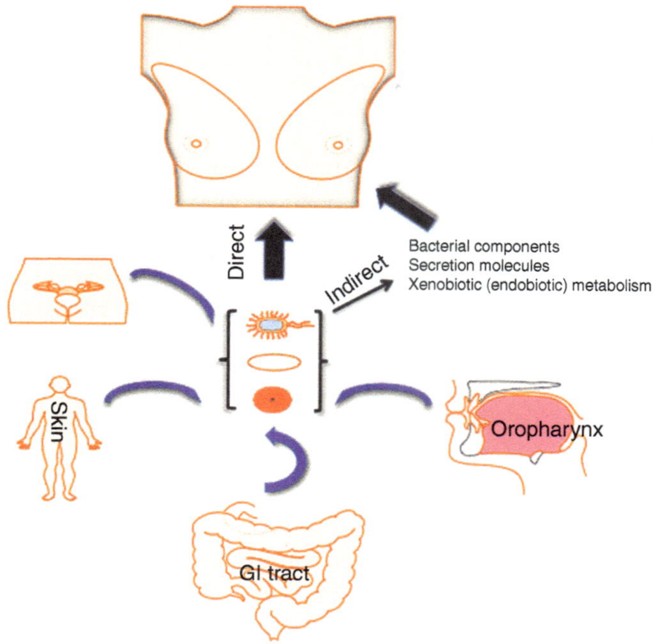

Fig. 1 Schematic of microbial influence on breast cancer. The direct effects implicate microbes on skin/breast tissue that directly influence breast tumor development via contact with breast tissue. Indirect mechanisms implicate structural and functional components of bacteria [from proximal or distal sources like skin, orpharynx, gastrointestinal (GI) tract], and/or secretion products (molecules) or metabolites derived from xenobiotic/endobiotic metabolism.

observed between blood sex hormones and breast cancer in postmenopausal women, circulating estrogens seem to be an appropriate marker of tissue exposure, as suggested by Hankinson and Eliassen.[12]

Circulating estrogens (and metabolites) are, in general, bioactive. In premenopausal women, estradiol is synthesized in the ovaries and is the most important estrogen. In postmenopausal women, estrogens are synthesized in peripheral tissues. Estradiols, estrones and estriols are produced via aromatization/dehydrogenases in peripheral tissues. These are then further subjected to hepatic phase II conjugation reactions via enzymes that add either a glucuronic acid (UDP-glucuronosyltransferases) and/or a sulfate (sulfotransferases) (the latter is a minor component of the metabolite pool) and released into bile. In both rodents and humans, these pathways are relatively conserved, in that only conjugated estrogens are available in bile and intestines. Conjugation is thus considered a detoxification reaction by which hormones either become water-soluble and are excreted in the urine or feces, or turn into a more lipophilic moiety with elevated half-lives.[10,13–15]

In the intestines, β-glucuronidases (BGUS) catalyze the hydrolysis of endogenous β-glucuronides produced in the liver and exogenous β-glucuronides found in the diet, such as complex carbohydrates. Many of these metabolites include steroid hormones, xenobiotics, and bile acids excreted into the intestinal tract by bile after hepatic glucuronidation. The removal of glucuronic acid from conjugated substrates by intestinal microbial β-glucuronidases promotes reabsorption of their respective aglycones into the enterohepatic circulation. There are distinct and well-characterized bacterial BGUS genes (*gus* genes) commonly found in commensal bacteria in both rodents and humans (reviewed in Ref. [16]). BGUS activity can be significantly modulated by diet and by bacterial context. Increased fecal BGUS activity has been reported in healthy humans consuming high fat and/or protein diets,[17,18] while fiber consumption decreases BGUS activity.[19,20] Indeed quorum sensing and bacterial communities determine BGUS expression while antibiotics totally eradicate bacteria and BGUS activity. Notably, antibiotics kill bacteria and can induce their own consequences to human health (e.g., drug resistance, reduced mucosal immunity).[16] Furthermore, antibiotic usage has revealed unexpected increased risk associations with breast cancer, which is controversial and attributed to poor study designs.[21] Selected probiotics (e.g., lactic acid bacteria) can suppress BGUS activity; however, its role in mechanisms and of chemoprevention has not been established or proven to be safe.[22] Until recently, there are no specific chemical inhibitors of microbial BGUS; however, nonspecific low-potency (mM) inhibitors of general BGUS activity (e.g., calcium glucarate or glucarolactone) have been described previously.[23–25] Dietary modulation is always a strategy to reduce weight and control dietary intake; however, this is much more difficult to implement as a primary prevention strategy alone.

Microbial BGUS activity is central in determining "the estrabolome," and thus the body estrogen burden.[21,26] Urinary estrone in women is highly correlated with functional activity of fecal BGUS and by contrast, fecal BGUS inversely correlates with total fecal estrogens. The authors of this human study conclude that "differences in systemic estrogens were attributable to differences in the intestinal microbiota" and "BGUS activity."[27] Similar results in rodents support the role of estrogen metabolism in breast carcinogenesis.[28] It is important to note that even small changes in estrogen uptake via the intestine daily can add over time to larger total estrogen burden (in humans this time scale is in the order of years) as there would be an expected daily small surge of estrogen levels over the previous days baseline. If we add exogenous estrogens, as well as environmental estrogens to this mix, there could be substantial accumulation of estrogen in the body over time.

Our laboratory in conjunction with Dr. Matt Redinbo's laboratory at UNC, Chapel Hill described a first-in-class, novel microbial specific inhibitor of BGUS.[29] Our lead compound described Inh1 as a nanomolar (nM) potency inhibitor of BGUS and did not have any influence on mammalian BGUS activity. The compound did not kill bacteria and was nontoxic to human and rodent cells in vitro and organs in vivo.[29–31] The inhibitor, Inh1, has low-systemic bioavailability and very successfully prevented the action of BGUS enzymes on xenobiotic glucuronides in vitro and in vivo. This has prevented the major intestinal side effects of these xenobiotic compounds.[16,32–35] The advantage of using a stealth microbial-specific BGUS inhibitor over the non-specific inhibitors available is that there is expected to be limited off-target inhibition of the mammalian enzyme. Theoretically, mammalian BGUS enzymes are important in regulating aglycones emanating from other physiologic compounds, such as cholesterol, bile acids, and other hormones. Furthermore, our approach will specifically probe and answer the speculation that microbial BGUS enzymes are important regulators of the estrabolome and breast cancer risk (Fig. 2).

The study of breast cancer prevention is best achieved in multistep models of breast carcinogenesis. While no model completely recapitulates human breast tumors, some are better at delineating the steps toward carcinoma, namely, hyperplastic lesions, adenomas, in situ lesions, and frank invasive carcinomas. Furthermore, to assess endogenous, as well as exogenous hormones as a trigger of breast tumors, models that are estrogen receptor positive, as a start, would be extremely valuable and pertinent. Thus, precision mouse modeling rather than a "one size fits all" approach is likely to yield to the best findings as a "proof-of-concept."[36] Taking into consideration this concept, Dr. Jeffery Pollard developed the Polyoma MT (PyMT) mouse model at our institution. In this model, the expression of the oncoprotein, polyoma middle T antigen (PyMT), is under the control of mouse mammary tumor virus LTR (MMTV LTR) and is therefore restricted to the mammary epithelium. In this model mammary hyperplasia is detected as early as 4 weeks in over 50%–60% of mice and most importantly there is a stepwise progression in the adenoma/carcinoma sequence. By 12–14 weeks the majority of mice develop carcinoma. The most interesting aspect of this model is that

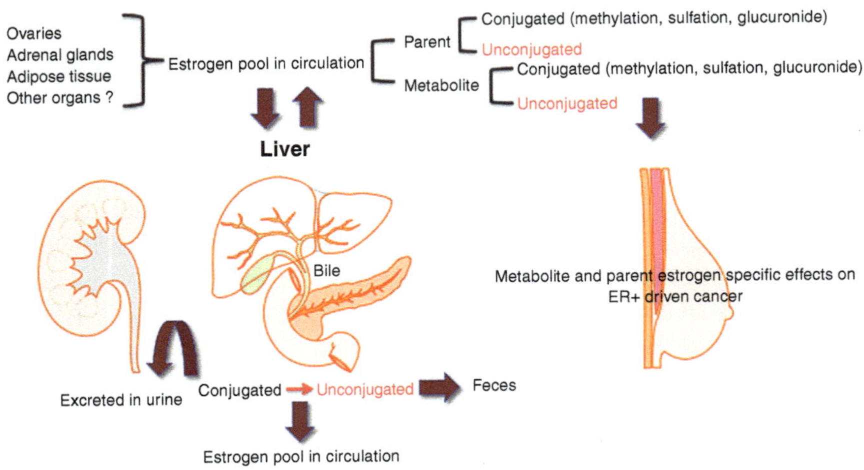

Fig. 2 The estrabolome and breast cancer. The ovaries/adrenals/adipose andnother tissues contribute to the total estrogen pool in the body. Liver is a major site for metabolite formation and conjugation/deconjugation reactions. These are critical for clearance of estrogen from the body (via urine and feces). In the intestines, microbial enzymes are implicated in the deconjugation of estrogens. The net aglycones are recirculated and contribute toward the total estrogen burden in the body. In circulation, the various conjugated and unconjugated estrogens/metabolites have differential effects on estrogen-receptor positive (ER+) breast cancer growth and development.

the early occurrence of tumors allows for rapid assessment of chemoprevention and that the early lesions are significantly dependent on estrogen receptor signaling.[37] This model has been extensively characterized for probing tumor biology as it relates to both rodent and human breast cancer, as well as experimental therapies.[38–46] From the perspective of this application, in which the estrogen burden effects are being evaluated, the PyMT mouse is tractable and has excellent breeding potential, short latency to tumor development, and estrogen-receptor positive (ER+)–associated early lesions. The important point is that for a proof-of-concept, more human-like models (e.g., longer latency) are not required as the principal goal is only to test if estrogen drives early breast tumor lesions given our interventions. Alternative human-like long latency (~>11 months) estrogen receptor (ER)–driven models (i.e., tTA/Tag/ERα mice) are more cumbersome and involve use of antibiotics like tetracycline that in our experimental design would pose problems with data interpretation in the control arm.[47,48] Non-GEM models would not provide us with the opportunity to study the multistep progression of malignancy and early cancer intervention effects.

The positive correlation between bioactive estrogen and breast cancer development in postmenopausal women is now well established. A *major* metabolic route of estrogen disposition (clearance) in rodents and humans is via intestinal/hepatic glucuronidation, secretion into the intestines, and recirculation of bioactive estrogen via microbial BGUS action in situ. An interesting but untested proposal is that blocking the action of microbial selective BGUS with our inhibitors will reduce the recirculation of bioactive estrogens, thus, curbing estrogen receptor (ER) positive breast carcinogenesis in an at-risk (PyMT) rodent model of breast cancer. These and related concepts in breast cancer are being tested at the Albert Einstein College of Medicine, Bronx, NY under the sponsorship of the United States Department of Defense.

3. BACTERIA AND BREAST CANCER TISSUE

Bacteria are abundantly present in breast cancer tissue.[49–51] The Shannon diversity index averages 3.6 and is comparable to the diversity seen in other microbial compartments (e.g., the intestines) but higher than, for example, that observed in the vagina.[49] Proteobacteria is the most abundant phylum represented in breast tissue. Other common phylum is the firmicutes. It is unclear at present if the distribution of bacteria differs at the species level between normal and adjacent tumor tissues. A more recent paper suggests that there is indeed a good correlation between breast cancer with enrichment of taxa of lower abundance that include genera *Fusobacterium, Atopobium, Gluconacetobacter, Hydrogenophaga*, and *Lactobacillus*.[50] In another paper, in women with breast cancer there was a higher relative abundance of *Bacillus*, *Enterobacteriaceae*, and *Staphylococcus*. They also showed that *Escherichia coli* (a member of the *Enterobacteriaceae* family) and *Staphylococcus epidermidis*, all isolated from breast cancer patients, induced-DNA double-stranded breaks in HeLa cells.[51] Nipple aspirates from breast cancer patients can also differ in bacterial community composition to those from "normal" or "healthy control" women. For example, the genus *Alistipes* was more common in aspirates of breast cancer cases while the unclassified genus of Sphingomonadaceae family was more prevalent in aspirates of healthy control women. Interestingly, the bacteria from breast cancer patients all shared a common enzyme expressed, β-glucuronidase.[52] Interestingly, quorum-sensing molecules present within the milieu of biofilms in breast tissue have been associated with changes in function of neutrophils, macrophages, endothelial,[53] and epithelial cells.[54,55] As an extension of these observations, Human Papilloma Virus also correlates with stat3 signaling and IL17 expression in breast cancer tissue.[56]

4. INTESTINAL BACTERIA AND BREAST CANCER

Microbial dysbiosis seems to be a hallmark of human breast cancer.[57,58] There is potential for alterations in the composition of the breast microbiota via the intestines—for example, oral lactobacilli administered to patients with mastitis has been isolated from the breast tissue and correlated with abrogation of inflammation in the breast.[59] Indeed, postmenopausal women with breast cancer have lower α-diversity but higher β-diversity, which seemed to be estrogen independent.[60] One view of how gut bacteria alter mammary tumor response is likely via alteration of neutrophil-associated inflammatory responses in the intestine and distal sites.[61] Indeed, compiling the studies, one can posit that estrogen-independent breast cancer–related conditions, such as adiposity, insulin resistance, dyslipidemia (metabolic syndrome), leukocytosis, and elevated level of C-reactive protein might well be independent predictors of microbiota-contributed factors toward breast carcinogenesis.[62,63] Further indication of a microbial association with breast cancer comes from a breast cancer staging study, in which patients with grade III cancer had a borderline significance higher absolute count of *Blautia* sp. as compared with patients with grade I breast cancer ($P = 0.048$). In addition, absolute counts of *Bifidobacterium* and *Blautia*, and of *Faecalibacterium prausnitzii* and *Blautia*, changed with clinical stage.[64,65] Certain bacterial orders overrepresented in breast cancer intestinal microbiome, for example, *Clostridiales* and *Bacteriodes*, might have estrogen metabolism potential and have been shown to coassociate with metabolite to parent estrogen ratios.[66] Other studies partly confirm associations of specific intestinal bacterial orders and breast cancer; however, the studies published to-date are small and cannot lead to any significant conclusions.[67] An alternative hypothesis has been proposed that estrogen-like molecules released from bacterial metabolism might propel certain cancers like leiomyomas and breast cancer.[68]

The intestinal microbiome has also been explored in animal models of breast cancer development. There is increased prevalence of breast tumors in C57BL/6 ApcMin Rag2-deficient female mice colonized with *Helicobacter hepaticus*.[69] Indeed, other models like the C3-1-Tag mice, a well-characterized mammary tumor model, support the role of intestinal microbiota in promoting breast tumors.[61,70] Indeed, lignin-metabolizing bacteria colonized mice have decreased breast tumor sizes compared with germ-free mice, which support that notion that populations of bacteria in the right diet milieu can be protective toward breast carcinogeneisis.[71]

The microbiome of the intestines may play dual roles in cancer—one that promotes malignant progression and the other that selectively uses specific subsets to promote antitumor immunity in the context of chemoimmunotherapy. For example, *Bacterioides fragilis* is a key player in the antitumor effectiveness of antibodies targeting cytotoxic T lymphocyte–associated protein 4 (CTLA-4).[72] Similar observations for Bifidobacteria and anti-PD1 therapy have been shown.[73] It has been conjectured that there might be differential protective properties to intestinal microbes in treatment-associated toxicity, while enhancing therapeutic efficacy.

5. CONCLUSIONS

The data summarized here provide a snapshot of the current research on the microbiome with respect to breast cancer development and treatment. The focus is to highlight the seminal features of the breast or intestinal microbiome that might be important for breast cancer development. Indeed, microbial taxa do segregate in patients with breast cancer when compared to those without breast cancer; however, at present these are simply association studies. Microbiome influences on immunotherapy show that perhaps other agents might follow similar associations. A considerable amount of future research is required to further explore if the microbiome is a major driver of breast cancer.

6. FUTURE PROSPECTS

The critical unknown prospect in all the studies published thus far are cause and effect. Like many early association studies, few are validated in larger scale studies. Thus, a future prospect would be to perform laboratory studies geared toward cause-and-effect. The second and more critical aspect of the microbiome is whether there is a network or connection between the intestinal and breast or skin microbiome, such that both act in concert to orchestrate the environment permissive toward breast cancer development. The research performed so far is piece-meal and warrants integration to get a better picture of how the microbiome might influence breast cancer development. Finally, mining of the microbiome for targets that might reduce estrogen metabolism and/or reduce drug toxicity is a promising step toward chemoprevention.

REFERENCES

1. Rudolph A, Chang-Claude J, Schmidt MK. Gene-environment interaction and risk of breast cancer. *Br J Cancer.* 2016;114(2):125–133.
2. Siegel R, Ma J, Zou Z, Jemal A. Cancer statistics, 2014. *Cancer J Clin.* 2014;64(1):9–29.
3. Huang WY, Newman B, Millikan RC, Schell MJ, Hulka BS, Moorman PG. Hormone-related factors and risk of breast cancer in relation to estrogen receptor and progesterone receptor status. *Am J Epidemiol.* 2000;151(7):703–714.
4. Hulka BS, Moorman PG. Breast cancer: hormones and other risk factors. *Maturitas.* 2001;38(1):103–113. [discussion 13-6].
5. Hulka BS, Moorman PG. Breast cancer: hormones and other risk factors. *Maturitas.* 2008;61(1–2):203–213. [discussion 13].
6. Althuis MD, Fergenbaum JH, Garcia-Closas M, Brinton LA, Madigan MP, Sherman ME. Etiology of hormone receptor-defined breast cancer: a systematic review of the literature. *Cancer Epidemiol Biomark Prevent.* 2004;13(10):1558–1568.
7. Hartman TJ, Sisti JS, Hankinson SE, Xu X, Eliassen AH, Ziegler R. Alcohol consumption and urinary estrogens and estrogen metabolites in premenopausal women. *Hormones Cancer.* 2016;7(1):65–74.
8. Simone V, D'Avenia M, Argentiero A, Felici C, Rizzo FM, De Pergola G, et al. Obesity and breast cancer: molecular interconnections and potential clinical applications. *Oncologist.* 2016;21(4):404–417.
9. Hankinson SE, Willett WC, Manson JE, Hunter DJ, Colditz GA, Stampfer MJ, et al. Alcohol, height, and adiposity in relation to estrogen and prolactin levels in postmenopausal women. *J Natl Cancer Inst.* 1995;87(17):1297–1302.
10. Samavat H, Kurzer MS. Estrogen metabolism and breast cancer. *Cancer Lett.* 2015;356 (2 Pt A):231–243.
11. Eliassen AH, Missmer SA, Tworoger SS, Hankinson SE. Endogenous steroid hormone concentrations and risk of breast cancer: does the association vary by a woman's predicted breast cancer risk? *J Clin Oncol.* 2006;24(12):1823–1830.
12. Hankinson SE, Eliassen AH. Endogenous estrogen, testosterone and progesterone levels in relation to breast cancer risk. *J Steroid Biochem Mol Biol.* 2007;106(1–5):24–30.
13. Lakhani NJ, Venitz J, Figg WD, Sparreboom A. Pharmacogenetics of estrogen metabolism and transport in relation to cancer. *Curr Drug Metab.* 2003;4(6):505–513.
14. Adlercreutz H, Martin F. Biliary excretion and intestinal metabolism of progesterone and estrogens in man. *J Steroid Biochem.* 1980;13(2):231–244.
15. Adlercreutz H, Martin F, Jarvenpaa P, Fotsis T. Steroid absorption and enterohepatic recycling. *Contraception.* 1979;20(3):201–223.
16. Mani S, Boelsterli UA, Redinbo MR. Understanding and modulating mammalian-microbial communication for improved human health. *Annu Rev Pharmacol Toxicol.* 2014;54:559–580.
17. Reddy BS, Hanson D, Mangat S, Mathews L, Sbaschnig M, Sharma C, et al. Effect of high-fat, high-beef diet and of mode of cooking of beef in the diet on fecal bacterial enzymes and fecal bile acids and neutral sterols. *J Nutr.* 1980;110(9):1880–1887.
18. Domellof L, Darby L, Hanson D, Mathews L, Simi B, Reddy BS. Fecal sterols and bacterial beta-glucuronidase activity: a preliminary metabolic epidemiology study of healthy volunteers from Umea, Sweden, and metropolitan New York. *Nutr Cancer.* 1982;4(2):120–127.

19. Reddy BS, Engle A, Simi B, Goldman M. Effect of dietary fiber on colonic bacterial enzymes and bile acids in relation to colon cancer. *Gastroenterology.* 1992;102(5):1475–1482.
20. Reddy BS, Engle A, Simi B, O'Brien LT, Barnard RJ, Pritikin N, et al. Effect of low-fat, high-carbohydrate, high-fiber diet on fecal bile acids and neutral sterols. *Prevent Med.* 1988;17(4):432–439.
21. Kwa M, Plottel CS, Blaser MJ, Adams S. The intestinal microbiome and estrogen receptor-positive female breast cancer. *J Natl Cancer Inst.* 2016;108(8).
22. Kumar M, Kumar A, Nagpal R, Mohania D, Behare P, Verma V, et al. Cancer-preventing attributes of probiotics: an update. *Int J Food Sci Nutr.* 2010;61(5):473–496.
23. Morita N, Walaszek Z, Kinjo T, Nishimaki T, Hanausek M, Slaga TJ, et al. Effects of synthetic and natural in vivo inhibitors of beta-glucuronidase on azoxymethane-induced colon carcinogenesis in rats. *Mol Med Rep.* 2008;1(5):741–746.
24. Oleson L, Court MH. Effect of the beta-glucuronidase inhibitor saccharolactone on glucuronidation by human tissue microsomes and recombinant UDP-glucuronosyltransferases. *J Pharm Pharmacol.* 2008;60(9):1175–1182.
25. Monograph, Calcium-D-glucarate. *Altern Med Rev.* 2002;7(4):336–339.
26. Dada OA, Martins OO. Drug effects on the intestinal absorption of estrogens. *J Steroid Biochem.* 1983;19(1c):821–825.
27. Flores R, Shi J, Fuhrman B, Xu X, Veenstra TD, Gail MH, et al. Fecal microbial determinants of fecal and systemic estrogens and estrogen metabolites: a cross-sectional study. *J Transl Med.* 2012;10:253.
28. Santen RJ, Yue W, Wang JP. Estrogen metabolites and breast cancer. *Steroids.* 2015;99 (Pt A):61–66.
29. Wallace BD, Wang H, Lane KT, Scott JE, Orans J, Koo JS, et al. Alleviating cancer drug toxicity by inhibiting a bacterial enzyme. *Science.* 2010;330(6005):831–835.
30. Roberts AB, Wallace BD, Venkatesh MK, Mani S, Redinbo MR. Molecular insights into microbial beta-glucuronidase inhibition to abrogate CPT-11 toxicity. *Mol Pharmacol.* 2013;84(2):208–217.
31. Wallace BD, Roberts AB, Pollet RM, Ingle JD, Biernat KA, Pellock SJ, et al. Structure and inhibition of microbiome beta-glucuronidases essential to the alleviation of cancer drug toxicity. *Chem Biol.* 2015;22(9):1238–1249.
32. Dou W, Mukherjee S, Li H, Venkatesh M, Wang H, Kortagere S, et al. Alleviation of gut inflammation by Cdx2/Pxr pathway in a mouse model of chemical colitis. *PLoS One.* 2012;7(7):e36075.
33. Boelsterli UA, Redinbo MR, Saitta KS. Multiple NSAID-induced hits injure the small intestine: underlying mechanisms and novel strategies. *Toxicol Sci.* 2013;131(2):654–667.
34. LoGuidice A, Wallace BD, Bendel L, Redinbo MR, Boelsterli UA. Pharmacologic targeting of bacterial beta-glucuronidase alleviates nonsteroidal anti-inflammatory drug-induced enteropathy in mice. *J Pharmacol Exp Ther.* 2012;341(2):447–454.
35. Saitta KS, Zhang C, Lee KK, Fujimoto K, Redinbo MR, Boelsterli UA. Bacterial beta-glucuronidase inhibition protects mice against enteropathy induced by indomethacin, ketoprofen or diclofenac: mode of action and pharmacokinetics. *Xenobiotica.* 2014;44 (1):28–35.
36. Le Magnen C, Dutta A, Abate-Shen C. Optimizing mouse models for precision cancer prevention. *Nat Rev Cancer.* 2016;16(3):187–196.
37. Lin EY, Jones JG, Li P, Zhu L, Whitney KD, Muller WJ, et al. Progression to malignancy in the polyoma middle T oncoprotein mouse breast cancer model provides a reliable model for human diseases. *Am J Pathol.* 2003;163(5):2113–2126.
38. Dadachova E, Nguyen A, Lin EY, Gnatovskiy L, Lu P, Pollard JW. Treatment with rhenium-188-perrhenate and iodine-131 of NIS-expressing mammary cancer in a mouse model remarkably inhibited tumor growth. *Nucl Med Biol.* 2005;32 (7):695–700.

39. Iyengar P, Espina V, Williams TW, Lin Y, Berry D, Jelicks LA, et al. Adipocyte-derived collagen VI affects early mammary tumor progression in vivo, demonstrating a critical interaction in the tumor/stroma microenvironment. *J Clin Invest.* 2005;115(5):1163–1176.
40. Lin EY, Li JF, Bricard G, Wang W, Deng Y, Sellers R, et al. Vascular endothelial growth factor restores delayed tumor progression in tumors depleted of macrophages. *Mol Oncol.* 2007;1(3):288–302.
41. Lin EY, Li JF, Gnatovskiy L, Deng Y, Zhu L, Grzesik DA, et al. Macrophages regulate the angiogenic switch in a mouse model of breast cancer. *Cancer Res.* 2006;66 (23):11238–11246.
42. Moadel RM, Nguyen AV, Lin EY, Lu P, Mani J, Blaufox MD, et al. Positron emission tomography agent 2-deoxy-2-[18F]fluoro-D-glucose has a therapeutic potential in breast cancer. *Breast Cancer Res.* 2003;5(6):R199–R205.
43. Lin EY, Pollard JW. Tumor-associated macrophages press the angiogenic switch in breast cancer. *Cancer Res.* 2007;67(11):5064–5066.
44. Wyckoff JB, Wang Y, Lin EY, Li JF, Goswami S, Stanley ER, et al. Direct visualization of macrophage-assisted tumor cell intravasation in mammary tumors. *Cancer Res.* 2007;67 (6):2649–2656.
45. Esbona K, Inman D, Saha S, Jeffery J, Schedin P, Wilke L, et al. COX-2 modulates mammary tumor progression in response to collagen density. *Breast Cancer Res.* 2016;18 (1):35.
46. Oren B, Urosevic J, Mertens C, Mora J, Guiu M, Gomis RR, et al. Tumour stroma-derived Lipocalin-2 promotes breast cancer metastasis. *J Pathol.* 2016;239:274–285.
47. Hruska KS, Tilli MT, Ren S, Cotarla I, Kwong T, Li M, et al. Conditional over-expression of estrogen receptor alpha in a transgenic mouse model. *Transgenic Res.* 2002;11(4):361–372.
48. Tilli MT, Frech MS, Steed ME, Hruska KS, Johnson MD, Flaws JA, et al. Introduction of estrogen receptor-alpha into the tTA/TAg conditional mouse model precipitates the development of estrogen-responsive mammary adenocarcinoma. *Am J Pathol.* 2003;163 (5):1713–1719.
49. Urbaniak C, Cummins J, Brackstone M, Macklaim JM, Gloor GB, Baban CK, et al. Microbiota of human breast tissue. *Appl Environ Microbiol.* 2014;80(10):3007–3014.
50. Hieken TJ, Chen J, Hoskin TL, Walther-Antonio M, Johnson S, Ramaker S, et al. The microbiome of aseptically collected human breast tissue in benign and malignant disease. *Sci Rep.* 2016;6:30751.
51. Urbaniak C, Gloor GB, Brackstone M, Scott L, Tangney M, Reid G. The microbiota of breast tissue and its association with breast cancer. *Appl Environ Microbiol.* 2016;82 (16):5039–5048.
52. Chan AA, Bashir M, Rivas MN, Duvall K, Sieling PA, Pieber TR, et al. Characterization of the microbiome of nipple aspirate fluid of breast cancer survivors. *Sci Rep.* 2016;6:28061.
53. De Spiegeleer B, Verbeke F, D'Hondt M, Hendrix A, Van De Wiele C, Burvenich C, et al. The quorum sensing peptides PhrG, CSP and EDF promote angiogenesis and invasion of breast cancer cells in vitro. *PLoS One.* 2015;10(3):e0119471.
54. Holm A, Vikstrom E. Quorum sensing communication between bacteria and human cells: signals, targets, and functions. *Front Plant Science.* 2014;5:309.
55. Reid G. Can breast microbiota provide protective effects against cancer? *Fut Microbiol.* 2016;11:987–989.
56. Zhang N, Ma ZP, Wang J, Bai HL, Li YX, Sun Q, et al. Human papillomavirus infection correlates with inflammatory Stat3 signaling activity and IL-17 expression in patients with breast cancer. *Am J Transl Res.* 2016;8(7):3214–3226.
57. Xuan C, Shamonki JM, Chung A, Dinome ML, Chung M, Sieling PA, et al. Microbial dysbiosis is associated with human breast cancer. *PLoS One.* 2014;9(1):e83744.

58. Shapira I, Sultan K, Lee A, Taioli E. Evolving concepts: how diet and the intestinal microbiome act as modulators of breast malignancy. *ISRN Oncol.* 2013;2013:693920.
59. Arroyo R, Martin V, Maldonado A, Jimenez E, Fernandez L, Rodriguez JM. Treatment of infectious mastitis during lactation: antibiotics versus oral administration of Lactobacilli isolated from breast milk. *Clinical Infect Dis.* 2010;50(12):1551–1558.
60. Goedert JJ, Jones G, Hua X, Xu X, Yu G, Flores R, et al. Investigation of the association between the fecal microbiota and breast cancer in postmenopausal women: a population-based case-control pilot study. *J Natl Cancer Inst.* 2015;107(8).
61. Lakritz JR, Poutahidis T, Mirabal S, Varian BJ, Levkovich T, Ibrahim YM, et al. Gut bacteria require neutrophils to promote mammary tumorigenesis. *Oncotarget.* 2015;6 (11):9387–9396.
62. Gunter MJ, Xie X, Xue X, Kabat GC, Rohan TE, Wassertheil-Smoller S, et al. Breast cancer risk in metabolically healthy but overweight postmenopausal women. *Cancer Res.* 2015;75(2):270–274.
63. Moore LL, Chadid S, Singer MR, Kreger BE, Denis GV. Metabolic health reduces risk of obesity-related cancer in framingham study adults. *Cancer Epidemiol Biomark Prevent.* 2014;23(10):2057–2065.
64. Bard J-M, Luu HT, Dravet F, Michel C, Moyon T, Pagniez A, et al. Relationship between intestinal microbiota and clinical characteristics of patients with early stage breast cancer. *FASEB J.* 2015;29(Suppl. 1).
65. Luu TH, Michel C, Bard JM, Dravet F, Nazih H, Bobin-Dubigeon C. Intestinal proportion of *Blautia* sp. is associated with clinical stage and histoprognostic grade in patients with early-stage breast cancer. *Nutr Cancer.* 2017;69(2):267–275.
66. Fuhrman BJ, Feigelson HS, Flores R, Gail MH, Xu X, Ravel J, et al. Associations of the fecal microbiome with urinary estrogens and estrogen metabolites in postmenopausal women. *J Clin Endocrinol Metab.* 2014;99(12):4632–4640.
67. Yang J, Tan Q, Fu Q, Zhou Y, Hu Y, Tang S, et al. Gastrointestinal microbiome and breast cancer: correlations, mechanisms and potential clinical implications. *Breast Cancer.* 2016;24:220–228.
68. Minelli EB, Beghini AM, Vesentini S, Marchiori L, Nardo G, Cerutti R, et al. Intestinal microflora as an alternative metabolic source of estrogens in women with uterine leiomyoma and breast cancer. *Ann N Y Acad Sci.* 1990;595(1):473–479.
69. Rao VP, Poutahidis T, Ge Z, Nambiar PR, Boussahmain C, Wang YY, et al. Innate immune inflammatory response against enteric bacteria *Helicobacter hepaticus* induces mammary adenocarcinoma in mice. *Cancer Res.* 2006;66(15):7395–7400.
70. Green JE, Shibata MA, Yoshidome K, Liu ML, Jorcyk C, Anver MR, et al. The C3 (1)/SV40 T-antigen transgenic mouse model of mammary cancer: ductal epithelial cell targeting with multistage progression to carcinoma. *Oncogene.* 2000;19(8):1020–1027.
71. Mabrok HB, Klopfleisch R, Ghanem KZ, Clavel T, Blaut M, Loh G. Lignan transformation by gut bacteria lowers tumor burden in a gnotobiotic rat model of breast cancer. *Carcinogenesis.* 2012;33(1):203–208.
72. Vetizou M, Pitt JM, Daillere R, Lepage P, Waldschmitt N, Flament C, et al. Anticancer immunotherapy by CTLA-4 blockade relies on the gut microbiota. *Science.* 2015;350 (6264):1079–1084.
73. Sivan A, Corrales L, Hubert N, Williams JB, Aquino-Michaels K, Earley ZM, et al. Commensal *Bifidobacterium* promotes antitumor immunity and facilitates anti-PD-L1 efficacy. *Science.* 2015;350(6264):1084–1089.

CHAPTER NINE

Complementary and Alternative Medicine and Breast Cancer

Ramadevi Subramani*, Rajkumar Lakshmanaswamy†,1

* Center of Emphasis in Cancer Research, Paul L. Foster School of Medicine, Texas Tech University Health Sciences Center El Paso, El Paso, TX, United States

† Center of Emphasis in Cancer Research, Paul L. Foster School of Medicine, Graduate School of Biomedical Sciences, Texas Tech University Health Sciences Center El Paso, El Paso, TX, United States

[1] Corresponding author. E-mail address: rajkumar.lakshmanaswamy@ttuhsc.edu

Contents

Abstract

Complementary and alternative medicine (CAM) has been in use among cancer patients for a long time. There are several types of CAM that are practiced in various parts of the world. For example, traditional medicinal practices followed in India and China are frequently used by cancer patients. CAM is broadly classified into five different categories: (1) traditional medicines, (2) mind–body interventions, (3) biology-based practices, (4) manipulative body-based practices, and (5) energy medicine. In this review, we have compiled data from the available literature regarding CAM use

Progress in Molecular Biology and Translational Science, Volume 151
ISSN 1877-1173
http://dx.doi.org/10.1016/bs.pmbts.2017.07.008

in breast cancer patients. We have highlighted the current concepts and the need for more structured studies to facilitate the implementation of CAM as a mainstream option for cancer patients.

1. INTRODUCTION

Breast cancer is the most frequently diagnosed type of cancer, and it is also the common cause of cancer-related death in women worldwide. A country's economic development has a huge impact on breast cancer incidence and mortality. Mortality rates are decreasing in the majority of high-income countries, although breast cancer incidence rates have been stable or slightly increasing. By contrast, in low-income countries, both breast cancer incidence rates and mortality rates are increasing. This trend predominantly maybe due to the acceptance of Western lifestyle components, such as diet, physical activity, and reproductive patterns in low-income and developing countries. Additionally, this trend also indicates a lack of awareness about both breast cancer and benefits of an early diagnosis, in addition to a lack of affordable high-quality treatment in low-income countries.[1]

2. TYPES OF BREAST CANCER TREATMENT

Many different life-saving treatments are available for breast cancer patients; the appropriate treatments are determined based on the type and stage of breast cancer. To a certain extent, other medical issues and personal preferences also determine the choice of treatment. The standard treatment of breast cancer can be classified into five types: surgery, radiation therapy, chemotherapy, endocrine therapy, and targeted therapy. Standard treatment options typically depend on: the size of the tumor, pathology scores, family medical history, patient's age when presenting breast carcinoma, and risk factors associated with environmental conditions. In addition to these standard therapies, several patients also use complementary and alternative treatments.

Complementary and alternative medicine (CAM) refers to medical products and practices that are used either exclusively or in addition to standard care, but are not part of standard treatments. CAM treatments have been widely used throughout the world for centuries. These treatments have not been subject to rigorous investigation in the way that standard treatments have been. While research on CAM is growing, there remains a lack of knowledge regarding its risks, benefits, and side effects, and how it interacts with standard care. Various types of CAM treatments hold a great value as cancer therapy. Recently, there has been a remarkable increase in recommending CAM therapies, along with standard treatments, to breast cancer patients. The combination of standard and CAM treatments is known as integrative medicine. Current research is attempting to understand the potential benefits of integrative healthcare under various conditions, for example, relief of symptoms in cancer patients and pain management for army soldiers and veterans.

3. CLASSIFICATION OF CAM

The use of CAM has dramatically increased among cancer patients; however, understanding the applications of CAM remains a challenge due to discrepancies in methodology across research studies. The National Center for Complementary and Alternative Medicine (NCCAM) defined CAM as a group of diverse medical systems, practices, and products that are not presently considered to be part of conventional medicine.[2] NCCAM classified CAM treatments into five major categories: (1) traditional medical practices, such as whole medical systems; (2) mind–body interventions; (3) biological substance–based practices; (4) manipulative and body-based practices; and (5) energy medicine[3] (Fig. 1).

3.1 Whole Medical Systems

Whole medical systems refer to complete systems of theory and practice that evolved independently from, or parallel to, conventional or allopathic medicine. These systems include multiple groups of practices or remedies. Many different types of traditional systems of medicine are practiced in different cultures across the world. The major Eastern whole medical systems are Indian Ayurvedic medicine and traditional Chinese medicine (TCM). From the Western hemisphere, the predominant whole medical systems are homeopathy and naturopathy. Other whole medical systems are practiced in Native American, African, Middle Eastern, Central and South American, and Tibetan cultures.

3.1.1 Ayurvedic Medicine (Ayurveda)

Ayurveda means "the science of life." It is one of the world's oldest holistic natural healing system that was developed over 3000 years ago in India.

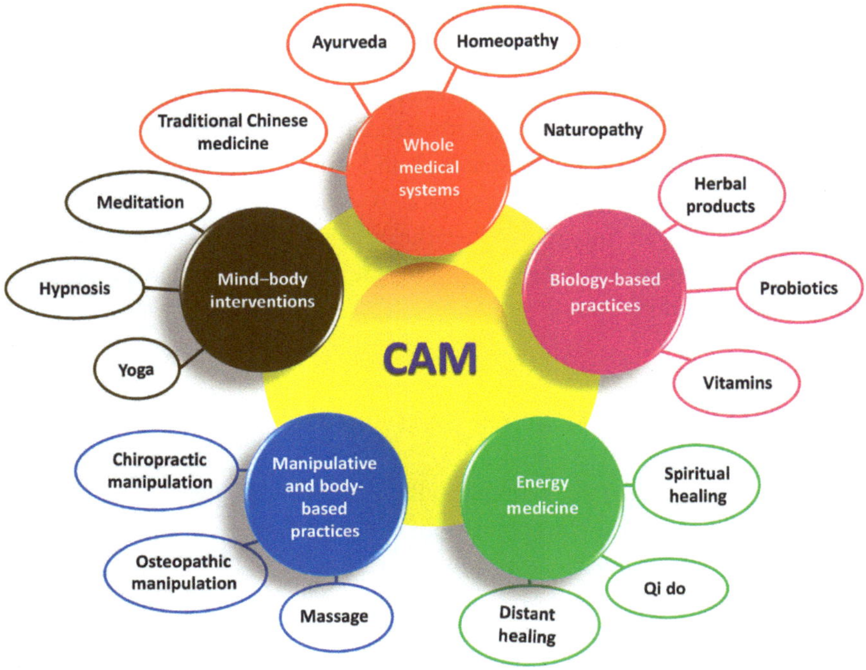

Fig. 1 Classification of complementary and alternative medicine (CAM) treatments into five major categories: (1) whole medical systems, (2) biology-based practices, (3) energy medicine, (4) manipulative and body-based practices, and (5) mind–body interventions.

Ayurveda is a comprehensive system of medicine that considers the body, the mind, and the spirit to be equally important in the restoration of the innate harmony of the individual. India's original systems of meditation and yoga are the foundations of this medical system. Its key intention is to promote good health, rather than fighting against disease. Nevertheless, treatments have been developed for various health complications, such as cancer, diabetes, cardiovascular conditions, and neurological disorders. Among the components of Ayurveda are diet, exercise, meditation, herbs, massage, exposure to sunlight, and controlled breathing. There are many Ayurvedic hers such as turmeric,[4,5] amla,[6] ashwagandha,[7] kumanjam,[8] black cumin,[9] rhubarb root, triphala, among others,[10] that possess medicinal qualities to treat a variety of cancers, including breast cancer. Nevertheless, extensive investigation with proper experimental design, including satisfactory sample sizes, randomization, and control groups, is required to establish the effectiveness of Ayurvedic treatments.[11]

3.1.2 Traditional Chinese Medicine

TCM originated in ancient China and has evolved over thousands of years. Similar systems also exist in Japan, Korea, and Vietnam. TCM is composed of several different practices, such as acupuncture, moxibustion, Chinese herbal medicine, tui na and t'ai chi massages, dietary therapy, and qi gong. The most common approaches are Chinese herbal medicine, acupuncture, and t'ai chi massages.

3.1.2.1 Chinese Herbal Medicine

Chinese medicinal herbs are used to restore the qi, or vital energy, and to balance the yin and yang forces. Yin represents the cold, slow, or passive energy, and yang represents the hot, excited, or active energy. In Chinese herbal medicine, botanicals or herbs are usually used to treat the underlying cause of the disease. They are also used as a preventative health measure by boosting the immune system to promote well-being before any disease occurs.

The Xihuang pill is a well-known traditional Chinese medication derived from the combination of *Bos taurus domesticus* Gmelin, *Moschus moschiferus* Linnaeus, *Boswellia carterii* bird, and *Commiphora myrrha* Engl. The aqueous extract of the Xihuang pill inhibits the cell viability, arrests the cell cycle at S phase, and induces intrinsic-mediated apoptosis in triple-negative breast cancer cell line Hs578T. Further clinical studies are needed to support the efficacy of this aqueous extract on breast cancer patients.[12]

Shikonin is a naphthoquinone extracted from *Lithospermum*, a traditional Chinese herb. The effect of this extract was tested against the exosome release in breast cancer cell line MCF-7. Shikonin inhibited proliferation of MCF-7 cells by reducing the release of tumor-derived exosomes.[12]

Bakuchiol, found in the Chinese herb Fructus Psoraleae, the dried ripe fruit of *Psoralea corylifolia* L., is a meroterpene with a partial terpenoid structure. It is shown to possess antimicrobial, antiinflammatory, antioxidative, antiosteoporotic, and antidepression or antistress activities.[13–17] The anticancer potential of bakuchiol has been noted; bakuchiol was shown to be potent against breast cancer cell growth.[18,19] The antibreast cancer effect of bakuchiol was accomplished by inducing S phase arrest and apoptosis via the intrinsic mitochondrial pathway. However, in a recent study, bakuchiol exhibited estrogenic activity both in vitro and in vivo systems. Further studies are warranted to prove the potential of bakuchiol as an antibreast cancer drug.[20]

3.1.2.2 Acupuncture

Acupuncture is extensively practiced by many acupuncturists, physicians, dentists, and other practitioners to relieve, prevent, or treat pain and other various health conditions. The practice originated in China over 2000 years ago. It is based on the theory of yin and yang in which the universe is viewed as dualistic forces that must be balanced to create harmony and health. Acupuncturists believe that there are 12 meridians that flow throughout the body and are related to various organs. Acupuncture can also alter brain chemistry by regulating the release of neurotransmitters and neurohormones, thereby affecting the parts of the central nervous system that are related to sensation and involuntary body functions. It has been proposed that acupuncture works by conducting electromagnetic signals at a rate higher than the norm. The increased rate of electromagnetic signaling facilitates the activity of pain-killing biochemicals.

Acupuncture is suggested to patients with breast cancer because of its effectiveness, affordability, low level of side effects, and potential to improve the quality of life in patients. The combination of acupuncture and moxibustion, the application of heat from the burning of the herb at the acupuncture point, is termed as warm acupuncture. Warm acupuncture was used on patients with breast cancer–related chronic lymphedema for 1 month, along with diosmin (a semisynthetic flavonoid) treatment, for the swelling of various parts in the upper arm. The results demonstrated that warm acupuncture was more effective than oral diosmin in reducing breast cancer–related chronic lymphedema. Warm acupuncture proved to be clinically safe and significantly improved patients' quality of life without any adverse effects during the treatment period. Further research should be conducted to find the appropriate acupoint locations for warm acupuncture and to determine the duration of treatment needed to achieve maximum effectiveness for patients' health.[21] Local and systematic therapies are used to treat breast cancer and are associated with short- and long-term side effects. Short-term side effects associated with conventional breast cancer therapies were effectively reduced or cured by acupuncture.[22]

Saam acupuncture is a Korean variation and is currently being used by Korean clinicians. In a pilot study, Saam acupuncture was tested on breast cancer patients as an alternative treatment. These patients received 18 sessions of acupuncture, 3 times a week, for 6 weeks. Patients were frequently evaluated to analyze the effects of this treatment. The study results demonstrated that Saam acupuncture has the potential to treat breast cancer patients with lymphedema. In addition, it was also suggested that it can be used as a complementary or alternative treatment to reduce lymphedema in breast cancer patients. Further research is needed with a randomized, controlled study with a large sample size to analyze the effectiveness and safety of Saam acupuncture.[23]

3.1.2.3 T'ai Chi

T'ai chi is a Chinese movement therapy that uses breathing and slow movements to promote healing. It is based on the same principles as acupuncture, acupressure, and qi gong. T'ai chi originated in China in the 13th century; it focuses on flexibility and strength. Like acupuncture and Chinese herbal medicine, it draws on the principles of yin and yang. T'ai chi aims to join together the internal and external forces, each movement having a purposeful meaning. Some of the benefits of t'ai chi are better circulation, balance, a reduction in stress and anxiety, enhanced aerobic capacity, and increased flexibility.

Several clinical studies have attested to the safety and health benefits of t'ai chi intervention. Randomized and nonrandomized clinical trials have addressed the efficacy of t'ai chi on various health conditions and diseases, such as cardiovascular disease,[24–26] osteoarthritis,[27,28] diabetes, schizophrenia, depression, balance control, and prevention of falls.[29–32] The effects of t'ai chi on breast cancer are well reported, and it is widely used by breast cancer survivors.[33–39] Metaanalysis data from nine randomized controlled trials demonstrated the effects of t'ai chi in relieving treatment-related side effects and quality of life in women with breast cancer. The effects of t'ai chi on 322 breast cancer patients were analyzed by studying the pathology (e.g., interleukin-6 and insulin-like growth factor 1), physical capacity [i.e., handgrip, limb physical fitness, and body mass index (BMI)], and well-being (i.e., physical, social, emotional, and general quality of life) of the patients. The overall data revealed that t'ai chi was positively correlated with improved handgrip strength and limb elbow flexion, but did not demonstrate any significant effect on pathology; BMI; physical, social, or emotional well-being; or general health-related quality of life.[35] However, a recent study by Larkey *et al.* reported that practicing t'ai chi has some advantage for BMI reduction and overall quality of life compared to nonmeditative gentle exercise in breast cancer survivors.[33] In another study, 3 months of t'ai chi was shown to reduce cellular inflammatory responses and decrease the expression of genes encoding proinflammatory mediators in breast cancer survivors with insomnia.[36] In contrast, a systematic review and metaanalysis with 5 randomized controlled trials involving 407 breast cancer survivors demonstrated that t'ai chi failed to improve the primary outcome, quality of life, as well as secondary outcomes, such as BMI, bone mineral density, and muscle strength.[40] However, no recent comprehensive systematic review has investigated the overall evidence of the effectiveness of t'ai chi in prevention, treatment, and rehabilitation for human health. There are broad variations of t'ai chi differing in style, intensity, duration, and learning and practicing methods; therefore, the documentation of t'ai chi intervention needs to be improved.

3.1.3 Naturopathy

Naturopathy, an intriguing branch of alternative medicine, primarily emphasizes the powerful nature of the human body's natural homeostatic processes. This system of medicine offers a treatment pathway that remedies the whole individual, restores the body's natural balances, targets the origins of disease, and prevents future illnesses. While many other branches of allopathic or alternative medicine provide treatments specific to a particular ailment, naturopathic physicians generally focus on acting as guides toward healthy lifestyle practices and ensuring the long-term well-being of their patients. Intriguingly, many naturopathy practitioners have begun to integrate the system's traditional principles with ambitious research aims, modern scientific design, and clinical applications. Current naturopathic treatments include physical therapy, nutritional supplements, homeopathy, hydrotherapy, hormone therapies, and individualized exercise regimens, all devoted to reducing the patient's physiological and psychological stress. Naturopathic physicians have also recently recommended TCM remedies.[41] A recent study examining 1817 visits to naturopathic physicians in the USA found that the patients and their naturopathic physicians were predominately female and Caucasian, and that the physicians primarily prescribed botanical medicines, vitamins, and minerals to treat chronic headaches, back pains, and fatigue.[42] According to a cross-sectional mail survey of American and Canadian naturopathic physicians, respondents treating patients with breast cancer monitored their patients through diagnostic imaging, physical examinations, and analyses of their patients' lifestyle behaviors. Of the remedies prescribed, 94% were suggested changes in an individual's diet, 88% were botanical medicines, and 84% were supplemental nutrients, including vitamin C and coenzyme Q-10 treatments.[43,44] While naturopathy demonstrates promising potential in ensuring patient health and provides an alternative pathway to wellness by focusing on an individual's natural healing processes, further clinical trials are required to confirm the replacement of conventional medicine with naturopathic remedies.

3.1.4 Homeopathy

Homeopathy is an alternative system of medicine developed by Samuel Christian Hahnemann, a 19th century German physician. Homeopathy is primarily governed by the principle of *similiasimilibuscurantur*, meaning "like cures like." According to Hahnemann, treatments for a particular ailment should generate side effects similar to those of the illness itself to incite a homeostatic, or secondary, response to remedy these bodily disturbances.[45] This branch of alternative medicine firmly adheres to a "law of minimum dose," where dosage quantities are inversely proportional to a substance's potency. Thus, while many homeopathic treatments contain active ingredients, the remedies undergo excessive dilution such that the resulting dosages contain small quantities of the active constituents. Hahnemann first began formulating homeopathic treatment plans by carefully observing his patient's physiological and psychological conditions and providing personalized treatment plans for the individual based on past trials. He would instruct healthy individuals to repeatedly ingest substances that produced the common symptoms of various ailments. Homeopathic remedies often include substances from various animals and plants, including cinchona bark and phytolacca, diluted and incorporated into ingestible pills and tablets or topical creams and ointments.[46] Recent research examined the effect of homeopathic dilutions of *Terminalia chebula* on the growth of MDA-MB-231 and MCF-7 breast cancer cells. The results have demonstrated that the remedy, particularly at a 1:10 dilution ratio, suppresses the growth of breast cancer cells.[47] While numerous trials on homeopathic treatments exist, and approximately 5 million adults and 1 million children in the United States employed homeopathic remedies in 2012, the effectiveness of this medical approach faces considerable skepticism in the scientific community.[48] Published studies on homeopathy focus on the effectiveness of homeopathic remedies for specific illnesses, the relationship between treatment dilutions and potency, and the differences between the effects of current homeopathic antidotes and placebos. Unfortunately, many of the published clinical trials lack the proper experimental design. Although a few reputable studies were found, the cumulative data is insufficient and unreliable.[49] Homeopathy demonstrates notable potential in treating innumerable ailments; however, the effectiveness and efficiency of the alternative pathway remains contested.[50]

Although the various whole medical systems differ notably in their philosophical methodologies to prevent or treat disease, they share several common elements. These whole medical systems are dependent on the faith that one's body has the ability to heal itself. Healing frequently involves the balancing of mind, body, and spirit. In these systems, treatment is often personalized and based on the specific symptoms of the patient. To better understand these whole medical systems, research efforts are needed to evaluate them, as they are commonly practiced with adequate experimental rationale and not as individual therapies.

3.2 Mind–Body Interventions

Mind–body therapies explore the relationships between the mind, body, spirit, and behavior upon the belief that the mind can regulate bodily functions and symptoms. Many of the techniques involved in these therapies to relax the mind and body are ultimately intended to promote good health and treat diseases. Mind–body interventions include therapies, such as meditation, breathing, yoga, hypnosis, biofeedback, t'ai chi, and visual imagery, which act as effective treatments for anxiety and depression in patients with cancer.[51] Interestingly, various types of art therapies, such as music therapy and dance and movement therapies, the latter being a part of mind–body intervention, showed beneficial effects on anxiety in patients with breast cancer.[52] However, the study with 13 trials using a total of 606 patients also demonstrated that the art therapy didn't have positive effects on depression in breast cancer patients.[52] Furthermore, the study concluded that music interventions may have beneficial effects on anxiety, pain, mood, and quality of life in cancer patients, and may have a minimal effect on heart rate, respiratory rate, and blood pressure.[52] Another article showed the beneficial effects of music intervention on cancer-related fatigue and pain.[53] In addition, other forms of mind–body interventions, such as meditation, relaxation, imagery and hypnosis, and cognitive behavioral therapies also alleviated three cancer-related symptoms: fatigue, pain, and sleep disturbance.[53] However, to date, no systemic review has analyzed the existing data on different types of art therapies especially in breast cancer women. Most patients prefer this intervention due to the low physical and emotional risk associated with it. Additionally, the treatment cost is relatively low, allowing more patients to actively participate in the treatment.

Mind–body interventions are generally suggested to treat vasomotor symptoms, such as hot flashes and night sweats in healthy menopausal women and breast cancer survivors. Moreover, 26 randomized controlled trials were conducted using mind–body interventions, such as yoga ($n = 5$), hypnosis ($n = 3$), mindfulness ($n = 2$), relaxation ($n = 7$), paced breathing ($n = 4$), reflexology ($n = 1$), and cognitive behavioral therapy ($n = 4$). A recent study reported that cognitive behavioral therapy and relaxation therapies were consistent in their efficacy for treating vasomotor symptoms in breast cancer patients. Breathing, relaxation techniques, and cognitive behavioral therapies can be beneficial in the treatment of vasomotor symptoms. Yoga has demonstrated positive impact on reducing fatigue in breast cancer patients and survivors.[54]

3.3 "Biology"-Based Practices

Biology-based practices refer to the use of ingredients available in nature, such as herbs, foods, vitamins, and other natural substances to strengthen, treat, and balance the human health. Common supplements used for biology-based practices are dietary supplements, such as vitamins, probiotics, prebiotics, herbal products, fatty acids, amino acids, and functional foods. Food and Drug Administration (FDA) approval is not required for natural remedies except in the case of a new dietary ingredient. Unlike pharmaceutical products, biology-based supplements are not subject to rigorous testing to prove their safety and effectiveness for treatment. Therefore, the efficacy of biology-based substances remains unsubstantiated.

3.4 Manipulative and Body-Based Practices

Manipulative and body-based practices stimulate the movement of the body to promote health and healing. Various types of manipulative and body-based practices are currently used, such as massage, reflexology, Rolfing, craniosacral therapy, the Alexander technique, the Trager approach, and chiropractic and osteopathic manipulation. Approximately 27% of breast cancer patients use manipulative and body-based practices to improve breast cancer–related mortality outcomes.

3.5 Energy Medicine

Energy medicine is a method to transmit healing energy to patient's body through the hands of a practitioner to restore or balance the body's energy field for better health. Energy healing therapy has been used to treat various health conditions.[48] There are several approaches to energy medicine, such as hands-on, hands-off, and distant healing, where the healer and patient are in different locations.[55] Interestingly, 57% of trials with distant healing intervention showed a positive treatment effect on any medical condition.[55] There are nine different schools of energy medicine: biofield energy healing, spiritual healing, contact healing, distant healing, qi do, therapeutic touch, Reiki, polarity therapy, and qi gong. Fatigue is a common side effect of cancer treatment. Biofield therapies, such as polarity therapy, Reiki, and therapeutic touch, are often used as adjuvant therapy for cancer-related fatigue. Clinical trials have confirmed the impact of both polarity therapy and Reiki intervention in mitigating cancer-related fatigue.[56,57] Reiki intervention was further found to improve quality of life. It is a spiritual and vibrational healing intervention that maintains physiological homeostasis through a subtle vibrational field, without the involvement of physical manipulation or the ingestion or application of any substances. This spiritual form of healing not only tends to the physical ailment but also addresses spiritual growth. A small, qualitative study demonstrated the positive influence of spiritual healing on the survivorship experiences of breast cancer patients who were in long-term therapy.[58]

Qi gong is an old Chinese art that combines gentle physical movement, mental focus, and deep breathing, each focused on specific area of the body. It is an economical, effective, and nonaggressive intervention that balances mind, body, and spirit. Qi gong has been shown to reduce the side effects of chemotherapy. It has proven useful in treating conditions, such as leukopenia; however, it did not show significant effect in the platelet and hemoglobin levels. It is suggested that additional scientific studies on qi gong intervention are needed before it can be implemented in clinical practice as a complementary therapy for cancer.[59]

4. PHYTOMEDICINE IN BREAST CANCER PREVENTION AND TREATMENT

The incidence of cancer is largely associated with lifestyle factors, of which diet is primary.[60] Accordingly, several natural nutritional products have been examined as complementary or alternative treatments for cancer. Nutritional supplements and herbal supplements are the most commonly used CAMs by cancer patients.[61] Therefore, it is essential to understand the impact of phytomedicine in breast cancer prevention and treatment. Herbal medicine is also known as phytomedicine, phytotherapy, or botanical medicine. Several types of herbal medicine have been in use around the word for centuries, such as Indian Ayurvedic medicine, Chinese herbal medicine, Tibetan herbal medicine, and Western herbal medicine. Herbs consist of several bioactive substances that may vary according to the geographic location, harvest season, postharvest processing, and storage. These factors also significantly influence the composition and concentration of the bioactive compounds of the herbs. Numerous modern drugs used in standard therapy for cancer are made from plants. Paclitaxel (Taxol), vinblastine, vincristine, topotecan, irinotecan, etoposide, and teniposide are among the classic examples of plant-derived drugs approved by the FDA for cancer therapy. Traditional healers use whole-plant or crude extract, or mixtures of plants, to maintain the delicate chemical balance between the compounds for the effective remedy. Conversely, industries isolate a herb's single bioactive compound needed for the treatment. In the last decade, several research studies have been conducted to identify and understand the anticancer effects of both the single active compounds and the crude extracts found in herbs. Phytochemicals are classified based on their biosynthetic origin. Flavonoids, carotenoids, phenolics, alkaloids, steroids, terpenoids, saponins, tannins, glycosides, and proteins have been used as single or combined therapeutic agents in breast cancer drug development[62] (Fig. 2). Phytochemicals from plants are more widely used for cancer treatment than other sources due to their safety, low toxicity, and general availability.[63,64] It is well known that tumorigenesis is a multistep process that begins with initiation followed by

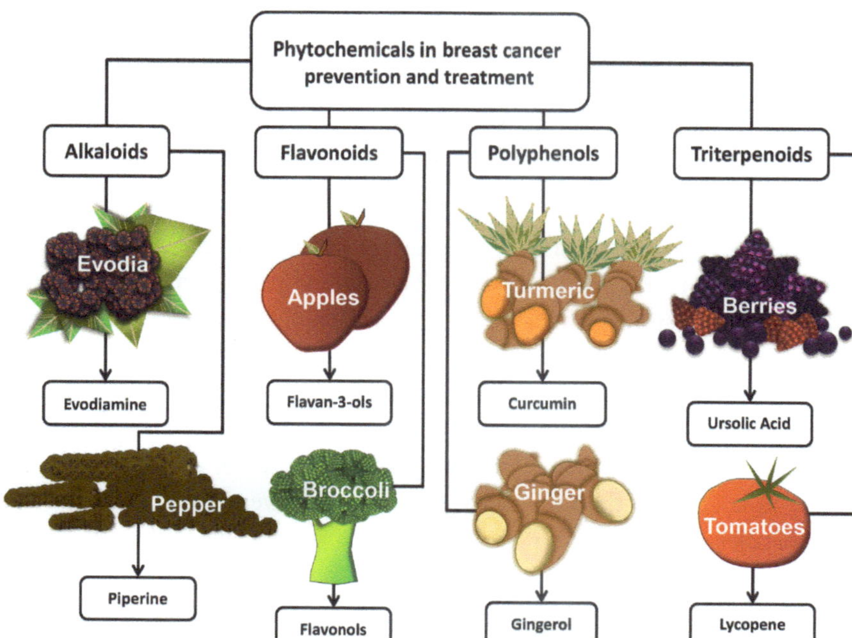

Fig. 2 Examples of phytochemicals (and their sources) used in breast cancer prevention and their treatment.

promotion, progression, and metastasis, involving various genetic and epigenetic alterations.[65] Phytochemicals exert antiangiogenic, antiinflammatory, antioxidative stress, antiproliferative, and proapoptotic effects by influencing various signaling pathways. This makes phytochemicals effective anticancer drugs that can inhibit cancer at multiple stages.

4.1 Alkaloids

In the early 19th century, the discovery, substantiation, and isolation of alkaloids from medicinal plants led to the beginning of scientific pharmacy.[66] Plants produce alkaloids as natural by-products. These alkaloid by-products are derived from secondary metabolic pathways that serve as an important defense mechanism for the plant against pathogenic organisms.[67] Alkaloids, such as camptothecin and vinblastine, are clinically used as drugs for anticancer treatment,[68] and have been successful in this capacity. Alkaloids are mostly present in higher plants that belong to the Ranunculaceae, Leguminosae, Papaveraceae, Menispermaceae, and Loganiaceae families. Alkaloids contain pharmacologically active nitrogenous compounds that have been of interest due to their various physiological applications for animals and humans.[67] For example, tetrandrine, a member of the bisbenzylisoquinoline alkaloids, a large family of alkaloids found in medicinal plants that have shown great potential for clinical application in countries, such as China, has been used pharmacologically to treat a variety of pathologies. Upon discovering an effective new method of delivery to specifically target breast cancer cells, researchers have now utilized tetrandrine's antiproliferative, antiinflammatory, and apoptotic-inducing properties to inhibit mammosphere formation of inflammatory and aggressive breast cancer–initiating cells (tumor-initiating cells) in vitro. Researchers have further used tetrandrine to have a similar effect on mammosphere formation in cells isolated from patient samples.[69,70] Studies have demonstrated the therapeutic potential of many other alkaloids that exercise antitumor properties in breast cancer cells, such as fangchinoline, which is derived from the dry roots of Menispermaceae, and mahanine, an isolate of the leaves of *Murraya koenigii* (curry tree).[70,71] Although much is to be discovered on alkaloids, the pharmacological effects of these compounds in treating breast cancer are promising.[67]

Other naturally derived alkaloids, such as berberine, evodiamine, matrine, piperine, sanguinarine, and tetrandrine, have also exhibited antiproliferative and antimetastatic effects on various types of cancers, including breast cancer, in preclinical models.[72–80] Evodiamine, a quinolone alkaloid and a constituent from the Chinese herb Evodia Fructus, inhibited the proliferation of adriamycin-resistant human breast cancer NCI/ADR-RES cells, both in in vitro and in vivo models. It also induced apoptosis and cell cycle arrest at G2/M phase. The antitumor activity of evodiamine in human multiple-drug resistant breast tumor xenograft was better than paclitaxel.[81] Piperine, an amide alkaloid, was isolated from the Indian spice black pepper (*Pipernigrum* Linn) and long pepper (*Piperlongum* Linn).[82] A study has shown that breast cancer stem cell self-renewal was significantly inhibited by piperine without causing any toxicity to differentiated cells.[83] Piperine induced apoptotic cell death and cell cycle blockage in 4T1 mouse mammary carcinoma cells, demonstrating its anticancer activity. It suppressed local primary tumor growth and occurrence of spontaneous metastases in 4T1-implanted BALB/c female mice.[84] Furthermore, piperine induced apoptotic cell death in human HER2-overexpressing breast cancer cells through the inhibition of expression of HER2, SREBP-1, FAS, MMP-9, and AP-1 via blockage of ERK1/2, p38MAPK, and AKT signaling.[85] It enhanced the tumor necrosis factor–related apoptosis by inhibiting survivin and p65 phosphorylation in human triple-negative breast cancer cells, such as MDA-MB-468, MDA-MB-231, and 4T1 orthotopic-inoculated 4T1-luc mouse model,[86] and it impaired the growth and motility of triple-negative breast cancer cells. Additionally, tumor growth of TNBC xenografts in immune-deficient mice was reduced by treatment of piperine.[76] Nanoparticle delivery of piperine, along with rapamycin, increased the bioavailability of the drug by four- to eightfold. In future, this promising approach could be beneficial for treating breast cancer patients.[87]

4.2 Polyphenols

Polyphenols are widely distributed in fruits, vegetables, spices, nuts, grains, tea, coffee, and wine.[88,89] Approximately 8000 types of polyphenols have been identified in the plant kingdom.[90] The new strategy of combining conventional agents in nanoparticles with polyphenols has increased the efficacy of therapeutics and reduced their related side effects. This novel approach also improves the bioavailability and intracellular stability of polyphenols due to the constant release of the cancer-fighting compounds.

The polyphenolic compound, genistein, found in soy, decreased the breast cancer risk in women who had soy food intake during their adolescence.[91] In a recent study, genistein was shown to inhibit the growth of triple-negative breast cancer cells by altering the phosphorylation of proteins involved in cell cycle regulation, DNA damage response, and other cellular processes.[92] It further restored the expression of estrogen receptor alpha (ERα) in clinically aggressive ERα-negative breast cancer, which in turn enhanced the tamoxifen-dependent antiestrogen sensitivity.[93] The chemo-preventive effect of genistein is well established in multiple stages of carcinogenesis: cancer initiation, promotion, and progression.[94,95] Genistein decreased cell proliferation and induced apoptotic cell death in MCF-7 breast cancer cells. It decreased the cancer stem–like cell population via inhibition of the Gli–Hedgehog signaling pathway.[96] Interestingly, genistein derivatives have exhibited anticancer effects in ERα-positive breast cancer cells at much lower concentrations (~10–15 time) than their parental compound, genistein.[97] Similarly, in another study, consumption of resveratrol from grapes reduced the risk of breast cancer, but resveratrol from wine failed to do so.[98] A study revealed that consumption of epigallocatechin-3-gallate (EGCG), an active ingredient from green tea extract, reduced the axillary lymph node metastases and the frequency of recurrence in premenopausal women with stage I and stage II breast cancer.[99] However, a contradictory result was found with Japanese women, due to lack of correlation between breast cancer risk and plasma levels of the tea polyphenols.[100] EGCG and paclitaxel combined with nanoparticle induced apoptosis and decreased cell survival, angiogenesis, metastasis, and NF-κB activation in breast cancer cell line MDA-MB-231. It also downregulated P-glycoprotein expression involved in development of multidrug resistance.[101] Similar to the nanoparticle combination, when EGCG was loaded into liposomes or chitosan-coated liposomes, it increased the stability of sustained release and intracellular concentration, which eventually decreased the cell viability and induced apoptosis.[102] In another procedure, curcumin was combined with poly(lactic-*co*-glycolic) acid nanoparticles to improve the bioavailability and effectiveness of polyphenols within the cells. This combination for 10 days in MCF-7 cells inhibited cell proliferation and induced apoptosis and cell cycle arrest at S and G2/M phases.[103] However, encapsulation of polyphenols into nanoparticles alone may not be sufficient. Further analysis is required to prove the efficacy of polyphenols in managing the side effects of conventional chemotherapy in breast cancer cells.

Curcumin, a low–molecular weight polyphenolic compound, is an active molecule from *Curcuma longa*, widely known as turmeric. Curcumin has garnered considerably more interest than several other chemopreventive agents due to its multitargeted biological and molecular effects on breast cancer. It inhibits cell motility, invasiveness, and epithelial–mesenchymal transition by downregulating the expression of vimentin and upregulating the expression of E-cadherin in MCF-7 and MDA-MB-231 breast cancer cells.[104] Wnt signaling involved in epithelial–mesenchymal transition and metastasis is significantly inhibited by curcumin.[83] Curcumin also induced reactive oxygen species (ROS) in MCF-7 and MDA-MB-231 breast cancer cell lines.[105] Contrary to its prooxidant effect, in another study, curcumin reduced nickel oxide nanoparticle–induced ROS in MCF-7 cells.[106] Curcumin's contradiction regarding pro- or antioxidant effects may be due to the differences in the cell and animal models, the amount of drug administration, or the experimental parameters. Curcumin[107] and other polyphenolic compounds, such as quercetin,[108] resveratrol,[109] genistein,[110] and apigenin,[111] induced the cell cycle arrest in the cells at G2/M phase. The involvement of curcumin in apoptosis induction was found in neu transgenic mice.[112] Furthermore, curcumin treatment induced both early and late apoptosis by nuclear accumulation of tumor suppressor p53 activity and cell cycle inhibitor p21.[107] It also decreased the elevated transcription of DNMT1, DNMT3a, and DNMT3b genes, and reduced the breast cancer risk.[113] Indeed, most polyphenols, such as curcumin, EGCG, genistein, and resveratrol, induced epigenetic modifications by decreasing DNA methyl transferase expression in breast cancer cells. Treatment with curcumin has an inhibitory effect on breast cancer stem cells; it inhibits mammosphere formation and increasing expression levels of epithelial markers, cytokeratin 18 and 19, while decreasing levels of cyclin D1, c-myc, vimentin, MMP-2 and -9, and the nuclear localization of β-catenin.[114] Curcumin inhibits the activation of carcinogen and activates carcinogen-detoxifying enzymes.[115] Curcumin influences signaling molecules, such as PI3K/AKT, MAPK, and NF-κB.[116] A recent study demonstrated that curcumin enhanced the quality of life in BALB-neuT transgenic mice by increasing the tumor-free survival and reducing tumor multiplicity with no toxicity.[112]

Grape seeds contain a phytoestrogen known as resveratrol, a selective estrogen receptor modulator. The molecule is known to serve as an antioxidant,[117–120] antiinflammatory,[121–123] antiangiogenic, antimetastatic,[124,125] and a proapoptotic[126–128] in breast cancer models for prevention and therapy. Resveratrol induced apoptosis by upregulating proapoptotic molecules, such as Bax, p53, and p21Cip1/WAF1 and downregulating antiapoptotic or survival molecules, such as survivin, Bcl-2, Bcl-xL, and cIAPS. This eventually resulted in reduced cyclin D1 and cyclin E cell cycle regulatory proteins. Furthermore, resveratrol significantly suppressed migration and invasion of breast cancer cells. Hippo/YAP signaling is vital for tumorigenesis and progression of several cancers. Resveratrol, either alone or in combination with other agents, attenuated breast cancer invasion through inactivation of Rho A, which led to activation of LATS and induction of YAP phosphorylation.[129] Similarly, it improved the efficacy of sorafenib-facilitated apoptosis via induction of ROS, caspase 3, PARP cleavage, and cell cycle inhibition.[117] It further inhibited the MMP, COX-2, VEGF, and interleukin genes involved in metastasis and angiogenesis.[130–132] Resveratrol has also been shown to play a pivotal role in modulating the activity of key molecular players such as MAPK, FAK, and PI3K/Akt in an ER-dependent and -independent way.[133–136] Resveratrol has antiestrogenic effects; it decreases tumor suppressor proteins p53 and steroid receptor ERα.[137] In vitro, xenograft, and carcinogen-inducible animal models demonstrated the chemopreventive activity of resveratrol.

Gingerol is primarily isolated from the medicinal plant ginger (*Zingiber officinale*). This pungent ingredient from the Zingiberaceae family is reported to possess several pharmacological and physiological, such as antiinflammatory, antitumorigenic, proapoptotic, antihepatotoxic, and cardiotonic effects.[138] Gingerol inhibits metastasis of MDA-MB-231 human breast cancer cell line by decreasing the invasion, motility, and activities of MMP-2 and MMP-9 in a dose-dependent manner.[138] It can alter β-catenin, PKC epsilon, and GSK-3β pathways, and regulate protein degradation. Shogaol is one of the biologically active phenylpropanoid-derived compounds from ginger, and is reported to have anticancer and antiinflammatory effects.[139–141] Shogaol is a potent inhibitor of breast cancer cell invasion. The antimetastatic effects of shogaol in MDA-MB-231 were demonstrated by the downregulation of MMP-9 transcription through the targeting of the ERK and NF-κB activation cascade.[142] In another study, 4-shogaol was shown to suppress breast cancer cell invasion, migration, NF-κB activation, and the induction of mesenchymal–epithelial transition, thus inhibiting metastasis of breast cancer cells. An animal study demonstrated that shogaol exhibits its antimetastatic effects through inhibition of IκB and the translocation of NF-κB/Snail in breast cancer cells in mice.[143] Therefore, shogaol holds the potential to be an innovative therapeutic agent for treatment of metastatic breast cancer.

Capsaicin is one of the active components of hot chili peppers, which has anticancer activity. It triggers apoptosis in breast cancer cells.[144–146] The apoptotic effect of capsaicin could be through decreased mitochondrial membrane potential, reduced procaspase-7 expression, and increased cleavage of PARP-1 in MCF-7 and BT-20 breast cancer cells. Capsaicin inhibited breast cancer cell growth by arresting cell cycle in the S phase.[146]

Furthermore, it is also capable of inducing apoptotic cell death of breast cancer stem cells by inhibiting the translocation of notch intracellular membrane domain into the nucleus.[147] Apoptosis inducing factor–mediated caspase-independent apoptosis was observed in capsaicin-induced apoptosis even in caspase-3–deficient MCF-7 cells.[145,146] Therefore, further clinical studies using capsaicin could prove its therapeutic efficacy for breast cancer.

4.3 Flavonoids

Flavonoids are rich in fruits and vegetables and are involved in various biological activities. Epidemiological data has demonstrated that flavonoids decrease the risk of breast cancer (Table 1).[148] Flavonoids and nonflavonoids are the two major groups of polyphenols.[88] Some of the subclasses of flavonoids include flavonols, flavones, flavan-3-ols, anthocyanidins, flavanones, and isoflavones. Nonflavonoids consist of phenolic acids (benzoic acids and cinnamic acids), stilbenes, lignans, tannins, and other polyphenols, such as curcumin, rosmarinic acid, and gingerol.[88,149] Flavonoids represent 60% of polyphenols, and phenolic acids represent 30%.[149,150]

Although flavonoids have protective effects due to their influence on various cancer-related signaling pathways, human clinical trials with flavonoid supplementation on disease prevention revealed that no natural flavonoid has been approved for breast cancer treatment. However, the structure of natural flavonoids has served as an integral model for the development of chemopreventive and therapeutic agents.[182] Tea, herbs, fruits, and vegetables are rich in flavonoids. Green tea and black tea contain are approximately 25% constituted by flavonoids.

Flavonoids have been reported to have an effect on breast cancers through multiple mechanisms.[183] They can block the bioactivation of procarcinogens;[184,185] inhibit estrogen-producing and -metabolizing enzymes;[186,187] and suppress the breast cancer resistance protein (BCRP), a key player in development of multidrug resistance.[188,189] In a study, a collection of 124 compounds and their derivatives from natural sources were tested to elucidate the inhibitory potential of these compounds against BCRP and multidrug resistance–associated protein 2 (MRP2); 36% of compounds were effective against BCRP but only 3.2% had an inhibitory effect on MRP2. The same study further established that specific flavones and flavonols had higher inhibitory effects against BCRP. Most BCRP inhibitors are found in all flavonoid subclasses. Structure–activity relationship analyses are required to understand the interaction between BCRP and their inhibitors.[190]

Table 1 List of Flavonoid Compounds Isolated From Plants and its Role in Breast Cancer.

Flavonoid Subclasses	Dietary Flavonoids	Plant Sources	Roles in Breast Cancer	References
Anthocyanidins	Cyanidin, delphinidin, malvidin, pelargonidin, peonidin, petunidin	Red, blue, and purple berries, red and purple grapes	Cyanidin, delphinidin, malvidin, and petunidin inhibited cell proliferation Cyanidin and delphinidin induced apoptosis and autophagy in breast cancer Peonidin has antitumor effects Pelargonidin has anticancer effects	151–154
Flavan-3-ols	*Monomers (catechins)*: (+)-catechin, (−)-epicatechin, (−)-epigallocatechin, (+)-gallocatechin, and their gallate derivatives	White, green, and oolong teas, cocoa-based products, grapes, berries, apples	EGCG decreased growth and invasion of SUM-149 and SUM-190 cells Reduced growth and mammospheres formation in SUM-149 cells Induced apoptosis in SUM-149 cells, along with increased ALDH activity EGCG analogs reduced cell proliferation, mammospheres formation, and mTOR pathway $CD44^{+}/CD24^{-}$ stem cell population was observed in MDA-MB-231 cells EGCG induced apoptosis and cell cycle arrest, and downregulated telomerase, vascular endothelial growth factor, and aromatase activity	155–163
	Dimers and polymers: proanthocyanidins	Apples, berries, cocoa-based products, red grapes, red rice, red wine	Grape seed proanthocyanidins inhibited vasculogenic mimicry in human triple-negative breast cancer cells Proanthocyanidins from grape seeds induced apoptosis and inhibited metastasis of breast cancer cells Proanthocyanidins from red rice inhibited MDA-MB-231 breast cancer cell invasion	164–166
	Theaflavins	Black tea	Theaflavins induced apoptosis and antimigratory effect by NF-κB via p53–ROS cross talk	167–169

Flavonols	Isorhamnetin, kaempferol, myricetin, quercetin	Onions, scallions, kale, broccoli, apples, berries, teas	Isorhamnetin inhibited cell proliferation and induced apoptosis in breast cancer Kaempferol suppressed metastatic behavior in breast cancer Myricetin suppresses cell viability in breast cancer Quercetin has antimalignant properties	170–173
Flavones	Apigenin, luteolin, baicalein, chrysin	Parsley, thyme, celery, hot peppers	Apigenin caused a reduction in tumor due to decrease proteasomal activity in breast cancer Luteolin inhibited cell viability in triple-negative breast cancer Baicalein inhibits epithelial–mesenchymal transition Chyrsin improved breast cancer cell cytotoxicity	174–176
Flavanones	Eriodictyol, hesperetin, naringenin	Citrus fruit and juices, for example, oranges, grapefruits, lemons	Eriodictyol and naringenin have antiproliferative effects in breast cancer Hesperetin induced apoptosis in breast cancer	177,178
Isoflavones	Daidzein, genistein, glycitein, biochanin A, formononetin	Soybeans, soy foods, legumes	Daidzein and genistein are involved in breast cancer chemoprevention High concentrations of glycitein can inhibit cell growth Biochanin A and formononetin have antitumorigenic effects	179–181

ALDH, Aldehyde dehydrogenase; EGCG, epigallocatechin-3-gallate; ROS, reactive oxygen species.

Chalcones (1,3-diaryl-2-propen-1-ones) are precursors of flavonoids known for their wide range of bioactivities, including therapeutic effects against cancer.[191] In a recent study, the antitumor activity of two chalcones, for example, *trans*-chalcone and licochalcone A, were tested on breast cancer cell line MCF7. The study results demonstrated that the two chalcones induced genotoxicity, growth arrest at G1 phase, and intrinsic-mediated apoptosis. Additionally, *trans*-chalcone also induced the expression of CIDEA gene and protein, which are involved in DNA fragmentation during apoptosis.[192]

Myricetin is a naturally occurring flavonoid from the bark of *Myricarubra*. It is also found in berries, fruits, vegetables, herbs, tea, and wine. It is known for its antioxidant properties and therapeutic effects in various cancers both in vitro and in vivo. Myricetin induced programmed cell death through extrinsic-[193] and intrinsic-mediated apoptotsis,[194] and it induced G2/M cell cycle arrest.[195] In an in vitro study, myricetin was shown to suppress the cell viability of MCF-7 breast cancer cells by altering PAK1/MEK/ERK/GSK3β/β-catenin/cyclin D1/PCNA/survivin/Bax–caspase-3 signaling pathways. Presently, the use of myricetin in a therapeutic context is hampered due to poor bioavailability, low metabolic stability, reduced cell uptake, and ineffective delivery to some of the cellular compartments, such as mitochondria. However, a small library of myricetin-derived new chemical entities was assessed in breast cancer cell lines to improve the redox potential, cell uptake, and mitochondrial targeting. Oncamex is one of these myricetin-derived flavonoids that showed antitumor activity by decreasing cell viability and inducing cytotoxicity and apoptosis in a preclinical breast cancer model. Research has further demonstrated that Oncamex is capable of targeting specific mitochondrial delivery and modulating redox potential. Therefore, this study suggests that Oncomex could be a potential anticancer agent.[196]

However, in a study, no significant correlation was observed between intake of flavonol (quercetin, kaempferol, and myricetin)- or flavone (apigenin and luteolin)-rich foods and incidence of total or site-specific cancers.[197]

4.4 Triterpenoids

More than 60% of natural products are terpenoids, also known as isoprenoids. Terpenoids play a major role in traditional medicine due to their aromatic qualities. Terpenoids are responsible for the scent of eucalyptus; the flavors of cinnamon, cloves, ginger; and the color of tomatoes and sunflowers. Terpenoids are broadly classified into monoterpenoids, diterpenoids, triterpenoids, tetraterpenoids, carotenoids, sesquiterpenoids, volatile oils, resins, oleoresins, oleogum resins, and balsams. Here, we discuss the potential biological activities of some of the natural triterpenoids in breast cancer.

Mounting evidence suggest that naturally occurring triterpenoids from the plant kingdom have beneficial effects against several types of human diseases, including breast cancer. The chemopreventive and anticancer potential of triterpenoids is established through their antiinflammatory, antiproliferative, proapoptotic, and antiangiogenic effects both in vitro and in vivo models. The subclasses of triterpenoids, such as lupane, oleanane, ursane (ursolic acid), and cucurbitacin, are also recognized for their beneficial effects, as demonstrated in preclinical and clinical investigations.[198]

Ursolic acid is a pentacyclic triterpenoid found in fruits and vegetables, such as apples, berries, pears, prunes, and natural medicinal plants (e.g., *Rhododendron hymenanthes* Makino, *Rosmarinus officinalis*, *Eriobotrya japonica*, *Calluna vulgaris*, *Ocimum sanctum*, and *Eugenia jambolana*).[199] Ursolic acid inhibited breast cancer cell proliferation through the induction of apoptosis and G1/G2 cell cycle arrest. It also scavenges free radicals and has strong antiinflammatory benefits.[200] Many preclinical and clinical studies provided ample evidence that naturally occurring ursolic acid and its derivatives have chemopreventive and therapeutic effects against cancer.[199] Ursolic acid modulates transcription factors, growth factor receptors, inflammatory cytokines, and other key intracellular molecular targets involved in various cellular functions, such as cancer cell proliferation, apoptosis, invasion, metastasis, and angiogenesis.

In another study, ursolic acid and its derivatives were reported to inhibit proliferation of breast cancer cells through the induction of apoptosis.[201,202] It also suppressed migration and invasion by altering c-Jun N-terminal kinase, Akt, and mammalian target of rapamycin signaling.[203] Further, it inhibited the transcription factor FoxM1, which is essential for the progression and proliferation of breast cancer cells.[204] Preclinical studies have proven the chemopreventive effects of ursolic acid using in vivo models. A recent study showed that ursolic acid inhibited tumor growth and size of syngeneic MMTV-Wnt-1 tumor cells injected to ovariectomized C57BL/6 postmenopausal breast cancer mice model. It also inhibited cell cycle progression and induced necrosis in MMTV-Wnt-1 tumor cell line.[205]

Novel triterpenoids from raisins exhibit the highest antiproliferative activity in MCF-7 cells by altering Ras/Raf/ERK protein tyrosine kinase signaling.[206] 3β,7β,25-Trihydroxycucurbita-5,23(E)-dien-19-al (TCD), a triterpenoid molecule isolated from bitter gourd, suppressed proliferation, histone deacetylases, and cytoprotective autophagy. TCD also induced apoptosis and increased ROS generation in MCF-7 and MDA-MB-231 breast cancer cells. These results suggest that TCD has translational potential as an anticancer drug for breast cancer treatment.[207]

In a study friedolanostane, a triterpenoid isolated from leaves of *Garcinia celebica*, was shown to induce apoptosis by inhibiting AKT.[208] *Maytenus* sp. root bark–derived triterpenoid, tingenin B, is cytotoxic against breast cancer stem cell–enriched population of MCF-7, as it induced apoptosis. Tingenin B decreased mitochondrial membrane potential and also increased endoplasmic reticulum stress to induce cell death. However, preclinical and clinical studies are warranted to elucidate and corroborate its possible use in breast cancer treatment.[209]

Lycopene is a compound that has displayed antioxidant and antiproliferative characteristics in numerous in vitro and animal studies.[210] This compound effectively inhibited the development and progression of breast cancer cells via its multifarious ability to arrest cell cycle, induce apoptosis, downregulate epidermal growth factor receptor, suppress estrogen-mediated signaling, and enhance gap junction communication.[211–219] Although more research is necessitated to prove these qualities definitively, lycopene has specifically shown the capacity to inhibit mammary carcinogenesis induced by dimethylbenz[a]anthracene through phase I enzyme inhibition, reinforcing its chemopreventive potential.[220] When used in combination with other anticancer compounds, such as quinacrine and tocopherol, lycopene complements their anticancer effects and improves efficacy through the synergistic relationships.[221,222] In a recent study involving breast and prostate cancer cells, lycopene was observed to significantly inhibit cell growth by influencing reduced NF-κB–responsive gene regulation.[223]

Withaferin A (WA) is commonly used in the Indian Ayurvedic system of medicine and is extracted from an Indian medicinal plant, *Withania somnifera*.[224] Recognized as a safe option to increase vitality and longevity in treating a variety of physical and mental conditions, WA is a steroidal lactone, verified by significant evidence, that possesses antiinflammatory, antitumor, antistress, hematopoietic, immunomodulatory, and rejuvenating properties.[225,226] WA serves as a pivotal biochemical constituent of *W. somnifera*, which is regularly recognized as ashwagandha, winter cherry, or Indian ginseng in the Indian traditional system of medicine, and has demonstrated anticancer activity in various types of cancers, including breast cancer.[224,227–237] A 2016 study confirmed WA as a promising chemopreventive phytochemical with the capacity to inhibit at least two different subtypes of breast cancer.[238]

In MDA-MB-231 and MCF-7 breast cancer cells, WA facilitated the downregulation of a oncogenic transcription factor, STAT3, which is typically activated in human malignancies.[234] Following treatment with WA, MDA-MB-231 and MCF-7 breast cancer cells underwent apoptosis and entered irrevocable arrest in the G_2/M phase of cell cycle.[236] In a mammary cancer chemopreventive study, treatment with WA blocked breast cancer stem cells through the inhibition of Kruppel-like factor 4 (KLF4).[239] Other mechanisms involving WA's multifactorial anticancer effects include FOXO3a- and Bim-dependent apoptosis, ROS production, and activation of the Bax/Bak cascade.[229,234,237] In a recent in vitro study conducted by Zhang et al.,[228] the proapoptotic and growth-inhibitory effects of WA were specifically facilitated by the downregulation of ERα, RET, and HSF1, and the upregulation of phospho-p38 MAPK, p53, and p21. These results further verified WA's detrimental influence on breast cancer cells.[227,228] Along with WA's proapoptotic and growth-inhibitory effects, further affirmation of its potential has been appreciated through its ability to act as an effective inhibitor of the transcription factor NF-κB, an influencer of genes controlling inflammation, antiapoptosis, and antimicrobial activity.[240,241] Collectively, WA has demonstrated commendable influence on the complex molecular pathways of numerous types of cancer cells.

Carotenoids, found commonly as a significant bioactive component in fruits and vegetables, are noteworthy chemopreventative agents that are able to inhibit cell proliferation and malignant transformation, while also promoting apoptosis in transformed cells.[211,242–246] Through the influence of biological processes, carotenoids enhance immunity, protect against cellular mutagenesis, inhibit tumorigenesis, and stimulate antioxidant activity through the entrapment of free radicals.[247–250] Specific types of carotenoids, such as α-, β-, γ-carotene, exist as provitamins and can be converted to active vitamin A, which exercises potent antioxidant properties.[212] Regarding breast cancer cell growth, several carotenoids and respective derivative products of β-carotene have been observed to inhibit both ER-positive and ER-negative breast cancer cell growth.[212,242]

β-Carotene is a highly pigmented, fat-soluble compound found in a variety of fruits, vegetables, oils, and grains. When treated with β-carotene, the growth of MCF-7 and MDA-MB-231 cells regressed substantially. In

MCF-7, β-carotene induced apoptosis, increased production of ROS, and upregulated peroxisome proliferator–activated receptor gamma (PPARγ) expression.[212,243] These observations were reinforced by another study where breast cancer cell lines were treated with β-carotene and lycopene, which increased apoptosis, inhibited proliferation, and arrested cell cycle.[251] One specific study reported an overall decrease in breast cancer risk (OR = 0.41; 95% CI, 0.22–0.79) among women in the highest quintile of serum β-carotene;[252] while there is an increase in breast cancer risk (OR = 2.21; 95% CI, 1.29–3.79) among women in the lowest quintile.[253] The Finnish Social Institution's Mobile Clinic Health Survey (FSIIMCHS) similarly reported a decreased risk of breast cancer among women with low β-carotene serum levels (OR = 0.3; 95% CI, 0.1–1.0).[254–256] However, there exists a notable discrepancy between epidemiological studies concerning β-carotene and its role in breast cancer treatment; several other studies have determined that no association exists between β-carotene levels and breast cancer risk.[257,258] Due to the differing results, more research is required to accurately assess the correlation, if any, between β-carotene and breast cancer risk. Still, it should be noted that β-carotene's chemical properties remain consistent in their hostility to breast cancer cells.

5. PREVALENCE OF CAM USE

Various studies revealed that between 30% and 50% of the adult population in industrialized countries use some type of CAM to prevent or treat a range of health-related complications.[259] Women tend to use CAM more frequently[260] almost 80% of the women use CAM for general purposes.[2] Between 36% and 80% of cancer patients have reported the use of CAM for relief of treatment-related symptoms.[61] A metaanalysis of articles on CAM published from 1998 to 2012, reported that approximately 46% of cancer patients in North America use CAM therapies.[261] Another study using multivariate analysis revealed that CAM users are more likely women, who are older, more educated, affluent, and have health insurance.[260,262,263] Interestingly, whites are more likely to use CAM compared to Asians, blacks, and Hispanics.[264] Self-reported analysis indicates that the use of CAM increases following a cancer diagnosis; breast cancer survivors are the maximum users among cancer survivors.[2] In a secondary analysis of the Women's Healthy Eating and Living Study, it was projected that 80% of breast cancer patients and survivors in the United States use some form of CAM.[265] Another study suggested that breast cancer survivors were 36% more likely to have used CAM than the general population.[266]

6. ADVANTAGES OF CAM IN BREAST CANCER PATIENTS

CAM is a nonhomogenous field; it is a common term that covers a wide range of over 100 therapeutic philosophies, approaches, and healing modalities. Even though CAM is more popular among breast cancer patients, it is essential to assess the ratio of potential benefit to risk in regard to CAM use.[267] For example, many breast cancer patients use complementary therapies to reduce the psychoemotional trauma of coping with the illness and recovery.[268–270] Overall, patients who participate in CAM display lower levels of anxiety and depression and are more hopeful about their chances of survival.[271]

Across many studies, CAM was shown to be correlated with advantages, such as boosting a patient's immune system, preventing the recurrence of cancer, controlling and treating side effects of accompanied therapies, and improving overall survival ability.[272,273] As an adjuvant therapy to chemotherapy in comparison with chemotherapy alone, breast cancer patients using Chinese herbal medicine reported enhanced tumor response, improved Karnofsky performance score, and alleviated toxicity induced by chemotherapy.[274] In a 2009 study exploring the "effect of yoga on symptom management in breast cancer patients," researchers observed a significant negative correlation between breast cancer patients who participated in yoga therapy and symptoms, such as fatigue, nausea, vomiting, and psychological distress. In addition to a reduction of symptoms, the study also concluded a significant improvement in the overall activity level in the group participating in yoga as a complementary therapy, as opposed to control group participants who did not partake in the yoga therapy.[275]

Compared to nonusers, breast cancer patients who participated in CAM less frequently reported a general deterioration of their health status, emphasizing an overall advantage for its ability to improve well-being.[268] Although future research is required to evaluate the applications and to test the effectiveness of CAM on breast cancer outcomes, the advantages of CAM therapies observed in many studies are extensive and noteworthy.

7. DISADVANTAGES OF CAM IN BREAST CANCER PATIENTS

Unlike conventional treatments, CAM treatments are not generally subjected to testing for undesired side effects. Yet, any form of CAM treatment that exhibits biological or psychological effects on patients could cause harmful side effects. To discredit this fact, the "appeal to nature fallacy" has often been employed under the justification that "nature cannot be harmful." However, the study demonstrated that patients with impaired hepatic or renal function are highly prone to experience adverse side effects from CAM treatments.[276,277]

Though breast cancer patients commonly use CAM, the effects of CAM on breast cancer survival have not been extensively studied. Neuhouser *et al.*[278] asserted that the use of CAM has no association with breast cancer–specific mortality or general mortality. The Healthy Eating, Activity, and Lifestyle (HEAL) study was conducted with breast cancer patients at different stages (I–IIIA) ($n = 707$). Of the participants, 60.2% were reported to use CAM postdiagnosis. The most commonly used CAM were natural products and plant-based supplements. However, no correlation was found between CAM and mortality outcomes. Therefore, the authors of the study suggest that randomized controlled trials are required to understand whether there is risk or benefit associated with the types of CAM tested in regard to breast cancer survival.[278]

In another study, allogeneic hematopoietic stem cell transplant survivors ($n = 432$) used a variety of CAM treatments: 30% vitamin therapy, 26% manipulative and body-based therapies, 17.2% spiritual or mind–body therapy, 13.6% dietary modification, 13.5% herbal supplements, 3.5% Chinese medicine, 3% Reiki, and 3% homeopathy. These study results demonstrate that documented therapies have neither benefit nor toxicity. Nevertheless, awareness of CAM use and any potential interactions with conventional treatment is critical for patient safety, particularly in the case of a potentially life-threatening disease, such as cancer.[2,279] Further studies are essential to understand the potency and safety of CAM treatments, given the growing number of users.[280]

8. CONCLUSIONS

The prevalence of CAM use in breast cancer patients demonstrates the extent of its incorporation into standard care. Nevertheless, CAM is not administered as part of integrative medicine alongside standard care due to the lack of knowledge regarding the risks, benefits, side effects of its use, and its interaction with standard care. Several types of CAM treatments hold great value as cancer therapy. Therefore, it is necessary to develop physicians' knowledge about CAM treatments, as they have a long history in traditional practice, as demonstrated by the use of phytotherapy. The intention of the healthcare system should be to provide care that is comprehensive in practice, both effective and safe for the patient, collaborative among disciplines, and inclusive of all appropriate interventions from various sources. In addition to the healthcare system, the involvement of policy makers and the advancement of research efforts are necessary to make decisions about the future of integrative medical practice to provide cost-effective treatments for breast cancer patients. Interventional trials may be required to understand the efficacy of specific CAM therapies and to determine which of them are best suited for combination with standard care. Several studies have demonstrated that phytomedicine is safe, feasible, clinically efficient, cost effective, and has the potential for health system integration. While the safety and advantages of CAM are causing its growth in popularity, 70% of patients reported their physicians did not take time to discuss it with them as an option and only 16% of patients believed the physician was well informed regarding it as a treatment option.[272] Therefore, it is imperative that further research be conducted to elaborate on CAM's effects on the overall survival and recurrence rates in breast cancer patients.[274]

REFERENCES

1. DeSantis CE, Bray F, Ferlay J, Lortet-Tieulent J, Anderson BO, Jemal A. International variation in female breast cancer incidence and mortality rates. *Cancer Epidemiol Biomarkers Prev.* 2015;24(10):1495–1506.
2. Saquib J, Madlensky L, Kealey S, et al. Classification of CAM use and its correlates in patients with early-stage breast cancer. *Integr CancerTher.* 2011;10(2):138–147.
3. Kim MY, Choi SD, Ryu A. Is complementary and alternative therapy effective for women in the climacteric period? *J Menopausal Med.* 2015;21(1):28–35.
4. Chang CC, Fu CF, Yang WT, Chen TY, Hsu YC. The cellular uptake and cytotoxic effect of curcuminoids on breast cancer cells. *Taiwan J Obstet Gynecol.* 2012;51 (3):368–374.
5. Sun XD, Liu XE, Huang DS. Curcumin induces apoptosis of triple-negative breast cancer cells by inhibition of EGFR expression. *Mol Med Rep.* 2012;6(6):1267–1270.
6. Ngamkitidechakul C, Jaijoy K, Hansakul P, Soonthornchareonnon N, Sireeratawong S. Antitumour effects of *Phyllanthus emblica* L.: induction of cancer cell apoptosis and inhibition of in vivo tumour promotion and in vitro invasion of human cancer cells. *Phytother Res.* 2010;24(9):1405–1413.
7. Llanos GG, Araujo LM, Jimenez IA, Moujir LM, Bazzocchi IL. Withaferin A-related steroids from *Withania aristata* exhibit potent antiproliferative activity by inducing apoptosis in human tumor cells. *Eur J Med Chem.* 2012;54:499–511.

8. Suhail MM, Wu W, Cao A, et al. Boswellia sacra essential oil induces tumor cell-specific apoptosis and suppresses tumor aggressiveness in cultured human breast cancer cells. *BMC Complement Altern Med.* 2011;11:129.
9. Sutton KM, Doucette CD, Hoskin DW. NADPH quinone oxidoreductase 1 mediates breast cancer cell resistance to thymoquinone-induced apoptosis. *Biochem Biophys Res Commun.* 2012;426(3):421–426.
10. Chang CY, Chan HL, Lin HY, et al. Rhein induces apoptosis in human breast cancer cells. *Evid Based Complement Alternat Med.* 2012;2012:952504.
11. Hardy ML. Research in Ayurveda: where do we go from here? *Altern Ther Health Med.* 2001;7(2):34–35.
12. Zheng W, Han S, Jiang S, et al. Multiple effects of Xihuang pill aqueous extract on the Hs578T triple-negative breast cancer cell line. *Biomed Rep.* 2016;5(5):559–566.
13. Lim SH, Ha TY, Kim SR, Ahn J, Park HJ, Kim S. Ethanol extract of *Psoralea corylifolia* L. and its main constituent, bakuchiol, reduce bone loss in ovariectomised Sprague-Dawley rats. *Br J Nutr.* 2009;101(7):1031–1039.
14. Choi SY, Lee S, Choi WH, Lee Y, Jo YO, Ha TY. Isolation and anti-inflammatory activity of bakuchiol from *Ulmus davidiana* var. *japonica*. *J Med Food.* 2010;13(4): 1019–1023.
15. Kim KA, Shim SH, Ahn HR, Jung SH. Protective effects of the compounds isolated from the seed of *Psoralea corylifolia* on oxidative stress-induced retinal damage. *Toxicol Appl Pharmacol.* 2013;269(2):109–120.
16. Huang Y, Liu X, Wu Y, Li Y, Guo F. Meroterpenes from *Psoralea corylifolia* against *Pyricularia oryzae*. *Planta Med.* 2014;80(15):1298–1303.
17. Mao H, Wang H, Ma S, et al. Bidirectional regulation of bakuchiol, an estrogenic-like compound, on catecholamine secretion. *Toxicol Appl Pharmacol.* 2014;274(1):180–189.
18. Chen HL, Feng HJ, Li YC. Vitro antitumor activity and synthesis of the key intermediate of bakuchiol. *Yao Xue Xue Bao.* 2010;45(4):467–470.
19. Xin D, Wang H, Yang J, et al. Phytoestrogens from psoralea corylifolia reveal estrogen receptor-subtype selectivity. *Phytomedicine.* 2010;17(2):126–131.
20. Li L, Chen X, Liu CC, Lee LS, Man C, Cheng SH. Phytoestrogen bakuchiol exhibits in vitro and in vivo anti-breast cancer effects by inducing S phase arrest and apoptosis. *Front Pharmacol.* 2016;7:128.
21. Yao C, Xu Y, Chen L, et al. Effects of warm acupuncture on breast cancer-related chronic lymphedema: a randomized controlled trial. *Curr Oncol.* 2016;23(1):e27–e34.
22. Hack CC, Voiss P, Lange S, et al. Local and systemic therapies for breast cancer patients: reducing short-term symptoms with the methods of integrative medicine. *Geburtshilfe Frauenheilkd.* 2015;75(7):675–682.
23. Jeong YJ, Kwon HJ, Park YS, Kwon OC, Shin IH, Park SH. Treatment of lymphedema with Saam acupuncture in patients with breast cancer: a pilot study. *Med Acupunct.* 2015;27(3):206–215.
24. Yeh GY, Wang C, Wayne PM, Phillips R. Tai chi exercise for patients with cardiovascular conditions and risk factors: a systematic review. *J Cardiopulm Rehabil Prev.* 2009;29 (3):152–160.
25. Yeh GY, Wang C, Wayne PM, Phillips RS. The effect of tai chi exercise on blood pressure: a systematic review. *Prev Cardiol.* 2008;11(2):82–89.
26. Pan L, Yan J, Guo Y, Yan J. Effects of tai chi training on exercise capacity and quality of life in patients with chronic heart failure: a meta-analysis. *Eur J Heart Fail.* 2013;15 (3):316–323.
27. Yan JH, Gu WJ, Sun J, Zhang WX, Li BW, Pan L. Efficacy of tai chi on pain, stiffness and function in patients with osteoarthritis: a meta-analysis. *PLoS One.* 2013;8(4):e61672.
28. Kang JW, Lee MS, Posadzki P, Ernst E. T'ai chi for the treatment of osteoarthritis: a systematic review and meta-analysis. *BMJ Open.* 2011;1(1):e000035.

29. Sjosten N, Vaapio S, Kivela SL. The effects of fall prevention trials on depressive symptoms and fear of falling among the aged: a systematic review. *Aging Ment Health.* 2008;12(1):30–46.
30. Maciaszek J, Osinski W. The effects of tai chi on body balance in elderly people—a review of studies from the early 21st century. *Am J Chin Med.* 2010;38(2):219–229.
31. Liu B, Liu ZH, Zhu HE, Mo JC, Cheng DH. Effects of tai chi on lower-limb myodynamia in the elderly people: a meta-analysis. *J Tradit Chin Med.* 2011;31(2):141–146.
32. Yang GY, Wang LQ, Ren J, et al. Evidence base of clinical studies on tai chi: a bibliometric analysis. *PLoS One.* 2015;10(3):e0120655.
33. Larkey LK, Roe DJ, Smith L, Millstine D. Exploratory outcome assessment of qigong/tai chi easy on breast cancer survivors. *Complement Ther Med.* 2016;29:196–203.
34. Larkey L, Huberty J, Pedersen M, Weihs K. Qigong/tai chi easy for fatigue in breast cancer survivors: rationale and design of a randomized clinical trial. *Contemp Clin Trials.* 2016;50:222–228.
35. Pan Y, Yang K, Shi X, Liang H, Zhang F, Lv Q. Tai chi chuan exercise for patients with breast cancer: a systematic review and meta-analysis. *Evid Based Complement Alternat Med.* 2015;2015:535237.
36. Irwin MR, Olmstead R, Breen EC, et al. Tai chi, cellular inflammation, and transcriptome dynamics in breast cancer survivors with insomnia: a randomized controlled trial. *J Natl Cancer Inst Monogr.* 2014;2014(50):295–301.
37. Larkey LK, Roe DJ, Weihs KL, et al. Randomized controlled trial of qigong/tai chi easy on cancer-related fatigue in breast cancer survivors. *Ann Behav Med.* 2015;49(2):165–176.
38. Galantino ML, Callens ML, Cardena GJ, Piela NL, Mao JJ. Tai chi for well-being of breast cancer survivors with aromatase inhibitor-associated arthralgias: a feasibility study. *Altern Ther Health Med.* 2013;19(6):38–44.
39. Fong SS, Ng SS, Luk WS, et al. Shoulder mobility, muscular strength, and quality of life in breast cancer survivors with and without tai chi qigong training. *Evid Based Complement Alternat Med.* 2013;2013:787169.
40. Yan JH, Pan L, Zhang XM, Sun CX, Cui GH. Lack of efficacy of tai chi in improving quality of life in breast cancer survivors: a systematic review and meta-analysis. *Asian Pac J Cancer Prev.* 2014;15(8):3715–3720.
41. Ritenbaugh C, Hammerschlag R, Calabrese C, et al. A pilot whole systems clinical trial of traditional Chinese medicine and naturopathic medicine for the treatment of temporomandibular disorders. *J Altern Complement Med.* 2008;14(5):475–487.
42. Boon HS, Cherkin DC, Erro J, et al. Practice patterns of naturopathic physicians: results from a random survey of licensed practitioners in two US states. *BMC Complement Altern Med.* 2004;4:14.
43. Standish LJ, Greene K, Greenlee H, Kim JG, Grosshans C. Complementary and alternative medical treatment of breast cancer: a survey of licensed North American naturopathic physicians. *Altern Ther Health Med.* 2002;8(5):68–70.
44. Standish LJ, Greene K, Greenlee H, Kim JG, Grosshans C. Complementary and alternative medical treatment of breast cancer: a survey of licensed north American naturopathic physicians. *Altern Ther Health Med.* 2002;8(4):74–76.
45. Teixeira MZ. New homeopathic medicines: use of modern drugs according to the principle of similitude. *Homeopathy.* 2011;100(4):244–252.
46. Castro M. Homeopathy. A theoretical framework and clinical application. *J Nurse Midwifery.* 1999;44(3):280–290.
47. Wani K, Shah N, Prabhune A, Jadhav A, Ranjekar P, Kaul-Ghanekar R. Evaluating the anticancer activity and nanoparticulate nature of homeopathic preparations of *Terminalia chebula. Homeopathy.* 2016;105(4):318–326.

48. Black LI, Clarke TC, Barnes PM, Stussman BJ, Nahin RL. Use of complementary health approaches among children aged 4-17 years in the united states: National Health Interview Survey, 2007-2012. *Natl Health Stat Report.* 2015;(78):1–19.
49. Cucherat M, Haugh MC, Gooch M, Boissel JP. Evidence of clinical efficacy of homeopathy. A meta-analysis of clinical trials. HMRAG. homeopathic medicines research advisory group. *Eur J Clin Pharmacol.* 2000;56(1):27–33.
50. Der Marderosian AH. Understanding homeopathy. *J Am Pharm Assoc.* 1996;NS36 (5):317–321.
51. Ando M, Morita T, Akechi T, et al. The efficacy of mindfulness-based meditation therapy on anxiety, depression, and spirituality in Japanese patients with cancer. *J Palliat Med.* 2009;12(12):1091–1094.
52. Boehm K, Cramer H, Staroszynski T, Ostermann T. Arts therapies for anxiety, depression, and quality of life in breast cancer patients: a systematic review and meta-analysis. *Evid Based Complement Alternat Med.* 2014;2014:103297.
53. Kwekkeboom KL, Cherwin CH, Lee JW, Wanta B. Mind-body treatments for the pain-fatigue-sleep disturbance symptom cluster in persons with cancer. *J Pain Symptom Manage.* 2010;39(1):126–138.
54. Cramer H, Lange S, Klose P, Paul A, Dobos G. Can yoga improve fatigue in breast cancer patients? A systematic review. *Acta Oncol.* 2012;51(4):559–560.
55. Astin JA, Harkness E, Ernst E. The efficacy of "distant healing": a systematic review of randomized trials. *Ann Intern Med.* 2000;132(11):903–910.
56. Mustian KM, Roscoe JA, Palesh OG, et al. Polarity therapy for cancer-related fatigue in patients with breast cancer receiving radiation therapy: a randomized controlled pilot study. *Integr Cancer Ther.* 2011;10(1):27–37.
57. Tsang KL, Carlson LE, Olson K. Pilot crossover trial of Reiki versus rest for treating cancer-related fatigue. *Integr Cancer Ther.* 2007;6(1):25–35.
58. Barlow F, Walker J, Lewith G. Effects of spiritual healing for women undergoing long-term hormone therapy for breast cancer: a qualitative investigation. *J Altern Complement Med.* 2013;19(3):211–216.
59. Yeh ML, Lee TI, Chen HH, Chao TY. The influences of chan-chuang qi-gong therapy on complete blood cell counts in breast cancer patients treated with chemotherapy. *Cancer Nurs.* 2006;29(2):149–155.
60. Amin AR, Kucuk O, Khuri FR, Shin DM. Perspectives for cancer prevention with natural compounds. *J Clin Oncol.* 2009;27(16):2712–2725.
61. Novak KL, Chapman GE. Oncologists' and naturopaths' nutrition beliefs and practices. *Cancer Pract.* 2001;9(3):141–146.
62. Nishino H, Tokuda H, Satomi Y, et al. Cancer chemoprevention by phytochemicals and their related compounds. *Asian Pac J Cancer Prev.* 2000;1(1):49–55.
63. Elkady AI, Abuzinadah OA, Baeshen NA, Rahmy TR. Differential control of growth, apoptotic activity, and gene expression in human breast cancer cells by extracts derived from medicinal herbs *Zingiber officinale*. *J Biomed Biotechnol.* 2012;2012:614356.
64. Ahmed SI, Hayat MQ, Tahir M, et al. Pharmacologically active flavonoids from the anticancer, antioxidant and antimicrobial extracts of *Cassia angustifolia* Vahl. *BMC Complement Altern Med.* 2016;16(1):460.
65. Shu L, Cheung KL, Khor TO, Chen C, Kong AN. Phytochemicals: cancer chemoprevention and suppression of tumor onset and metastasis. *Cancer Metastasis Rev.* 2010;29(3):483–502.
66. Petrovska BB. Historical review of medicinal plants' usage. *Pharmacogn Rev.* 2012;6 (11):1–5.
67. Ghosh B. Polyamines and plant alkaloids. *Indian J Exp Biol.* 2000;38(11):1086–1091.
68. Lu JJ, Bao JL, Chen XP, Huang M, Wang YT. Alkaloids isolated from natural herbs as the anticancer agents. *Evid Based Complement Alternat Med.* 2012;2012:485042.

69. Xu W, Debeb BG, Lacerda L, Li J, Woodward WA. Tetrandrine, a compound common in Chinese traditional medicine, preferentially kills breast cancer tumor initiating cells (TICs) in vitro. *Cancers.* 2011;3(2):2274–2285.
70. Liu T, Liu X, Li W. Tetrandrine, a Chinese plant-derived alkaloid, is a potential candidate for cancer chemotherapy. *Oncotarget.* 2016;7(26):40800–40815.
71. Ismail A, Noolu B, Gogulothu R, Perugu S, Rajanna A, Babu SK. Cytotoxicity and proteasome inhibition by alkaloid extract from *Murraya koenigii* leaves in breast cancer cells-molecular docking studies. *J Med Food.* 2016;19(12):1155–1165.
72. Sefidabi R, Mortazavi P, Hosseini S. Antiproliferative effect of berberine on canine mammary gland cancer cell culture. *Biomed Rep.* 2017;6(1):95–98.
73. Su K, Hu P, Wang X, et al. Tumor suppressor berberine binds VASP to inhibit cell migration in basal-like breast cancer. *Oncotarget.* 2016;7(29):45849–45862.
74. Wang KL, Hsia SM, Yeh JY, Cheng SC, Wang PS, Wang SW. Anti-proliferative effects of evodiamine on human breast cancer cells. *PLoS One.* 2013;8(6):e67297.
75. Du J, Wang XF, Zhou QM, et al. Evodiamine induces apoptosis and inhibits metastasis in MDAMB-231 human breast cancer cells in vitro and in vivo. *Oncol Rep.* 2013;30 (2):685–694.
76. Greenshields AL, Doucette CD, Sutton KM, et al. Piperine inhibits the growth and motility of triple-negative breast cancer cells. *Cancer Lett.* 2015;357(1):129–140.
77. Yu P, Liu Q, Liu K, Yagasaki K, Wu E, Zhang G. Matrine suppresses breast cancer cell proliferation and invasion via VEGF-akt-NF-kappaB signaling. *Cytotechnology.* 2009;59 (3):219–229.
78. Choi YH, Choi WY, Hong SH, et al. Anti-invasive activity of sanguinarine through modulation of tight junctions and matrix metalloproteinase activities in MDA-MB-231 human breast carcinoma cells. *Chem Biol Interact.* 2009;179(2–3):185–191.
79. Kalogris C, Garulli C, Pietrella L, et al. Sanguinarine suppresses basal-like breast cancer growth through dihydrofolate reductase inhibition. *Biochem Pharmacol.* 2014;90(3): 226–234.
80. Gao JL, Ji X, He TC, et al. Tetrandrine suppresses cancer angiogenesis and metastasis in 4T1 tumor bearing mice. *Evid Based Complement Alternat Med.* 2013;2013:265061.
81. Liao CH, Pan SL, Guh JH, et al. Antitumor mechanism of evodiamine, a constituent from chinese herb Evodiae Fructus, in human multiple-drug resistant breast cancer NCI/ADR-RES cells in vitro and in vivo. *Carcinogenesis.* 2005;26(5):968–975.
82. Szallasi A. Piperine: researchers discover new flavor in an ancient spice. *Trends Pharmacol Sci.* 2005;26(9):437–439.
83. Kakarala M, Brenner DE, Korkaya H, et al. Targeting breast stem cells with the cancer preventive compounds curcumin and piperine. *Breast Cancer Res Treat.* 2010;122 (3):777–785.
84. Lai LH, Fu QH, Liu Y, et al. Piperine suppresses tumor growth and metastasis in vitro and in vivo in a 4T1 murine breast cancer model. *Acta Pharmacol Sin.* 2012;33 (4):523–530.
85. Do MT, Kim HG, Choi JH, et al. Antitumor efficacy of piperine in the treatment of human HER2-overexpressing breast cancer cells. *Food Chem.* 2013;141(3):2591–2599.
86. Abdelhamed S, Yokoyama S, Refaat A, et al. Piperine enhances the efficacy of TRAIL-based therapy for triple-negative breast cancer cells. *Anticancer Res.* 2014;34 (4):1893–1899.
87. Katiyar SS, Muntimadugu E, Rafeeqi TA, Domb AJ, Khan W. Co-delivery of rapamycin- and piperine-loaded polymeric nanoparticles for breast cancer treatment. *Drug Deliv.* 2016;23(7):2608–2616.
88. Fantini M, Benvenuto M, Masuelli L, et al. In vitro and in vivo antitumoral effects of combinations of polyphenols, or polyphenols and anticancer drugs: perspectives on cancer treatment. *Int J Mol Sci.* 2015;16(5):9236–9282.

89. Manach C, Scalbert A, Morand C, Remesy C, Jimenez L. Polyphenols: food sources and bioavailability. *Am J Clin Nutr.* 2004;79(5):727–747.
90. Pandey KB, Rizvi SI. Plant polyphenols as dietary antioxidants in human health and disease. *Oxid Med Cell Longev.* 2009;2(5):270–278.
91. Shu XO, Jin F, Dai Q, et al. Soyfood intake during adolescence and subsequent risk of breast cancer among Chinese women. *Cancer Epidemiol Biomarkers Prev.* 2001;10 (5):483–488.
92. Fang Y, Zhang Q, Wang X, et al. Quantitative phosphoproteomics reveals genistein as a modulator of cell cycle and DNA damage response pathways in triple-negative breast cancer cells. *Int J Oncol.* 2016;48(3):1016–1028.
93. Li Y, Meeran SM, Patel SN, Chen H, Hardy TM, Tollefsbol TO. Epigenetic reactivation of estrogen receptor-alpha (ERalpha) by genistein enhances hormonal therapy sensitivity in ERalpha-negative breast cancer. *Mol Cancer.* 2013;12:9.
94. Chan HY, Leung LK. A potential protective mechanism of soya isoflavones against 7,12-dimethylbenz[a]anthracene tumour initiation. *Br J Nutr.* 2003;90(2): 457–465.
95. Taylor CK, Levy RM, Elliott JC, Burnett BP. The effect of genistein aglycone on cancer and cancer risk: a review of in vitro, preclinical, and clinical studies. *Nutr Rev.* 2009;67 (7):398–415.
96. Fan P, Fan S, Wang H, et al. Genistein decreases the breast cancer stem-like cell population through hedgehog pathway. *Stem Cell Res Ther.* 2013;4(6):146.
97. Marik R, Allu M, Anchoori R, Stearns V, Umbricht CB, Khan S. Potent genistein derivatives as inhibitors of estrogen receptor alpha-positive breast cancer. *Cancer Biol Ther.* 2011;11(10):883–892.
98. Levi F, Pasche C, Lucchini F, Ghidoni R, Ferraroni M, La Vecchia C. Resveratrol and breast cancer risk. *Eur J Cancer Prev.* 2005;14(2):139–142.
99. Nakachi K, Suemasu K, Suga K, Takeo T, Imai K, Higashi Y. Influence of drinking green tea on breast cancer malignancy among Japanese patients. *Jpn J Cancer Res.* 1998;89 (3):254–261.
100. Iwasaki M, Inoue M, Sasazuki S, et al. Plasma tea polyphenol levels and subsequent risk of breast cancer among japanese women: a nested case-control study. *Breast Cancer Res Treat.* 2010;124(3):827–834.
101. Narayanan S, Mony U, Vijaykumar DK, Koyakutty M, Paul-Prasanth B, Menon D. Sequential release of epigallocatechin gallate and paclitaxel from PLGA-casein core/shell nanoparticles sensitizes drug-resistant breast cancer cells. *Nanomedicine.* 2015;11 (6):1399–1406.
102. de Pace RC, Liu X, Sun M, et al. Anticancer activities of (−)-epigallocatechin-3-gallate encapsulated nanoliposomes in MCF7 breast cancer cells. *J Liposome Res.* 2013;23 (3):187–196.
103. Verderio P, Bonetti P, Colombo M, Pandolfi L, Prosperi D. Intracellular drug release from curcumin-loaded PLGA nanoparticles induces G2/M block in breast cancer cells. *Biomacromolecules.* 2013;14(3):672–682.
104. Huang T, Chen Z, Fang L. Curcumin inhibits LPS-induced EMT through downregulation of NF-kappaB-snail signaling in breast cancer cells. *Oncol Rep.* 2013;29 (1):117–124.
105. Singh N, Zaidi D, Shyam H, Sharma R, Balapure AK. Polyphenols sensitization potentiates susceptibility of MCF-7 and MDA MB-231 cells to centchroman. *PLoS One.* 2012;7(6):e37736.
106. Siddiqui MA, Ahamed M, Ahmad J, et al. Nickel oxide nanoparticles induce cytotoxicity, oxidative stress and apoptosis in cultured human cells that is abrogated by the dietary antioxidant curcumin. *Food Chem Toxicol.* 2012;50(3–4): 641–647.

107. Banerjee M, Singh P, Panda D. Curcumin suppresses the dynamic instability of microtubules, activates the mitotic checkpoint and induces apoptosis in MCF-7 cells. *FEBS J.* 2010;277(16):3437–3448.
108. Li J, Zhu F, Lubet RA, et al. Quercetin-3-methyl ether inhibits lapatinib-sensitive and -resistant breast cancer cell growth by inducing G(2)/M arrest and apoptosis. *Mol Carcinog.* 2013;52(2):134–143.
109. Hsieh TC, Wong C, John Bennett D, Wu JM. Regulation of p53 and cell proliferation by resveratrol and its derivatives in breast cancer cells: an in silico and biochemical approach targeting integrin alphavbeta3. *Int J Cancer.* 2011;129(11):2732–2743.
110. Tominaga Y, Wang A, Wang RH, Wang X, Cao L, Deng CX. Genistein inhibits Brca1 mutant tumor growth through activation of DNA damage checkpoints, cell cycle arrest, and mitotic catastrophe. *Cell Death Differ.* 2007;14(3):472–479.
111. Choi EJ, Kim GH. Apigenin causes G(2)/M arrest associated with the modulation of p21(Cip1) and Cdc2 and activates p53-dependent apoptosis pathway in human breast cancer SK-BR-3 cells. *J Nutr Biochem.* 2009;20(4):285–290.
112. Masuelli L, Benvenuto M, Fantini M, et al. Curcumin induces apoptosis in breast cancer cell lines and delays the growth of mammary tumors in neu transgenic mice. *J Biol Regul Homeost Agents.* 2013;27(1):105–119.
113. Khan SI, Aumsuwan P, Khan IA, Walker LA, Dasmahapatra AK. Epigenetic events associated with breast cancer and their prevention by dietary components targeting the epigenome. *Chem Res Toxicol.* 2012;25(1):61–73.
114. Mukherjee S, Mazumdar M, Chakraborty S, et al. Curcumin inhibits breast cancer stem cell migration by amplifying the E-cadherin/beta-catenin negative feedback loop. *Stem Cell Res Ther.* 2014;5(5):116.
115. Basnet P, Skalko-Basnet N. Curcumin: an anti-inflammatory molecule from a curry spice on the path to cancer treatment. *Molecules.* 2011;16(6):4567–4598.
116. Shehzad A, Wahid F, Lee YS. Curcumin in cancer chemoprevention: molecular targets, pharmacokinetics, bioavailability, and clinical trials. *Arch Pharm.* 2010;343(9): 489–499.
117. Mondal A, Bennett LL. Resveratrol enhances the efficacy of sorafenib mediated apoptosis in human breast cancer MCF7 cells through ROS, cell cycle inhibition, caspase 3 and PARP cleavage. *Biomed Pharmacother.* 2016;84:1906–1914.
118. Alosi JA, McDonald DE, Schneider JS, Privette AR, McFadden DW. Pterostilbene inhibits breast cancer in vitro through mitochondrial depolarization and induction of caspase-dependent apoptosis. *J Surg Res.* 2010;161(2):195–201.
119. Ergun O, Ergun G, Oktem G, et al. Enteral resveratrol supplementation attenuates intestinal epithelial inducible nitric oxide synthase activity and mucosal damage in experimental necrotizing enterocolitis. *J Pediatr Surg.* 2007;42(10):1687–1694.
120. Siddiqui MA, Saquib Q, Ahamed M, et al. Effect of *trans*-resveratrol on rotenone-induced cytotoxicity in human breast adenocarcinoma cells. *Toxicol Int.* 2011;18 (2):105–110.
121. Le Corre L, Chalabi N, Delort L, Bignon YJ, Bernard-Gallon DJ. Resveratrol and breast cancer chemoprevention: molecular mechanisms. *Mol Nutr Food Res.* 2005;49 (5):462–471.
122. Subbaramaiah K, Chung WJ, Michaluart P, et al. Resveratrol inhibits cyclooxygenase-2 transcription and activity in phorbol ester-treated human mammary epithelial cells. *J Biol Chem.* 1998;273(34):21875–21882.
123. Subbaramaiah K, Michaluart P, Chung WJ, Tanabe T, Telang N, Dannenberg AJ. Resveratrol inhibits cyclooxygenase-2 transcription in human mammary epithelial cells. *Ann NY Acad Sci.* 1999;889:214–223.
124. Tang FY, Su YC, Chen NC, Hsieh HS, Chen KS. Resveratrol inhibits migration and invasion of human breast-cancer cells. *Mol Nutr Food Res.* 2008;52(6):683–691.

125. Azios NG, Dharmawardhane SF. Resveratrol and estradiol exert disparate effects on cell migration, cell surface actin structures, and focal adhesion assembly in MDA-MB-231 human breast cancer cells. *Neoplasia.* 2005;7(2):128–140.
126. Aiyar SE, Park H, Aldo PB, et al. TMS, a chemically modified herbal derivative of resveratrol, induces cell death by targeting bax. *Breast Cancer Res Treat.* 2010;124(1): 265–277.
127. Filomeni G, Graziani I, Rotilio G, Ciriolo MR. *trans*-Resveratrol induces apoptosis in human breast cancer cells MCF-7 by the activation of MAP kinases pathways. *Genes Nutr.* 2007;2(3):295–305.
128. Singh N, Nigam M, Ranjan V, Sharma R, Balapure AK, Rath SK. Caspase mediated enhanced apoptotic action of cyclophosphamide- and resveratrol-treated MCF-7 cells. *J Pharmacol Sci.* 2009;109(4):473–485.
129. Kim YN, Choe SR, Cho KH, et al. Resveratrol suppresses breast cancer cell invasion by inactivating a RhoA/YAP signaling axis. *Exp Mol Med.* 2017;49(2):e296.
130. Bishayee A, Dhir N. Resveratrol-mediated chemoprevention of diethylnitrosamine-initiated hepatocarcinogenesis: inhibition of cell proliferation and induction of apoptosis. *Chem Biol Interact.* 2009;179(2–3):131–144.
131. Lanzilli G, Fuggetta MP, Tricarico M, et al. Resveratrol down-regulates the growth and telomerase activity of breast cancer cells in vitro. *Int J Oncol.* 2006;28(3):641–648.
132. Li Y, Liu J, Liu X, et al. Resveratrol-induced cell inhibition of growth and apoptosis in MCF7 human breast cancer cells are associated with modulation of phosphorylated akt and caspase-9. *Appl Biochem Biotechnol.* 2006;135(3):181–192.
133. Alkhalaf M. Resveratrol-induced growth inhibition in MDA-MB-231 breast cancer cells is associated with mitogen-activated protein kinase signaling and protein translation. *Eur J Cancer Prev.* 2007;16(4):334–341.
134. He X, Wang Y, Zhu J, Orloff M, Eng C. Resveratrol enhances the anti-tumor activity of the mTOR inhibitor rapamycin in multiple breast cancer cell lines mainly by suppressing rapamycin-induced AKT signaling. *Cancer Lett.* 2011;301(2):168–176.
135. Tang FY, Chiang EP, Sun YC. Resveratrol inhibits heregulin-beta1-mediated matrix metalloproteinase-9 expression and cell invasion in human breast cancer cells. *J Nutr Biochem.* 2008;19(5):287–294.
136. Waite KA, Sinden MR, Eng C. Phytoestrogen exposure elevates PTEN levels. *Hum Mol Genet.* 2005;14(11):1457–1463.
137. Saluzzo J, Hallman KM, Aleck K, et al. The regulation of tumor suppressor protein, p53, and estrogen receptor (ERalpha) by resveratrol in breast cancer cells. *Genes Cancer.* 2016;7(11–12):414–425.
138. Lee HS, Seo EY, Kang NE, Kim WK. 6-Gingerol inhibits metastasis of MDA-MB-231 human breast cancer cells. *J Nutr Biochem.* 2008;19(5):313–319.
139. Kundu JK, Na HK, Surh YJ. Ginger-derived phenolic substances with cancer preventive and therapeutic potential. *Forum Nutr.* 2009;61:182–192.
140. Pan MH, Hsieh MC, Kuo JM, et al. 6-Shogaol induces apoptosis in human colorectal carcinoma cells via ROS production, caspase activation, and GADD 153 expression. *Mol Nutr Food Res.* 2008;52(5):527–537.
141. Pan MH, Hsieh MC, Hsu PC, et al. 6-Shogaol suppressed lipopolysaccharide-induced up-expression of iNOS and COX-2 in murine macrophages. *Mol Nutr Food Res.* 2008;52(12):1467–1477.
142. Ling H, Yang H, Tan SH, Chui WK, Chew EH. 6-Shogaol, an active constituent of ginger, inhibits breast cancer cell invasion by reducing matrix metalloproteinase-9 expression via blockade of nuclear factor-kappaB activation. *Br J Pharmacol.* 2010;161 (8):1763–1777.
143. Hsu YL, Chen CY, Lin IP, Tsai EM, Kuo PL, Hou MF. 4-Shogaol, an active constituent of dietary ginger, inhibits metastasis of MDA-MB-231 human breast adenocarcinoma

cells by decreasing the repression of NF-kappaB/snail on RKIP. *J Agric Food Chem.* 2012;60(3):852–861.

144. Thoennissen NH, O'Kelly J, Lu D, et al. Capsaicin causes cell-cycle arrest and apoptosis in ER-positive and -negative breast cancer cells by modulating the EGFR/HER-2 pathway. *Oncogene.* 2010;29(2):285–296.
145. Lee MJ, Kee KH, Suh CH, Lim SC, Oh SH. Capsaicin-induced apoptosis is regulated by endoplasmic reticulum stress- and calpain-mediated mitochondrial cell death pathways. *Toxicology.* 2009;264(3):205–214.
146. Chang HC, Chen ST, Chien SY, Kuo SJ, Tsai HT, Chen DR. Capsaicin may induce breast cancer cell death through apoptosis-inducing factor involving mitochondrial dysfunction. *Hum Exp Toxicol.* 2011;30(10):1657–1665.
147. Shim Y, Song JM. Quantum dot nanoprobe-based high-content monitoring of notch pathway inhibition of breast cancer stem cell by capsaicin. *Mol Cell Probes.* 2015;29 (6):376–381.
148. Fink BN, Steck SE, Wolff MS, et al. Dietary flavonoid intake and breast cancer risk among women on long island. *Am J Epidemiol.* 2007;165(5):514–523.
149. Lall RK, Syed DN, Adhami VM, Khan MI, Mukhtar H. Dietary polyphenols in prevention and treatment of prostate cancer. *Int J Mol Sci.* 2015;16(2):3350–3376.
150. Ramos S. Effects of dietary flavonoids on apoptotic pathways related to cancer chemoprevention. *J Nutr Biochem.* 2007;18(7):427–442.
151. Zhang Y, Vareed SK, Nair MG. Human tumor cell growth inhibition by nontoxic anthocyanidins, the pigments in fruits and vegetables. *Life Sci.* 2005;76(13):1465–1472.
152. Aiyer HS, Warri AM, Woode DR, Hilakivi-Clarke L, Clarke R. Influence of berry polyphenols on receptor signaling and cell-death pathways: implications for breast cancer prevention. *J Agric Food Chem.* 2012;60(23):5693–5708.
153. Liu W, Xu J, Wu S, et al. Selective anti-proliferation of HER2-positive breast cancer cells by anthocyanins identified by high-throughput screening. *PLoS One.* 2013;8(12): e81586.
154. Long N, Suzuki S, Sato S, et al. Purple corn color inhibition of prostate carcinogenesis by targeting cell growth pathways. *Cancer Sci.* 2013;104(3):298–303.
155. Mineva ND, Paulson KE, Naber SP, Yee AS, Sonenshein GE. Epigallocatechin-3-gallate inhibits stem-like inflammatory breast cancer cells. *PLoS One.* 2013;8(9):e73464.
156. Chen D, Pamu S, Cui Q, Chan TH, Dou QP. Novel epigallocatechin gallate (EGCG) analogs activate AMP-activated protein kinase pathway and target cancer stem cells. *Bioorg Med Chem.* 2012;20(9):3031–3037.
157. Baliga MS, Meleth S, Katiyar SK. Growth inhibitory and antimetastatic effect of green tea polyphenols on metastasis-specific mouse mammary carcinoma 4T1 cells in vitro and in vivo systems. *Clin Cancer Res.* 2005;11(5):1918–1927.
158. Roy AM, Baliga MS, Katiyar SK. Epigallocatechin-3-gallate induces apoptosis in estrogen receptor-negative human breast carcinoma cells via modulation in protein expression of p53 and bax and caspase-3 activation. *Mol Cancer Ther.* 2005;4(1):81–90.
159. Mittal A, Pate MS, Wylie RC, Tollefsbol TO, Katiyar SK. EGCG down-regulates telomerase in human breast carcinoma MCF-7 cells, leading to suppression of cell viability and induction of apoptosis. *Int J Oncol.* 2004;24(3):703–710.
160. Yeh CW, Chen WJ, Chiang CT, Lin-Shiau SY, Lin JK. Suppression of fatty acid synthase in MCF-7 breast cancer cells by tea and tea polyphenols: a possible mechanism for their hypolipidemic effects. *Pharmacogenomics J.* 2003;3(5):267–276.
161. Deguchi H, Fujii T, Nakagawa S, Koga T, Shirouzu K. Analysis of cell growth inhibitory effects of catechin through MAPK in human breast cancer cell line T47D. *Int J Oncol.* 2002;21(6):1301–1305.
162. Sartippour MR, Shao ZM, Heber D, et al. Green tea inhibits vascular endothelial growth factor (VEGF) induction in human breast cancer cells. *J Nutr.* 2002;132(8):2307–2311.

163. Pianetti S, Guo S, Kavanagh KT, Sonenshein GE. Green tea polyphenol epigallocatechin-3 gallate inhibits her-2/neu signaling, proliferation, and transformed phenotype of breast cancer cells. *Cancer Res.* 2002;62(3):652–655.
164. Luan YY, Liu ZM, Zhong JY, Yao RY, Yu HS. Effect of grape seed proanthocyanidins on tumor vasculogenic mimicry in human triple-negative breast cancer cells. *Asian Pac J Cancer Prev.* 2015;16(2):531–535.
165. Mantena SK, Baliga MS, Katiyar SK. Grape seed proanthocyanidins induce apoptosis and inhibit metastasis of highly metastatic breast carcinoma cells. *Carcinogenesis.* 2006;27 (8):1682–1691.
166. Pintha K, Yodkeeree S, Limtrakul P. Proanthocyanidin in red rice inhibits MDA-MB-231 breast cancer cell invasion via the expression control of invasive proteins. *Biol Pharm Bull.* 2015;38(4):571–581.
167. Adhikary A, Mohanty S, Lahiry L, Hossain DM, Chakraborty S, Das T. Theaflavins retard human breast cancer cell migration by inhibiting NF-kappaB via p53-ROS cross-talk. *FEBS Lett.* 2010;584(1):7–14.
168. Lahiry L, Saha B, Chakraborty J, et al. Theaflavins target fas/caspase-8 and akt/pBad pathways to induce apoptosis in p53-mutated human breast cancer cells. *Carcinogenesis.* 2010;31(2):259–268.
169. Lahiry L, Saha B, Chakraborty J, et al. Contribution of p53-mediated bax transactivation in theaflavin-induced mammary epithelial carcinoma cell apoptosis. *Apoptosis.* 2008;13(6):771–781.
170. Hu S, Huang L, Meng L, Sun H, Zhang W, Xu Y. Isorhamnetin inhibits cell proliferation and induces apoptosis in breast cancer via akt and mitogenactivated protein kinase kinase signaling pathways. *Mol Med Rep.* 2015;12(5):6745–6751.
171. Lee GA, Choi KC, Hwang KA. Kaempferol, a phytoestrogen, suppressed triclosan-induced epithelial-mesenchymal transition and metastatic-related behaviors of MCF-7 breast cancer cells. *Environ Toxicol Pharmacol.* 2017;49:48–57.
172. Jiao D, Zhang XD. Myricetin suppresses p21-activated kinase 1 in human breast cancer MCF-7 cells through downstream signaling of the beta-catenin pathway. *Oncol Rep.* 2016;36(1):342–348.
173. Balakrishnan S, Bhat FA, Raja Singh P, et al. Gold nanoparticle-conjugated quercetin inhibits epithelial-mesenchymal transition, angiogenesis and invasiveness via EGFR/VEGFR-2-mediated pathway in breast cancer. *Cell Prolif.* 2016;49(6):678–697.
174. Cook MT, Liang Y, Besch-Williford C, Hyder SM. Luteolin inhibits lung metastasis, cell migration, and viability of triple-negative breast cancer cells. *Breast Cancer.* 2016;9:9–19.
175. Nguyen LT, Song YW, Cho SK. Baicalein inhibits epithelial to mesenchymal transition via downregulation of Cyr61 and LOXL-2 in MDA-MB231 breast cancer cells. *Mol Cells.* 2016;39(12):909–914.
176. Eatemadi A, Daraee H, Aiyelabegan HT, Negahdari B, Rajeian B, Zarghami N. Synthesis and characterization of chrysin-loaded PCL-PEG-PCL nanoparticle and its effect on breast cancer cell line. *Biomed Pharmacother.* 2016;84:1915–1922.
177. Wang H, Li YL, She WZ, et al. Studies on antiproliferative effect of flavones compounds isolated from yao herb medicines. *Zhong Yao Cai.* 2007;30(8):980–984.
178. Palit S, Kar S, Sharma G, Das PK. Hesperetin induces apoptosis in breast carcinoma by triggering accumulation of ROS and activation of ASK1/JNK pathway. *J Cell Physiol.* 2015;230(8):1729–1739.
179. Uifalean A, Schneider S, Gierok P, Ionescu C, Iuga CA, Lalk M. The impact of soy isoflavones on MCF-7 and MDA-MB-231 breast cancer cells using a global metabolomic approach. *Int J Mol Sci.* 2016;17(9):1443.
180. Zhang B, Su JP, Bai Y, Li J, Liu YH. Inhibitory effects of O-methylated isoflavone glycitein on human breast cancer SKBR-3 cells. *Int J Clin Exp Pathol.* 2015;8(7): 7809–7817.

181. Chen J, Ge B, Wang Y, Ye Y, Zeng S, Huang Z. Biochanin A promotes proliferation that involves a feedback loop of microRNA-375 and estrogen receptor alpha in breast cancer cells. *Cell Physiol Biochem.* 2015;35(2):639–646.
182. Magne Nde CB, Zingue S, Winter E, et al. Flavonoids, breast cancer chemopreventive and/or chemotherapeutic agents. *Curr Med Chem.* 2015;22(30):3434–3446.
183. Martinez-Perez C, Ward C, Cook G, et al. Novel flavonoids as anti-cancer agents: mechanisms of action and promise for their potential application in breast cancer. *Biochem Soc Trans.* 2014;42(4):1017–1023.
184. Ciolino HP, Yeh GC. Inhibition of aryl hydrocarbon-induced cytochrome P-450 1A1 enzyme activity and CYP1A1 expression by resveratrol. *Mol Pharmacol.* 1999;56 (4):760–767.
185. Moon YJ, Wang X, Morris ME. Dietary flavonoids: effects on xenobiotic and carcinogen metabolism. *Toxicol In Vitro.* 2006;20(2):187–210.
186. Sanderson JT, Hordijk J, Denison MS, Springsteel MF, Nantz MH, van den Berg M. Induction and inhibition of aromatase (CYP19) activity by natural and synthetic flavonoid compounds in H295R human adrenocortical carcinoma cells. *Toxicol Sci.* 2004;82 (1):70–79.
187. Rice S, Whitehead SA. Phytoestrogens and breast cancer—promoters or protectors? *Endocr Relat Cancer.* 2006;13(4):995–1015.
188. Katayama K, Masuyama K, Yoshioka S, Hasegawa H, Mitsuhashi J, Sugimoto Y. Flavonoids inhibit breast cancer resistance protein-mediated drug resistance: transporter specificity and structure-activity relationship. *Cancer Chemother Pharmacol.* 2007;60 (6):789–797.
189. Tan KW, Li Y, Paxton JW, Birch NP, Scheepens A. Identification of novel dietary phytochemicals inhibiting the efflux transporter breast cancer resistance protein (BCRP/ABCG2). *Food Chem.* 2013;138(4):2267–2274.
190. Sjostedt N, Holvikari K, Tammela P, Kidron H. Inhibition of breast cancer resistance protein and multidrug resistance associated protein 2 by natural compounds and their derivatives. *Mol Pharm.* 2017;14(1):135–146.
191. Sahu NK, Balbhadra SS, Choudhary J, Kohli DV. Exploring pharmacological significance of chalcone scaffold: a review. *Curr Med Chem.* 2012;19(2):209–225.
192. Bortolotto LF, Barbosa FR, Silva G, et al. Cytotoxicity of trans-chalcone and licochalcone A against breast cancer cells is due to apoptosis induction and cell cycle arrest. *Biomed Pharmacother.* 2017;85:425–433.
193. Siegelin MD, Gaiser T, Habel A, Siegelin Y. Myricetin sensitizes malignant glioma cells to TRAIL-mediated apoptosis by down-regulation of the short isoform of FLIP and bcl-2. *Cancer Lett.* 2009;283(2):230–238.
194. Ko CH, Shen SC, Hsu CS, Chen YC. Mitochondrial-dependent, reactive oxygen species-independent apoptosis by myricetin: roles of protein kinase C, cytochrome c, and caspase cascade. *Biochem Pharmacol.* 2005;69(6):913–927.
195. Zhang Q, Zhao XH, Wang ZJ. Flavones and flavonols exert cytotoxic effects on a human oesophageal adenocarcinoma cell line (OE33) by causing G2/M arrest and inducing apoptosis. *Food Chem Toxicol.* 2008;46(6):2042–2053.
196. Martinez-Perez C, Ward C, Turnbull AK, et al. Antitumour activity of the novel flavonoid Oncamex in preclinical breast cancer models. *Br J Cancer.* 2016;114(8):905–916.
197. Wang L, Lee IM, Zhang SM, Blumberg JB, Buring JE, Sesso HD. Dietary intake of selected flavonols, flavones, and flavonoid-rich foods and risk of cancer in middle-aged and older women. *Am J Clin Nutr.* 2009;89(3):905–912.
198. Patlolla JM, Rao CV. Triterpenoids for cancer prevention and treatment: current status and future prospects. *Curr Pharm Biotechnol.* 2012;13(1):147–155.
199. Ngo SN, Williams DB, Head RJ. Rosemary and cancer prevention: preclinical perspectives. *Crit Rev Food Sci Nutr.* 2011;51(10):946–954.

200. Yin R, Li T, Tian J, Xi P, Liu RH. Ursolic acid, a potential anticancer compound for breast cancer therapy. *Crit Rev Food Sci Nutr.* 2016.
201. Shan JZ, Xuan YY, Ruan SQ, Sun M. Proliferation-inhibiting and apoptosis-inducing effects of ursolic acid and oleanolic acid on multi-drug resistance cancer cells in vitro. *Chin J Integr Med.* 2011;17(8):607–611.
202. Kim KH, Seo HS, Choi HS, Choi I, Shin YC, Ko SG. Induction of apoptotic cell death by ursolic acid through mitochondrial death pathway and extrinsic death receptor pathway in MDA-MB-231 cells. *Arch Pharm Res.* 2011;34(8):1363–1372.
203. Yeh CT, Wu CH, Yen GC. Ursolic acid, a naturally occurring triterpenoid, suppresses migration and invasion of human breast cancer cells by modulating c-jun N-terminal kinase, akt and mammalian target of rapamycin signaling. *Mol Nutr Food Res.* 2010;54 (9):1285–1295.
204. Wang JS, Ren TN, Xi T. Ursolic acid induces apoptosis by suppressing the expression of FoxM1 in MCF-7 human breast cancer cells. *Med Oncol.* 2012;29(1):10–15.
205. De Angel RE, Smith SM, Glickman RD, Perkins SN, Hursting SD. Antitumor effects of ursolic acid in a mouse model of postmenopausal breast cancer. *Nutr Cancer.* 2010;62 (8):1074–1086.
206. Liu J, Wang Y, Liu RH, He X. Novel triterpenoids isolated from raisins exert potent antiproliferative activities by targeting mitochondrial and ras/raf/ERK signaling in human breast cancer cells. *Food Funct.* 2016;7(7):3244–3251.
207. Bai LY, Chiu CF, Chu PC, Lin WY, Chiu SJ, Weng JR. A triterpenoid from wild bitter gourd inhibits breast cancer cells. *Sci Rep.* 2016;6:22419.
208. Subarnas A, Diantini A, Abdulah R, et al. Apoptosis-mediated antiproliferative activity of friedolanostane triterpenoid isolated from the leaves of *Garcinia celebica* against MCF-7 human breast cancer cell lines. *Biomed Rep.* 2016;4(1):79–82.
209. Cevatemre B, Botta B, Mori M, Berardozzi S, Ingallina C, Ulukaya E. The plant-derived triterpenoid tingenin B is a potent anticancer agent due to its cytotoxic activity on cancer stem cells of breast cancer in vitro. *Chem Biol Interact.* 2016;260:248–255.
210. Levy J, Bosin E, Feldman B, et al. Lycopene is a more potent inhibitor of human cancer cell proliferation than either alpha-carotene or beta-carotene. *Nutr Cancer.* 1995;24 (3):257–266.
211. Hirsch K, Atzmon A, Danilenko M, Levy J, Sharoni Y. Lycopene and other carotenoids inhibit estrogenic activity of 17beta-estradiol and genistein in cancer cells. *Breast Cancer Res Treat.* 2007;104(2):221–230.
212. Prakash P, Russell RM, Krinsky NI. In vitro inhibition of proliferation of estrogen-dependent and estrogen-independent human breast cancer cells treated with carotenoids or retinoids. *J Nutr.* 2001;131(5):1574–1580.
213. Dorf PA. State laws regulating the practice of acupuncture. *J Leg Med.* 1975;3(3):39–45.
214. Salman H, Bergman M, Djaldetti M, Bessler H. Lycopene affects proliferation and apoptosis of four malignant cell lines. *Biomed Pharmacother.* 2007;61(6):366–369.
215. Chalabi N, Delort L, Le Corre L, Satih S, Bignon YJ, Bernard-Gallon D. Gene signature of breast cancer cell lines treated with lycopene. *Pharmacogenomics.* 2006;7 (5):663–672.
216. Molnar J, Gyemant N, Mucsi I, et al. Modulation of multidrug resistance and apoptosis of cancer cells by selected carotenoids. *In Vivo.* 2004;18(2):237–244.
217. Zhang LX, Cooney RV, Bertram JS. Carotenoids enhance gap junctional communication and inhibit lipid peroxidation in C3H/10T1/2 cells: relationship to their cancer chemopreventive action. *Carcinogenesis.* 1991;12(11):2109–2114.
218. Muto Y, Fujii J, Shidoji Y, Moriwaki H, Kawaguchi T, Noda T. Growth retardation in human cervical dysplasia-derived cell lines by beta-carotene through down-regulation of epidermal growth factor receptor. *Am J Clin Nutr.* 1995;62(6 suppl): 1535S–1540S.

219. Bertram JS, Pung A, Churley M, Kappock TJ, Wilkins LR, Cooney RV. Diverse carotenoids protect against chemically induced neoplastic transformation. *Carcinogenesis.* 1991;12(4):671–678.
220. Moselhy SS, Al mslmani MA. Chemopreventive effect of lycopene alone or with melatonin against the genesis of oxidative stress and mammary tumors induced by 7,12 dimethyl(a)benzanthracene in Sprague-Dawely female rats. *Mol Cell Biochem.* 2008;319(1–2):175–180.
221. Preet R, Mohapatra P, Das D, et al. Lycopene synergistically enhances quinacrine action to inhibit wnt-TCF signaling in breast cancer cells through APC. *Carcinogenesis.* 2013;34(2):277–286.
222. Al-Malki AL, Moselhy SS, Refai MY. Synergistic effect of lycopene and tocopherol against oxidative stress and mammary tumorigenesis induced by 7,12-dimethyl[a]benz-anthracene in female rats. *Toxicol Ind Health.* 2012;28(6):542–548.
223. Assar EA, Vidalle MC, Chopra M, Hafizi S. Lycopene acts through inhibition of IkappaB kinase to suppress NF-kappaB signaling in human prostate and breast cancer cells. *Tumour Biol.* 2016;37(7):9375–9385.
224. Mishra LC, Singh BB, Dagenais S. Scientific basis for the therapeutic use of *Withania somnifera* (ashwagandha): a review. *Altern Med Rev.* 2000;5(4):334–346.
225. Kulkarni SK, Dhir A. *Withania somnifera*: an Indian ginseng. *Prog Neuropsychopharmacol Biol Psychiatry.* 2008;32(5):1093–1105.
226. Winters M. Ancient medicine, modern use: *Withania somnifera* and its potential role in integrative oncology. *Altern Med Rev.* 2006;11(4):269–277.
227. Zhang X, Samadi AK, Roby KF, Timmermann B, Cohen MS. Inhibition of cell growth and induction of apoptosis in ovarian carcinoma cell lines CaOV3 and SKOV3 by natural withanolide withaferin A. *Gynecol Oncol.* 2012;124(3):606–612.
228. Zhang X, Mukerji R, Samadi AK, Cohen MS. Down-regulation of estrogen receptor-alpha and rearranged during transfection tyrosine kinase is associated with withaferin A-induced apoptosis in MCF-7 breast cancer cells. *BMC Complement Altern Med.* 2011;11:84.
229. Hahm ER, Moura MB, Kelley EE, Van Houten B, Shiva S, Singh SV. Withaferin A-induced apoptosis in human breast cancer cells is mediated by reactive oxygen species. *PLoS One.* 2011;6(8):e23354.
230. Thaiparambil JT, Bender L, Ganesh T, et al. Withaferin A inhibits breast cancer invasion and metastasis at sub-cytotoxic doses by inducing vimentin disassembly and serine 56 phosphorylation. *Int J Cancer.* 2011;129(11):2744–2755.
231. Wen Z, Wang Z, Wang S, et al. Discovery of molecular mechanisms of traditional chinese medicinal formula si-wu-tang using gene expression microarray and connectivity map. *PLoS One.* 2011;6(3):e18278.
232. Hahm ER, Lee J, Huang Y, Singh SV. Withaferin A suppresses estrogen receptor-alpha expression in human breast cancer cells. *Mol Carcinog.* 2011;50(8):614–624.
233. Yadav VR, Sung B, Prasad S, et al. Celastrol suppresses invasion of colon and pancreatic cancer cells through the downregulation of expression of CXCR4 chemokine receptor. *J Mol Med.* 2010;88(12):1243–1253.
234. Lee J, Hahm ER, Singh SV. Withaferin A inhibits activation of signal transducer and activator of transcription 3 in human breast cancer cells. *Carcinogenesis.* 2010;31(11):1991–1998.
235. Ndlovu MN, Van Lint C, Van Wesemael K, et al. Hyperactivated NF-kappaB and AP-1 transcription factors promote highly accessible chromatin and constitutive transcription across the interleukin-6 gene promoter in metastatic breast cancer cells. *Mol Cell Biol.* 2009;29(20):5488–5504.
236. Stan SD, Zeng Y, Singh SV. Ayurvedic medicine constituent withaferin A causes G2 and M phase cell cycle arrest in human breast cancer cells. *Nutr Cancer.* 2008;60(suppl 1):51–60.

237. Stan SD, Hahm ER, Warin R, Singh SV. Withaferin A causes FOXO3a- and bim-dependent apoptosis and inhibits growth of human breast cancer cells in vivo. *Cancer Res.* 2008;68(18):7661–7669.
238. Samanta SK, Sehrawat A, Kim SH, et al. Disease subtype-independent biomarkers of breast cancer chemoprevention by the ayurvedic medicine phytochemical withaferin A. *J Natl Cancer Inst.* 2016;109(6):djw293.
239. Kim SH, Singh SV. Mammary cancer chemoprevention by withaferin A is accompanied by in vivo suppression of self-renewal of cancer stem cells. *Cancer Prev Res.* 2014;7(7):738–747.
240. Ichikawa H, Takada Y, Shishodia S, Jayaprakasam B, Nair MG, Aggarwal BB. Withanolides potentiate apoptosis, inhibit invasion, and abolish osteoclastogenesis through suppression of nuclear factor-kappaB (NF-kappaB) activation and NF-kappaB-regulated gene expression. *Mol CancerTher.* 2006;5(6):1434–1445.
241. Kaileh M, Vanden Berghe W, Heyerick A, et al. Withaferin A strongly elicits IkappaB kinase beta hyperphosphorylation concomitant with potent inhibition of its kinase activity. *J Biol Chem.* 2007;282(7):4253–4264.
242. Tibaduiza EC, Fleet JC, Russell RM, Krinsky NI. Excentric cleavage products of beta-carotene inhibit estrogen receptor positive and negative breast tumor cell growth in vitro and inhibit activator protein-1-mediated transcriptional activation. *J Nutr.* 2002;132(6):1368–1375.
243. Cui Y, Lu Z, Bai L, Shi Z, Zhao WE, Zhao B. Beta-carotene induces apoptosis and up-regulates peroxisome proliferator-activated receptor gamma expression and reactive oxygen species production in MCF-7 cancer cells. *Eur J Cancer.* 2007;43(17):2590–2601.
244. Rock CL. Carotenoids: biology and treatment. *PharmacolTher.* 1997;75(3):185–197.
245. Wyss A. Carotene oxygenases: a new family of double bond cleavage enzymes. *J Nutr.* 2004;134(1):246S–250S.
246. Nagao A. Oxidative conversion of carotenoids to retinoids and other products. *J Nutr.* 2004;134(1):237S–240S.
247. Bendich A. Carotenoids and the immune response. *J Nutr.* 1989;119(1):112–115.
248. Krinsky NI. Antioxidant functions of carotenoids. *Free Radic Biol Med.* 1989;7(6):617–635.
249. Shultz TD, Chew BP, Seaman WR, Luedecke LO. Inhibitory effect of conjugated dienoic derivatives of linoleic acid and beta-carotene on the in vitro growth of human cancer cells. *Cancer Lett.* 1992;63(2):125–133.
250. Burton GW, Ingold KU. Beta-carotene: an unusual type of lipid antioxidant. *Science.* 1984;224(4649):569–573.
251. Gloria NF, Soares N, Brand C, Oliveira FL, Borojevic R, Teodoro AJ. Lycopene and beta-carotene induce cell-cycle arrest and apoptosis in human breast cancer cell lines. *Anticancer Res.* 2014;34(3):1377–1386.
252. Sato R, Helzlsouer KJ, Alberg AJ, Hoffman SC, Norkus EP, Comstock GW. Prospective study of carotenoids, tocopherols, and retinoid concentrations and the risk of breast cancer. *Cancer Epidemiol Biomarkers Prev.* 2002;11(5):451–457.
253. Toniolo P, Van Kappel AL, Akhmedkhanov A, et al. Serum carotenoids and breast cancer. *Am J Epidemiol.* 2001;153(12):1142–1147.
254. Knekt P, Aromaa A, Maatela J, et al. Serum vitamin A and subsequent risk of cancer: cancer incidence follow-up of the Finnish mobile clinic health examination survey. *Am J Epidemiol.* 1990;132(5):857–870.
255. Knekt P, Albanes D, Seppanen R, et al. Dietary fat and risk of breast cancer. *Am J Clin Nutr.* 1990;52(5):903–908.
256. Knekt P, Aromaa A, Maatela J, et al. Serum selenium and subsequent risk of cancer among Finnish men and women. *J Natl Cancer Inst.* 1990;82(10):864–868.

257. Cui Y, Shikany JM, Liu S, Shagufta Y, Rohan TE. Selected antioxidants and risk of hormone receptor-defined invasive breast cancers among postmenopausal women in the women's health initiative observational study. *Am J Clin Nutr.* 2008;87(4): 1009–1018.
258. Nagel G, Linseisen J, van Gils CH, et al. Dietary beta-carotene, vitamin C and E intake and breast cancer risk in the European Prospective Investigation into Cancer and Nutrition (EPIC). *Breast Cancer Res Treat.* 2010;119(3):753–765.
259. Astin JA, Marie A, Pelletier KR, Hansen E, Haskell WL. A review of the incorporation of complementary and alternative medicine by mainstream physicians. *Arch Intern Med.* 1998;158(21):2303–2310.
260. Eisenberg DM, Davis RB, Ettner SL, et al. Trends in alternative medicine use in the united states, 1990-1997: results of a follow-up national survey. *JAMA.* 1998;280 (18):1569–1575.
261. Horneber M, Bueschel G, Dennert G, Less D, Ritter E, Zwahlen M. How many cancer patients use complementary and alternative medicine: a systematic review and meta-analysis. *Integr Cancer Ther.* 2012;11(3):187–203.
262. Astin JA. Why patients use alternative medicine: results of a national study. *JAMA.* 1998;279(19):1548–1553.
263. Paramore LC. Use of alternative therapies: estimates from the 1994 Robert Wood Johnson Foundation National Access to Care Survey. *J Pain Symptom Manage.* 1997;13 (2):83–89.
264. Upchurch DM, Chyu L. Use of complementary and alternative medicine among American women. *Womens Health Issues.* 2005;15(1):5–13.
265. Saquib J, Parker BA, Natarajan L, et al. Prognosis following the use of complementary and alternative medicine in women diagnosed with breast cancer. *Complement Ther Med.* 2012;20(5):283–290.
266. Blaes AH, Kreitzer MJ, Torkelson C, Haddad T. Nonpharmacologic complementary therapies in symptom management for breast cancer survivors. *Semin Oncol.* 2011;38 (3):394–402.
267. Robotin MC, Penman AG. Integrating complementary therapies into mainstream cancer care: which way forward? *Med J Aust.* 2006;185(7):377–379.
268. Fasching PA, Thiel F, Nicolaisen-Murmann K, et al. Association of complementary methods with quality of life and life satisfaction in patients with gynecologic and breast malignancies. *Support Care Cancer.* 2007;15(11):1277–1284.
269. Shaharudin SH, Sulaiman S, Emran NA, Shahril MR, Hussain SN. The use of complementary and alternative medicine among Malay breast cancer survivors. *Altern Ther Health Med.* 2011;17(1):50–56.
270. Carlsson M, Arman M, Backman M, Hamrin E. Coping in women with breast cancer in complementary and conventional care over 5 years measured by the mental adjustment to cancer scale. *J Altern Complement Med.* 2005;11(3):441–447.
271. Helyer LK, Chin S, Chui BK, et al. The use of complementary and alternative medicines among patients with locally advanced breast cancer—a descriptive study. *BMC Cancer.* 2006;6:39.
272. Huebner J, Muenstedt K, Prott FJ, et al. Online survey of patients with breast cancer on complementary and alternative medicine. *Breast Care.* 2014;9(1):60–63.
273. Nahleh Z, Tabbara IA. Complementary and alternative medicine in breast cancer patients. *Palliat Support Care.* 2003;1(3):267–273.
274. Zhu L, Li L, Li Y, Wang J, Wang Q. Chinese herbal medicine as an adjunctive therapy for breast cancer: a systematic review and meta-analysis. *Evid Based Complement Alternat Med.* 2016;2016:9469276.
275. Vadiraja SH, Rao MR, Nagendra RH, et al. Effects of yoga on symptom management in breast cancer patients: a randomized controlled trial. *Int J Yoga.* 2009;2(2):73–79.

276. Roozbeh J, Hashempur MH, Heydari M. Use of herbal remedies among patients undergoing hemodialysis. *Iran J Kidney Dis.* 2013;7(6):492–495.
277. Del Prete A, Scalera A, Iadevaia MD, et al. Herbal products: benefits, limits, and applications in chronic liver disease. *Evid Based Complement Alternat Med.* 2012;2012:837939.
278. Neuhouser ML, Smith AW, George SM, et al. Use of complementary and alternative medicine and breast cancer survival in the health, eating, activity, and lifestyle study. *Breast Cancer Res Treat.* 2016;160(3):539–546.
279. Saxe GA, Madlensky L, Kealey S, Wu DP, Freeman KL, Pierce JP. Disclosure to physicians of CAM use by breast cancer patients: findings from the women's healthy eating and living study. *Integr Cancer Ther.* 2008;7(3):122–129.
280. Lindsay J, Kabir M, Gilroy N, et al. Epidemiology of complementary and alternative medicine therapy use in allogeneic hematopoietic stem cell transplant survivorship patients in Australia. *Cancer Med.* 2016;5(12):3606–3614.

INDEX

M

N

R

S

T

FSC
www.fsc.org
MIX
Papier | Fördert
gute Waldnutzung
FSC® C083411

Kolibri 360 GmbH
Ferdinand-Jühlke-Straße 7
99095 Erfurt, Deutschland
produktsicherheit@kolibri360.de